GE
197
.T48
1999

Thiele, Leslie Paul.

Environmentalism for
a new millennium.

27776

$35.00

DATE			

ENVIRONMENTALISM
FOR A NEW MILLENNIUM

ENVIRONMENTALISM
FOR A NEW MILLENNIUM

The Challenge of Coevolution

LESLIE PAUL THIELE

New York   Oxford

Oxford University Press

1999

Oxford University Press

Oxford   New York
Athens   Auckland   Bangkok   Bogotá   Buenos Aires   Calcutta
Cape Town   Chennai   Dar es Salaam   Delhi   Florence   Hong Kong   Istanbul
Karachi   Kuala Lumpur   Madrid   Melbourne   Mexico City   Mumbai
Nairobi   Paris   São Paulo   Singapore   Taipei   Tokyo   Toronto   Warsaw

and associated companies in
Berlin   Ibadan

Published by Oxford University Press, Inc.
198 Madison Avenue, New York, New York 10016

Oxford is a registered trademark of Oxford University Press

Library of Congress Cataloging-in-Publication Data
Thiele, Leslie Paul.
Environmentalism for a new millennium : the challenge of
coevolution / by Leslie Paul Thiele.
p.   cm.
Includes bibliographical references and index.
ISBN 0-19-512410-3
1. Environmentalism—United States.   I. Title.
GE197.T48   1999
363.7'005'0973—dc21   98-7722

1 2 3 4 5 6 7 8 9

Printed in the United States of America
on recycled acid-free paper

This book is dedicated to Jacob and Jonah.
May they inherit a world whose beauty and fertility
continue to astound and inspire.

Acknowledgments

I thank Albert Matheny, Arun Agrawal, Luke Garrott, Parakh Hoon, Kirk Jensen, Steven Sanderson, Judy Shapiro, Ken Wald, Donna Waller, and especially Paul Wapner for carefully reading and thoughtfully criticizing various iterations of the manuscript. This book would be much less convincing and much more confused had I not benefited from their insights and patience. I also thank Morton and Elinor Wapner for their ceaseless support and for providing a summer residence at the Jersey shore in 1994, where the writing of this book got its start. The research for this work was assisted by a Social Science Research Council–MacArthur Foundation Fellowship on Peace and Security in a Changing World. I am grateful not only for the financial aid but also for the intellectual fellowship that this afforded me. Finally, thanks are due to the editors of *Environmental Ethics* for permission to make selective use in chapter 5 of material on the discussion of biocentric and anthropocentric environmentalism found in Leslie Paul Thiele, "Nature and Freedom: A Heideggerian Critique of Biocentric and Sociocentric Environmentalism," *Environmental Ethics* 17 (Summer 1995): 171–90.

Preface

Environmentalists are inevitably saddled with a sense of crisis. Urgency pervades their work. I consider myself to be an environmentalist—more committed than some who would make a similar declaration, certainly less active than others. Like most environmentalists, I worry that time is running out on our efforts to stem the degradation of our land, air, and water, the alteration of our atmosphere, and the destruction of other species. It is a painful, chronic worry.

During a recent excursion to southern Utah's portion of the Colorado River, however, all sense of urgency dissolved. With the midday sun beating down from a relentlessly blue sky, my raft drifted down the silt-saturated waters. Layer upon layer of sedimented rock loomed high on both sides of the river, exposing the agedness of the earth and, for one with better credentials in geology than I, offering a well-kept notebook of eons past. With a quick glance from the meandering craft, I could encompass millions of years of the earth's history. Countless wedges of petrified time lay one upon the other, mutely testifying to an inhuman patience.

If human beings were suddenly to vanish from the earth, I asked myself that afternoon, what would visitors to the planet observe were they to explore its rivers and valleys a few million years from now? How would the earth appear once it had spent an eon in the absence of the human species? All our artifacts would eventually be reabsorbed. Vast cities and suburbs, huge industrial complexes, millions of miles of asphalt highways, and innumerable piles of waste would molder ever so slowly, undergo conversion by biotic and geologic processes, and be redigested by the earth. It struck me that despite humanity's 10,000 years of agricultural existence, 300 years of industrial existence, and 50 years of atomic existence, our geologic legacy might not even be noticeable to future river travelers eyeing such walls of stone. At most, compressed remnants of our tinkerings would litter a thin layer of rock slowly sinking into the earth's mantle. Someone struck with a similar thought speculated that the best evidence of humanity's stay on earth millions of years hence might be the relics of space equipment left on the moon.[1] Only on this celestial satellite would our presence endure, for the moon is devoid of the biotic and geologic transformations that inevitably make all earthly things fade in time.

Humanity's stay on Earth may be quite temporary. We are, like all of the planet's flora and fauna, ephemeral organisms situated on a currently hospitable but quite unsentimental planet. Natural history demonstrates that the earth's evolutionary cycles of life and death do not tolerate exceptions. Vertebrate species, fossil records indicate, usually arise and expire inside a ten million year span—even without human hands to accelerate the process. Were we to return to the earth after an absence of a hundred million years to survey its life forms, we would find few of the species that presently populate it. Perhaps only some ants and beetles scurrying about their business would look vaguely familiar. Yet we would be astounded by the many novel forms of life that had developed—organisms whose own pilgrimage on Earth had just begun or had not yet ended.

The earth is a tremendously fruitful place. Its fecundity, however, is currently being undermined at an accelerating rate. Conservation biologists observe that the earth's history of speciation—the evolution of new forms of life—may be coming to a grinding halt. Humanity's overbearing presence on the planet is the primary cause of the termination of vast evolutionary legacies, particularly among higher vertebrates. Unwilling to stem the growth of its own population, limit its consumption of resources, or reduce its production of pollution and poisons, humanity is endangering the earth's evolutionary potential itself.

With this in mind, environmentalists implore us to rescue the planet. The public response has been heartening. A how-to book entitled *50 Simple Things You Can Do to Save the Earth* even became a best-seller, with millions of copies sold.[2] The book's recommendations are mostly sound ones, and the world would be a better place if more people adopted them. Unfortunately, its endorsement of "green lifestyles" largely neglects the importance of political action and institutional change. Its message is troubling for another reason as well. On the one hand, saving a planet surely is asking too much of us as individuals, or even as a species. How can we save an entire planet when many of us cannot organize our own sock drawers and many millions more, far from worrying about sock drawers or planet-saving, do not have adequate nutrition, health care, shelter, or access to education? On the other hand, perhaps saving the planet is asking too little of us. After nearly four billion years of trials and tribulations, life on the third planet of the solar system is likely to abide a while longer, despite any carelessness or shortsightedness on the part of its human occupants. Certainly the earth does not need to be saved from its own natural processes—of speciation and extinction and slow but inevitable geologic transformation. That is simply nature taking its course. In any case, these planetary processes may well outlast the human species.

What, then, do environmentalists want us to save? The earth as a whole will take care of itself—given enough time. While any particular species, including *Homo sapiens,* might not be preserved, the earth itself will rebound from almost any catastrophe that befalls it. Were we to obliterate ourselves and most higher forms of life tomorrow with our nuclear weapons or industrial poisons, the earth would inevitably give birth to new forms of life. While these novel species might take millions or tens of millions of years to develop, this would constitute a relatively brief setback in the earth's four billion years of geologic and evolutionary history.

The problem with this point of view is that it operates with an inappropriate time scale for human concerns. We cannot save the planet in any literal sense. Our stirrings are no match for its patience. But there is much work to be done. What needs saving are the ecological legacies—some billions of years old—that we have come to know and love as well as many others we have yet fully to discover. We cannot save the planet or fully arrest its processes of change. Yet we should do everything in our power to establish the socioeconomic and ecological conditions that will allow the greatest diversity of the earth's plant and animal species—including the human species—to flourish and evolve.

In the span of a few thousand years humanity has spread itself and its artifacts across the globe, becoming at once the planet's most powerful, most widespread, and most dangerous form of life. Though constituting only one out of countless millions of species that currently make the earth a home, humans directly appropriate or otherwise redirect or destroy well over a third of the solar energy that is biologically fixed on the earth's landmasses. In gaining this ever-increasing prominence, we have caused the demise of thousands and perhaps millions of other species. Environmentalists lament the selfishness and shortsightedness of this planetary usurpation.

Other species are not the only casualties of our environmental malfeasance. Population growth, unrestrained resource consumption, and the careless generation of pollution and waste strip health, welfare, and joy from the lives of hundreds of millions of humans and undermine the prospects for their progeny. The environmental degradation that human beings visit upon the earth gets directly translated into human suffering: human disease from polluted air, land, and water, human hunger and want from depleted natural resources, and human physical and mental misery produced by overcrowded living conditions, a degraded natural environment, and the absence of wilderness.

Rhetoric about planet-saving aside, the real test for the environmental movement is whether it can sufficiently transform attitudes and behavior such that a high quality and dynamic diversity of life on Earth can be sustained. The task is to create and maintain the social, economic, and political conditions that will allow both humans and the greatest variety of other species to flourish and evolve. That is the challenge of coevolution. For many thousands of years now, the human aspiration to rise above nature has been largely achieved by trampling nature under. This aspiration, undeniably, has generated much economic, technological, and cultural development—what we now call civilization. Yet the time for a more grateful, self-consciously interdependent relationship has come. It is fitting that the challenge of coevolution faces us at the onset of a new millennium.

Contents

Lovers of our own lands, we are citizens of the world, conscious partakers in the sacrament of all human life or, more truly, of all sentient life.

Emily Balch, "Our Call"

We have frequently printed the word Democracy. Yet I cannot too often repeat that it is a word the real gist of which still sleeps, quite unawaken'd, notwithstanding the resonance and the many angry tempests out of which its syllables have come, from pen or tongue. It is a great word, whose history, I suppose, remains unwritten, because that history has yet to be enacted. It is, in some sort, younger brother of another great and often-used word, Nature, whose history also waits unwritten.

Walt Whitman, *Democratic Vistas*

Introduction

Environmentalism may constitute the most enduring and important social movement of the twentieth century, a movement whose importance,
in all probability, will increase over the coming decades and centuries.
How should we evaluate its historical development, current status, and
future trajectory at the onset of the new millennium? How should we
conceptually and practically account for this complex, diverse, and protean phenomenon? This book investigates environmentalism in an attempt to answer these questions. Rather than simply surveying or celebrating the various achievements of environmentalists—though these
achievements certainly merit an account and, many of them, a celebration—the following pages situate contemporary environmental struggles
in a historical and conceptual framework. By means of this framework,
the primary challenges that confront environmentalists today are presented.

A study of the current status of environmentalism ideally would be
both historically and geographically comprehensive. It would provide an
account of the development of environmental concerns over time, potentially leading back to the dawn of history. It would also be global in
scope. I do not attempt either of these tasks here.

This book is meant to describe the most salient features of contemporary environmentalism. While I believe that these features display
themselves to greater or lesser extent across the globe, the evidence for
their saliency is chiefly taken from the words and deeds of American

environmentalists. The brief historical analysis offered in chapter 1 is also primarily restricted to developments in the United States. The growth of environmentalism in Western Europe bears striking parallels to, and some significant differences from, its development in the United States.[1] The history of environmentalism in other countries displays greater variation. It is an undeniable shortcoming that the research for this book has been largely restricted to the United States, a shortcoming largely shared, if seldom admitted, by most U.S.-based publications in the field of environmental thought and practice. My own justification is pragmatic. The project was already vast in scope and I did not have the means at my disposal to extend its range.

At the same time, to speak of a *national* environmentalism today is to utter something of an oxymoron. Contemporary environmentalists (as I argue in chapter 4) cultivate a global sensibility. Though my research focuses primarily on the words and deeds of American environmentalists, these words and deeds are increasingly planetary in orientation, even when they remain local in application. This book, in any case, is not offered to the reader as the final word on the global environmental movement. It is a contribution to an ever-broadening and still-shifting conversation. My hope is that other scholars and activists with interests in and experience of environmentalism in different lands will be enticed to contribute to that conversation.

In chapter 1, I prepare the ground for the investigation of contemporary environmentalism with a much abbreviated history of the movement. Three "waves" of environmental concern are identified. These waves characterize the first century and a half of its development. The first and longest wave, from the mid-1800s to the 1960s, inaugurated a period of nature *conservation*. A small but significant number of citizens and government officials publicly advocated for the caretaking of nature and the husbandry of her resources. They championed the rational management of public lands and the preservation of wilderness and wildlife. This period saw the establishment of the first national parks and the first government agencies mandated with the stewardship of the nation's forests.

The second wave carried through to the early 1980s. It marked a period of *containment*. At this time, increasing numbers of citizens perceived the need to restrain the growth of human population and abate pollution. They focused on the environmental hazards of advanced industrial life. The point was not to undo industrial society, but to contain the spread of its environmentally destructive growth. Government and

the media responded to and heightened this concern about the environmental degradation that modern economies and lifestyles produced. A
mass environmental movement was born, though it remained in many
respects on the fringe of the mainstream.

The third wave of environmentalism, beginning in the early 1980s
and carrying through the early 1990s, was characterized by *co-optation.*
Conservation and containment describe the goals of environmentalists
in the first two waves. Co-optation describes the form rather than the
substance of environmental concern during the third wave. The word
co-optation signifies the mainstreaming of environmentalism. The political establishment, business interests, and the general public increasingly embraced environmental values during this period, in word if not
in deed. The movement's dissemination of environmental values, in turn,
was tempered to fit the needs of its vastly expanding base of patrons.
Not only did environmentalism become popular, it also became popularized. Environmental concern became a standard feature of life in
America. Yet life in America was itself marked by conflict and tension—
between businesses and consumers, corporations and communities, politicians and constituents, professionals and lay-people, classes and races.
Consequently, the environmental movement also became subject to division and discord. The movement's co-optation by and of various political, economic, and social groups produced strife within the movement
itself.

In chapters 2–5, I describe the key features and animating principles
of the fourth wave of environmentalism. This fourth wave, which has
only begun to crest above the third wave in the last few years, inaugu-
rates a period of *coevolution.* The viability and flourishing of the human
species, fourth-wave environmentalists insist, depend on our success in
preserving the life-sustaining capacities of a diverse biosphere in which
the "options for. . . .evolutionary change" remain intact.[2] Fourth-wave
environmentalists argue, in turn, that the successful protection of nature
depends on the opportunity for humans everywhere to safeguard their
lives and livelihoods.

In taking on the challenge of coevolution, environmental groups engage in various forms of education, advocacy, hands-on environmental
protection, preservation or restoration, and research. The large national
groups educate primarily through the publication of newsletters, magazines, and books and the production of videos, television specials, and
information packets for schools. They are joined by local and regional
groups in educating the public through public speeches, newspaper op-

ed pieces and letters to the editor, the organization of nature outings, ecology camps and workshops, and one-on-one conversations. The chief forms of environmental advocacy consist of lobbying lawmakers, mobilizing phone banks, writing letters and conducting political action campaigns, engaging in direct action and its accompanying media relations, disseminating policy papers, training volunteer leaders and organizing grassroots activists, influencing elections, organizing coalitions, and placing initiatives or referenda on ballots. Some groups engage directly in environmental protection and ecological preservation or restoration. They may work directly with industry to limit pollution, engage in litigation to stop environmentally destructive practices or to hold government agencies accountable for environmental protection, acquire land by purchase and manage it as a habitat preserve or wildlife sanctuary, or organize volunteers to care for public lands. Most of the previously mentioned activities are grounded in various levels of in-house research carried out in the natural and social sciences. This may include monitoring the flora and fauna on public and private lands as well as monitoring government agencies, industries, and the general public to record and analyze everything from public opinion to levels of waste production.

No one environmental organization does all these things, or at least none does them all well. Most specialize in particular methods of environmental advocacy and are equally focused, geographically or topically, in their selection of tasks. Some groups limit themselves to the protection of particular species of wildlife while others monitor the effects of hazardous waste disposal on particular human communities. To some extent, this represents a useful division of labor. But things are not that simple. Human resources are scarce. Because everything cannot be done well or at once, priorities need to be established. Disagreement among environmentalists as to the most important tasks at hand, the best tactics and strategies to employ, and the compromises that should or should not be made on the way, is rife and often rancorous.

Environmentalists have always differed widely, and often passionately, in their attitudes and opinions—each of the first three waves of the environmental movement was characterized by a plurality of approaches and concerns. The fourth wave is no different. It is decidedly not a homogeneous social or political force. Environmentalists today argue about whether other species should be protected for their own sakes or for the long-term benefits that they provide humankind. Their opinions diverge on whether big business is a potential friend of the environment or an incorrigible foe. They endlessly dispute whether the primary focus of action should be the curbing of human numbers (mostly in developing nations) or the controlling of human consumption (mostly in developed nations). Environmental journals are filled with vociferous debates that range from the abstractly philosophical to the very practical. And, of course, environmentalists on the ground disagree on a daily basis about concrete policy issues. They do not see eye to eye about the best way to protect dolphins from the tuna fishing industry. They dispute whether harnessing the renewable energy of the wind with turbines (whose whirling blades tend to kill birds and make noise) is good or bad. They argue about the ethics and efficiency of using market mechanisms to reduce pollution. The list could go on indefinitely. Disagreements regarding specific economic policies can be particularly vehement. During the negotiations for the North American Free Trade Agreement (NAFTA) in early 1990s, for instance, the environmental community and its national organizations were sharply split. Mainstream groups such as the Environmental Defense Fund, the Natural Resources Defense Council, and the World Wildlife Fund endorsed the treaty as an important if imperfect effort to globalize environmental protection. They hoped to strengthen the treaty's environmental provisions. Other organizations such as Friends of the Earth, Greenpeace, and the Sierra Club categorically opposed the agreement as a blank check for the expansion of corporate power and environmental neglect. Still other organizations refused to take a stand so as not to alienate an ambivalent public.

Unity, let alone uniformity, is simply not a feature of the contemporary environmental movement. As Carl Pope, executive director of the Sierra Club, stated, "Put ten environmentalists in a room and you'll get ten different opinions on the topic of your choice, from the reality of global warming to the best use—if any—for a cow. These differences are often quite pronounced, making a mockery of any effort to portray us as a monolithic 'special interest,' all cut from the same green cloth."[3] The environmental movement remains fractured along numerous fault lines. Observers have identified animated debates between shallow environmentalists and deep ecologists, biocentrists and anthropocentrists, eco-feminist and redneck naturalists, social ecologists and wildlife preservationists, eco-anarchists and authoritarian survivalists, environmental modernizers and green romantics, rationalists and transpersonalists, radicals and reformists, professional establishmentarians and grassroots populists. Even particular constituencies within the environmental movement that are typically portrayed as unified and homogeneous groups, such as the deep ecologists, display "distinctive and opposite political wings" not to mention divergent moral viewpoints and strategic orientations.[4] Indeed, factional strife within single organizations is relatively common.[5] The diversity of the environmental movement is sufficiently pronounced that any attempt to capture a singular essence easily distorts as much as it describes.[6] Perhaps we simply should not speak of the environmental movement, but refer instead to environmental movements.

Diversity poses its problems. Internal strife may weaken the environmental community in the face of opposition. Coalition formation and the coordination of activities are often hampered. But diversity also has its benefits. It makes environmentalists well equipped to respond to a wide range of problems in novel ways, and allows for much cross-fertilization of ideas and the sharing of tactics. In an increasingly complex world, a social movement must be multifaceted to be effective and to endure. For these reasons, it is widely accepted that the unification of environmental organizations would be "anathema to a healthy movement."[7] The most prevalent counsel is that environmentalists "should seek not unity but community."[8] Diversity facilitates evolutionary resilience. This holds true for the environmental community no less than for communities of other species. With this ecological truth in mind, contemporary environmentalists affirm their sense of community without aspiring to a homogeneous, conflict-free unity. By embracing their own diversity, environmentalists are effectively practicing the ecological lessons they preach.

Can we even speak of an environmental community, given the multiplicity of environmentalists' goals, tactics, and disagreements? "If there's anything upon which conservationists can all agree," Roger Kennedy, director of the National Park Service, observed, "it is that there are not enough of us."[9] Given the wide-range of problems that environmentalists face, it is easy to understand why they might agree about the deficiency of their own numbers. Perhaps their sense of community is purely a function of the opposition that they collectively confront.

This book suggests that there is more to the environmental community than a set of common obstacles and opponents. It offers an account of the values and sensibilities that bring a palpable coherence to a vast and inherently diverse social movement. Environmentalists constitute a social and political force too large, too varied, and too protean for their efforts ever to become symphonic. Yet their debates, exchanges, and collaborations are increasingly structured by a common theme. What defines the environmental community today is its development of coevolutionary attitudes and practices. This is chiefly manifested in the affirmation of interdependence and the struggle for sustainable development.

To embrace coevolution is to acknowledge, as Wendell Berry has suggested, that "the definitive relationships in the universe are. . . .interdependent."[10] It is to affirm the interdependence within and between nature and society. Environmentalists affirm this interdependence in three dimensions. They acknowledge and embrace our interdependence with (past and) future generations of human beings, with human neighbors both local and global, and with other life forms. Their threefold affirmation of interdependence marks an expansion of the moral universe. It denotes the extension of ethical relations across time, across space, and across species.

The growing awareness, acknowledgment, and practical affirmation of our generational, social, and ecological interdependencies are chiefly manifested in three related struggles carried on by environmentalists today. These are, respectively, the quest for environmental sustainability, the quest for environmental justice, and the quest for environmental integrity. These struggles may be understood as efforts to achieve intergenerational, social, and ecological justice.

Ecology is the study of the interdependencies displayed by networks of life. Environmentalists incorporate an appreciation of the dynamic features of ecological interdependence into human affairs. The economic (and military) interdependence of the world's peoples became increasingly apparent in the wake of the Second World War. Only within the last decade, however, has social and ecological interdependence been widely acknowledged. Cross-cutting linkages between nation states and (international) nongovernmental organizations have led international relations specialists, for more than a decade now, to focus their attention on the global phenomenon of "complex interdependence."[11] This interdependence has been called "the new reality of the century." More specifically for environmentalists, the new reality is the dynamic "intermeshing" of the world's social, economic, and ecological systems.[12]

The realization that all things are interdependent, historian Donald Worster writes, is the upshot of all our knowledge of human and natural ecology, particularly that which has come to light in the last few decades. The "interdependency principle," Worster maintains, is as fundamental and as objective a truth as we are capable of attaining. While human and natural systems may evidence changes in their patterns of interdependence, there is no exception to "the reality or extent of the interdependency itself."[13] Michael Oppenheimer of the Environmental Defense Fund wrote that "[o]ne of the most significant changes [in the perception of the environment and environmental problems over the last 25 years] is an increasing emphasis on connectivity. . . . Connectedness is not a new idea in the environmental arena. . . . In the future, however, the concept of connectedness will expand beyond relations among environmental problems to relations between environmental problems and issues previously viewed as social and economic."[14] Deb Callahan, president of the League of Conservation Voters, likewise has observed that "[a]s environmentalists, we know that understanding the interconnectedness of our world is basic to our work."[15] Fourth-wave environmentalists affirm and seek to safeguard social and ecological interdependencies.

Another defining feature of contemporary environmentalism is receptivity to change. Scientifically grounded knowledge of evolutionary change has been translated into an endorsement of social transformation and adaptation. Importantly, environmentalists' receptivity to change and their affirmation of interdependence prove mutually reinforcing. Evolutionary change within nature and society is produced by the conflict and cooperation that characterize interdependent relationships. In turn, evolutionary change produces new forms of interdependence.

The affirmation of dynamic interdependence is a salutary development. If humanity's social, economic, political, cultural, and technological activities cannot be made to coexist benignly with nature's evolutionary processes, much social and natural evolution may come to a grinding halt. This is the upshot of the "truth" of interdependence. Incorporating this truth into our daily lives and institutions is the task that fourth-wave environmentalists assume. The effort to meet the challenge of coevolution will define environmentalism in the new millennium.

There is, then, something of a convergence of orientations within the environmental movement. This convergence has manifested itself most concretely in the struggle for sustainable development, which has been called the "dominant global discourse of ecological concern."[16] The reason is clear. The discourse and practice of sustainable development bring together the threefold expansion of the moral universe that chiefly characterizes fourth-wave environmentalism. The struggle for sustainability identifies the effort to extend care and moral concern across time. The extension of care and moral concern across space is subsumed under the rubric of (social and economic) development. In turn, a defining feature of sustainable development is its mandate to preserve ecological health and diversity. Sustainable development thus entails the extension of care and moral concern to other species. It is in this sense that the struggle for sustainable development constitutes an effort to realize the promise of dynamic interdependence.

Victor Hugo famously observed that there is one thing stronger than any army and that is an idea whose time has come.[17] The waves of environmentalism outlined in this book represent the concrete effects of ideas whose times eventually arrived. These ideas germinated long before they were ripe enough to stimulate concerted action or become of widespread concern. In 1864, George Perkins Marsh, a key figure of environmentalism's first wave, anticipated second-wave efforts at containment. He also anticipated the contemporary insight that the greatest destruction of nature comes from the unintended effects of our actions. Likewise, in the late 1800s, Gifford Pinchot characterized the conservation movement as a struggle for democracy and equality, not unlike the third- and fourth-wave advocates of environmental justice. John Muir's own preservationist orientation (which together with Pinchot's resource conservationism largely defined the first wave of environmentalism) became a popular social force only a century later, with the efforts of deep ecologists. Employing concepts that are now commonplace among fourth-wave environmentalists, Fairfield Osborn wrote in 1948, before the first wave had crested, that we must harmonize human needs with natural processes in a way that recognizes the "interdependence of all the elements in the creative machinery of nature."[18] In short, the waves of environmentalism described in this book refer not so much to the genesis of particular ideas as to the widespread social enactment of ideas whose times had come.

The ideas underlying environmental care and protection have a long and checkered history. One might imagine their genesis with the first Pleistocene dweller to think twice about messing up his or her own cave. Only in time, however, did these ideas achieve sufficient internal coherence and sufficient reinforcement from material developments, such as the rapid growth of human populations and industry, to become stimulants of public action and governmental regulation. Only at particular historical junctures, in other words, did particular environmental ideas and commitments become effective social and political forces.

I identify these historical junctures as the four waves of environmentalism. The boundary dates set for these waves, though far from arbitrary, are not meant to be exact. There are few, if any, true watersheds in the history of the environmental movement. Its development has been the product of cultural drift rather than revolutionary upheaval. As the metaphor of waves in motion suggests, borders are fluid.

I cannot emphasize too strongly that these waves do not represent isolated or unrelated periods. They mark an evolutionary transformation of public sensibilities and concerns. They signify an expanding repertoire of practices and tactics. Earlier waves become the supportive undercurrents of those that follow. Each wave incorporates, amplifies, and redirects its predecessors' conceptual development, strategic approach, and organizational framework. First-wave (Muirian) preservationism and (Pinchotian) resource conservationism, for example, remain with us today. But these orientations have been supplemented by second-wave efforts to ensure the containment of industrial society. In turn, first-wave and second-wave concerns, concepts, and social legacies have been overlaid by the co-optation of environmentalism by political, corporate, and popular interests that chiefly defines the third wave. Fourth-wave environmentalism also marks a synthesis, reworking its own inheritance while responding to a changing world.

On his retirement after a ten-year term as president of the National Audubon Society in 1995, Peter Berle reflected on recent developments within the environmental movement. He informed members that sustainable development had become a central concern and that this concern was grounded on our appreciation of the interdependent diversity of life. Berle stated:

> In the past decade, our understanding of environmental problems has evolved. Ten years ago the concept of sustainability was not a driving force. Today we recognize that minimizing pollution and resource damage is not good enough. The real objective must be to meet human needs—both physical and spiritual—without limiting opportunity for future generations. . . . During the same period our concern for endangered species has broadened to encompass the recognition that biodiversity and maintaining all life within an ecosystem must be our objective. . . . While specific issues come and go. . . . all humankind needs to understand several basic principles: that the biosphere and all its components are interdependent; that humanity is part of nature; that nature in and of itself has intrinsic value; that biological and cultural diversity are fundamental characteristics of nature; and that we all have a responsibility to care for the community of life. Our continuing challenge is to make this understanding a universal guide for human conduct.[19]

None of the concerns that Berle outlines is in itself new. Some find their first articulations among the nineteenth-century resource conservationists and nature preservationists. Others emerged during the period of containment or the subsequent period of co-optation. What is new, and what chiefly identifies the fourth wave of environmentalism, is that each of these concerns has become a "driving force" within the movement. Each has become integrated into a relatively coherent agenda that animates most of the work that environmentalists engage in today.

This claim is based on the examination of thousands of direct-mail solicitations, letters to members, newsletters, journals, magazines, and other publications of environmental organizations; on hundreds of primary and secondary scholarly works, some of which are grounded in extensive survey data; and on scores of personal interviews with volunteer activists, professional staff, and executive officers of environmental groups (excerpts from which appear in the text as unnoted quotations). When describing what environmentalists believe, say, or do in the following pages, I take the liberty of employing phrases such as "fourth-wave environmentalists maintain that . . ." or "contemporary environmentalists speak of . . ." or "environmentalists struggle to. . . ." These are shorthand locutions. I do not mean to suggest that *all* environmentalists, or even an overwhelming majority of environmentalists, believe, say, or do any one thing. Rather, I am referring to what the more vocal and active elements within the environmental movement today believe, say, and do and what orientations and initiatives have, consequently, become driving forces within the movement. In the appendix, I explain the costs and benefits associated with this use of "ideal types" to describe the development and trajectory of environmentalism.

Environmentalism began as a movement to protect something beautiful and valuable that was "out there." In America, it rose into prominent view over a century ago with the efforts of nature lovers and outdoors-people. In the ensuing century, environmentalists became increasingly concerned with human welfare, waging prominent campaigns against pollution, resource depletion, and overpopulation. Attending to human welfare became part and parcel of environmental protection. The current focus on sustainable development marks the most expansive form of this linkage. The protection of human lives and livelihoods across the globe and across time is now intrinsic to environmental caretaking.

One interpretation of this development is that "the environment" has become increasingly humanized. A more appropriate interpretation,

I believe, is that human life has become increasingly ecologized. The environment is no longer something "out there" in need of protection. Rather, human beings are part of an environment that demands caretaking, an environment defined by complex relations of interdependence within a diverse biosphere.

In substantiating this account of contemporary environmentalism, I challenge a number of recent claims about it. First, the environmental movement is not in jeopardy of self-destructing from its internal divisions or fatally alienating its grassroots base owing to its mounting professionalism.[20] Second, while displaying a significant convergence, the environmental movement is far from unified. Moreover, its members do not perceive unity as necessary or even beneficial.[21] Third, while the movement itself remains a viable social force and is likely to grow, it is far from victorious. Significant achievements in the propagation of environmental values have not been matched by a similar success in fostering environmentally responsible behavior.[22] While there is much reason for hope, there is no justification for blind optimism about the state of the environment or the ultimate victory of the environmental movement. Sustainable development is much debated in theory, and remains largely undefined in practice. Interdependence is widely applauded, but rather narrowly embraced. Today and for the foreseeable future, coevolution will face us as a formidable challenge.

ENVIRONMENTALISM
FOR A NEW MILLENNIUM

From Conservation to Coevolution

First Wave: The Genesis of Conservation

Faced with nearly two billion acres of land available for westward expansion, the early settlers of America eyed a vast resource awaiting exploitation. The rapid development of much of this land was to be expected. Within relatively short order, signs of widespread degradation were evident. By the mid-1700s, a Swedish naturalist traveling in America could complain bitterly about the depletion of the soil's fertility and the lack of concern for natural and agricultural science among the colonists. He observed that "since the arrival of great crowds of Europeans, things are greatly changed; the country is well peopled, and the woods are cut down. The people, increasing in this country, have by hunting and shooting in part extirpated the birds, in part frightened them away. In spring the people steal eggs, mothers and young indifferently, because no regulations are made to the contrary. And if any had been made, the spirit of freedom which prevails in the country would not suffer them to be obeyed." The naturalist concluded his account: "I found everywhere the wisdom and goodness of the Creator; but too seldom saw any inclination to make use of them or adequate estimation of them among men."[1] The problem, evidently, was not simply that the European colonists of America cared less for the land than did the native peoples. Even relative to European standards, American settlers acted irresponsibly in their relationship to the land.

A "frontier" mentality had developed in the New World, and it proved ecologically disastrous. Nature was singled out by the pioneers as the malign source of their troubles. Unsettled America, an early colonist wrote, was a "hideous and desolate wilderness, full of wild beasts and wild men."[2] But wilderness—that which the frontier demarcated from "settled" land—was viewed not only as a obstacle but also as an opportunity. By means of its conquest and subjugation, natural America could become the source of potentially boundless wealth. The frontier mentality encouraged the belief that the land's natural bounty was undepletable. "The inexhaustibility of resources," Roderick Nash wrote in explaining Americans' lack of concern for the protection of nature, "was the dominant American myth. . . . [C]onservation seemed unnecessary. . . . Even people critical of resource exploitation could not escape the feeling that there was, after all, plenty of room for people *and* nature in the New World."[3] As time went on, the distance between ecological reality and frontier mythology grew. Nature, along with the native peoples of America, increasingly fell into the gap.

In 1890, the Census Bureau officially declared the American frontier to be closed. Nonetheless, the frontier mentality long persisted, as did its environmentally pernicious effects. Yet a keen desire to preserve the beauty and bounty of nature was not wholly unknown to the early colonists and settlers. Many poeticized their love of nature or extolled her sacredness in books, lectures, and sermons. Careful husbandry of the land and its resources had its exemplars—though it was not widely practiced. The small but significant number of citizens who struggled against the ecological hazards of America's frontier mentality initiated the first wave of environmentalism.

This was the genesis of conservation. The effort to conserve nature and natural resources achieved neither the level of sophistication nor the widespread acceptance in America that it had in Europe during the early and mid-1800s. Toward the end of the century, however, conservation in America began to surpass the standard European effort to manage natural resources efficiently. In 1864, for instance, Congress transferred the Yosemite Valley to the state of California on the condition that it be held in perpetuity for public recreation. With the preservation of Yosemite as a state park, the protection of nondomesticated animals and their habitats for purposes other than hunting and harvesting saw its first institutionalization. Six years later, in 1872, President Grant signed an act designating over two million acres of the Yellowstone area of northwestern Wyoming to be preserved as a national park. It was the

world's first. The creation of other state and national parks soon followed.

Though little concerned with the preservation of nature for its own
sake or for its aesthetic value, forester Gifford Pinchot (1865–1946)
played a key role in American conservation at the turn of the nineteenth
century. Schooled in German and French forestry practices, Pinchot advocated the scientific management of natural resources. Often considered
the "father of conservation" in America, Pinchot ran the Bureau of Forestry in the Department of Agriculture under President McKinley. Later,
during Theodore Roosevelt's second term, Pinchot headed the new Forest Service. The agency had come to control increasing amounts of public
lands newly established as national forests under the 1891 Forest Reserve Act. Outlining his philosophy of democratic utilitarianism, Pinchot
wrote: "The central thing for which Conservation stands is to make this
country the best possible place to live in, both for us and for our descendants. It stands against the waste of natural resources which cannot
be renewed, such as coal and iron; it stands for the perpetuation of the
resources which can be renewed, such as the food-producing soils and
the forests; and most of all its stands for an equal opportunity for every
American citizen to get his fair share of benefit from these resources,
both now and hereafter."[4] Conservation, Pinchot maintained, entailed a
"wise use of the earth" with the goal of attaining "the greatest good of
the greatest number for the longest time."[5] In his efforts to establish an
ethic of conservation, Pinchot battled against the "boomers" and "land-
grabbers" of his day, men who recklessly plundered western lands for
their mineral wealth and ravaged its forests for timber.

Working both in league with and in opposition to early resource
conservationists was a smaller group of individuals chiefly concerned
with safeguarding wilderness. Their hero was John Muir (1838–1914),
a Scottish-born immigrant who spent much of his life walking, hiking,
and climbing in America's wilds. Muir was less concerned with the utility of natural resources *for* humans than with protecting wildlife and
wildlands *from* humans. The preservation of nature, not the efficient
management of her resources, was his cause. Muir and his followers
came to be known as the "preservationist wing" of the early conservation movement.[6] Preservationists advocated the conservation of wilderness for its own sake and for the aesthetic or spiritual benefits it accorded
humanity.

For Muir, America's mountains and forests were "fountains of life,"
not simply "fountains of timber and irrigating rivers." To help preserve

this life-giving nature, Muir founded the Sierra Club in 1892. He remained its president until his death. The Sierra Club's initial efforts were directed at promoting the enjoyment and preservation of the forests, canyons, and mountains of California's Sierra Nevada. At the time, the primary threat to these mountains and valleys came from logging and the grazing of sheep and cattle. It did not take long before Muir vehemently opposed the more business-friendly efforts of his one-time friend, Gifford Pinchot.

The debate between Muirian nature preservationism and Pinchotian resource conservationism still reverberates within the environmental community. Resource conservationists maintain an essentially *anthropocentric* attitude toward nature while nature preservationists aspire to a *biocentric* (or ecocentric) point of view. Both groups are conservationists in the sense that they aim to conserve nature for the long term. The difference is whether nature is to be conserved as a stock of resources for human use or as a sacrosanct vessel of creation that bears its own inalienable rights. Pinchot, for example, viewed forests as "crops." He maintained that the economic "development" of natural resources was the first principle of conservation. He set forth the ideal of "a land subdued and controlled for the service of the people."[7] Nature preservationists, in contrast, embrace nature as a sacred realm in need of protection, regardless of the material benefits that might redound to human beings from its exploitation.

Resource conservationists had the upper hand in the first wave of American environmentalism. They formed the backbone of the early conservation movement. Yet the preservationist wing was not without influence. Moreover, there was a preservationist tendency within the practice, if not the philosophy, of much resource conservationism. Even the early hunter-conservationists of the first wave, men who viewed animals foremost as potential trophies to be mounted on their study walls or as food for the table, fought to preserve unspoiled natural habitats. They wanted to stalk their abundant prey in the wild rather than visit the remaining specimens in zoos. Sport fishermen, as well, aimed to preserve streams and rivers in a pristine state for their own and posterity's use. In other words, the resources that early conservationists fought to conserve often included wildland and wildlife. Their policies, if not their principles, frequently dovetailed with the efforts of preservationists. For the most part, then, the membership-based conservation organizations of this period assumed as their primary purpose the promotion and protection of the recreational interests of their members—primarily hunting, fishing, birdwatching, and hiking. Their goal was to coax government to pay heed at least as much if not more to these recreational interests than to the interests of extractive industries and ranchers.

A paradigmatic figure of this period was Aldo Leopold (1887–1948). Leopold personally exemplified the tension between anthropocentric and biocentric perspectives within the early conservation movement. Obtaining his master's degree from the Yale School of Forestry, which had been endowed by the Pinchot family, Leopold eventually worked under Pinchot in the Forest Service. He began his career as a resource conservationist. Various personal experiences, however, moved Leopold in a preservationist direction.

One such experience was his involvement with certain well-intentioned but nonetheless counterproductive efforts to manage forests and wildlife for human ends. Beginning in 1914, a federal predator extirpation campaign began on all public lands. Under the presumption that predators deprived deer hunters of their prey and threatened farm and ranch stock, wolves, mountain lions, and coyotes were systematically shot or poisoned. The extermination program was combined with efforts to increase the deer population by clearing forests so that meadows would form where deer might feed. As a public servant in charge of national forests in New Mexico and Arizona, Leopold was party to these policies for two decades. He eagerly worked to have wolves, mountain lions, and coyotes "cleaned out." Whenever possible, he shot predators and directed others to do the same.

By 1935, wolves were virtually exterminated in the contiguous United States. This did lead to the anticipated rise in deer populations. In many cases, however, the increased populations went well beyond the carrying capacity of the lands they occupied. At that point, either the deer had to be shot in droves, or mass starvation and disease would bring their populations well below the level that would have been maintained had the predators remained to winnow the herds. For Leopold, human management of nature had demonstrated its limitations. When it came to deer and wolf relations, he concluded, nature seemed to know best.

As if pulled between Pinchot and Muir, Leopold struggled to find a balance between social demands and environmental responsibilities. In the end, he came to view the human economy as a subset of an overarching ecological balance. In 1933, two years before co-founding the Wilderness Society, Leopold developed the concept of an ecological ethic. He would refine this ethic in *A Sand County Almanac*, posthumously published in 1949. In this, his most famous work, Leopold proposed that humanity transform its role from "conqueror of the land-community to plain member and citizen of it."[8] Ecological standards, he maintained, should structure morality and behavior. He suggested that people quit thinking about their relation to the land solely as an economic phenomenon. Instead, one should "[e]xamine each question in terms of what is ethically and esthetically right, as well as what is economically expedient. A thing is right when it tends to preserve the integrity, stability, and beauty of the biotic community. It is wrong when it tends otherwise."[9] To this day, Leopold's "land ethic" remains a standard for environmentalists.

Leopold's work, one historian has stated, "signaled the arrival of the Age of Ecology."[10] In point of fact, the age of ecology would be noticeably delayed. Sales of Leopold's book were slow and unimpressive upon its release. Only a few thousand copies were purchased over the following decade. The time for Leopold's idea had not yet come.

Second Wave: The Imperative of Containment

The first wave of environmentalism was characterized by the heroic deeds of a handful of individuals and small groups. Their efforts were aimed at the conservation of prized natural resources against the backdrop of a disappearing frontier. Beginning in the 1960s, and more markedly by the early 1970s, a second wave of environmentalism arose and gained increasing public support. This support was stimulated by signs of widespread environmental degradations and planetary limits. Rather than engaging in patchwork efforts of conservation, environmentalists took a broader and deeper view of the problem at hand. The struggle for containment describes this more radical vision.

The productive tension between resource conservation and nature preservation established in the first wave of environmentalism was carried on into the second wave. As the environment became a topic of concern for a growing number of Americans, however, the salient features of the problem changed. The worry was not only that human beings were degrading or destroying pristine wilderness and inefficiently managing natural resources. With mounting unease, Americans perceived that mother nature was starting to hit back. Humans were becoming the victims of their own environmental abuse.

Previously, those involved in environmental protection—both resource conservationists and nature preservationists—placed their activities under the rubric of conservation. The word *environment*, understood in terms of the natural environment, was not in common use and the term "environmentalist" had not been invented.[11] During the early 1970s, the currency of "environmentalism" came into general circulation. By the mid-1970s, talk of the environment was widespread in the nation's schools, on the streets, and in the halls of Congress. The efforts of environmentalists to combat the hazards associated with industrial life and human population growth were everywhere in evidence.

Like the earlier resource conservationists, second-wave environmentalists were troubled by the wasteful use of nature's resources. They also feared that much of nature was being made permanently unusable. Second-wave environmentalists were concerned with the efficient management of natural resources to satisfy human needs. But they also worried about the unending growth of these needs and the ecological costs of satisfying them. Second-wave environmentalists developed a sensitivity to the planetary limits of economic expansion and rising human populations. They also cultivated an ecosystemic perspective that was increasingly global in orientation. Conserving resources, diminishing or eliminating pollution, restraining population growth, and preserving wilderness and green space became common cause in the budding effort to transform, slow down, halt, or reverse the juggernaut of industrial expansion and population growth lest it undermine its own ecological foundations. Boundaries needed to be set. The job of containment had begun.

Observing the difference between the earlier conservationists and second-wave environmentalists, historian Samuel Hays wrote: "The conservation movement was an effort on the part of leaders in science, technology, and government to bring about more efficient development of physical resources. The environmental movement, on the other hand, was far more widespread and popular, involving public values that stressed the quality of human experience and hence of the human environment. Conservation was an aspect of the history of production that stressed efficiency, whereas environment was a part of the history of consumption that stressed new aspects of the American standard of living."[12] Hays's characterization highlights second-wave environmentalists' concern with pollution and the other effects of industrial expansion that were impairing human lives. Yet, Hays ignores the important role played by nature preservationists within the early conservationist movement and second-wave environmentalism. Admittedly, preservationists remained a minority during both of these periods. But they were a vocal and active minority and their strength grew in the 1960s and 1970s.

The rapid rise of second-wave concerns was sparked by Rachel Carson's *Silent Spring*, which first appeared in serial form in *The New Yorker* in 1962 and was published as a book later that year. In scrupulous detail, Carson documented the widespread use of pesticides and their devastating effects on bird populations. Americans, Carson predicted, might soon face a silent spring, a spring wholly deprived of their beloved songbirds. The publication of *Silent Spring*, one environmental philosopher wrote, "made explicit the underlying tension in American society between untrammeled economic growth and a growing concern for health and the quality of life."[13] For the sake of profit, the public learned, the chemical industry was manufacturing the death of nature and playing dice with their own lives. The book struck the American psyche like a bomb. It remained on the best-seller list for over thirty weeks.

With the publication of *Silent Spring*, the task of containment was begun in earnest. Isolated efforts to combat pollution and toxic contamination certainly arose before the 1960s. In the wake of Carson's work, however, public concern for the problem and public support for a solution reached unprecedented heights.

National dismay at the egregious environmental degradation caused by pesticides was accompanied by growing anxiety about the other byproducts of America's industrial, commercial, and agricultural success. Consumerism and the mass production of goods yielded a tremendous increase in litter. Waste disposal and energy resource problems (the latter heightened by the oil crisis of the early 1970s) mounted. Rapid suburban development led to the paving over of green spaces. Urban air quality was noticeably deteriorating, with deadly air inversions occurring in major cities. Oceans were becoming dumping grounds. Streams and rivers were contaminated with effluent that made their waters undrinkable and frequently unfit for swimming and fishing.

In the late 1960s, the Gemini and Apollo space flights brought home the first photographs of the Earth. The planet appeared as a blue-green gem floating in an ocean of night. Two decades earlier, astronomer Fred Hoyle predicted that "[o]nce a photograph of the Earth, taken from the outside is available . . . a new idea as powerful as any other in history will be let loose."[14] Hoyle well anticipated the environmental consciousness that emerged in the late 1960s. For many people, a personal relationship to the planet was formed. For the first time, the earth could be observed as a closed system with distinct and quite visible boundaries. It no longer presented itself as a vast world defying conquest. Rather, the earth seemed almost fragile in its limits. It appeared as a common home in need of caretaking.

That same year, 1968, Paul Ehrlich published *The Population Bomb*. It quickly became a best-seller. Americans were spellbound by Ehrlich's dire analyses. Ehrlich himself became a celebrity of sorts, appearing frequently on popular talk shows. He argued that overpopulation was the chief obstacle to solving many of the world's most pressing economic and ecological woes. Human reproduction, not only its industrial production, was taking a heavy toll on the environment. The problem, in Ehrlich's eyes, was straightforward: "Too many people . . . Too little food . . . A dying planet."[15] Ehrlich did not mince words. He spoke of population growth as a cancer that must be cut out. In the absence of drastic measures, he prophesied that humankind would quickly breed itself into oblivion.

Many of Ehrlich's predictions proved exaggerated. To his credit, however, Ehrlich almost single-handedly placed the issue of overpopulation on the environmental agenda. For many Americans, he globalized the concern for containment. Taking care of one's own home or one's own nation could no longer guarantee a healthy and happy future. The fates of people all over the globe were being ever more tightly tied together. After publishing his book, Ehrlich founded Zero Population Growth (ZPG), an organization with a mission of stemming the tide of human numbers. Within two years, ZPG had amassed 33,000 members in 380 chapters across the country.[16]

A Mass Movement Is Born

Widespread concerns about environmental decay in the United States and about the growth in the world's population burst into a true mass movement in 1970. On April 22 of that year, the first Earth Day was celebrated. An estimated twenty million people participated nationwide, with 10,000 grade schools, 2,000 colleges and 1,000 communities getting involved. The event was deemed "the largest one-day outpouring of public support for any social cause in American history."[17]

Earth Day was the brainchild of Democratic Senator Gaylord Nelson (who would go on to become chairman of the Wilderness Society). Nelson's purpose in organizing Earth Day was to allow latent public interest in environmental affairs a chance to express itself in the form of a nationwide "teach-in." By most accounts, Earth Day was a tremendous success. As Samuel Hays has noted, however, "Earth Day was as much a result as a cause."[18] Earth Day did not so much create concern for the environment as give it vent. It prodded anxiety into action and galvanized the public expression of nascent values. In turn, Earth Day gave environmentalism something it had hitherto lacked: a loud, national voice that was impossible for the country's leaders to ignore.

In large part, the success of Earth Day must be attributed to its motherhood and apple pie demeanor. Who, after all, could be against a clean, livable environment with sufficient resources for all? Amidst the divisive struggles of this era, namely, the war in Indochina and bitter race relations, most Americans could agree that a healthy environment was worth having. Organizations such as the John Birch Society and the Daughters of the American Revolution declared Earth Day a subversive, communist plot and speculated that its being held on April 22 was a "thinly veiled attempt to honor Lenin's birthday."[19] For the public at large, however, Earth Day was a music-filled celebration of life itself.

Some worried that it was little more. Journalist I. F. Stone deemed Earth Day a "gigantic snowjob." In Stone's eyes, it was a diversion. "Just as the Caesars once used bread and circuses," Stone wrote, "so ours use rock and roll, idealism, and noninflammatory social issues to turn the youth off from more urgent concerns which might really threaten our power structure."[20] Stone was thinking of the war in Indochina. Given that the environment was identified as "Nixon's New Issue," Stone may have had a point.[21] In retrospect, however, environmentalism mobilized at least as much criticism of "the establishment" as it dissipated. President Nixon, it should be noted, did not get the wheels of Earth Day in motion. He simply jumped on the environmental bandwagon after it was well on its way. It would quickly leave him behind. By 1972, Congress would be forced to override Nixon's veto of the Clean Water Act.

In the year following the inaugural Earth Day, biologist Barry Commoner accomplished for technology what Paul Ehrlich had achieved for population, namely, noteriety. Known in the late 1950s for his efforts to end the atmospheric testing of nuclear weapons, Commoner's ecological writings and speeches made him a national figure by the late 1960s. *Time* magazine dubbed him "the Paul Revere of Ecology." In *The Closing Circle*, published in 1971, Commoner outlined the sizable role played by technology in fostering the nation's and the world's environmental problems. Along with Lewis Mumford, another popular writer of the time, Commoner raised concerns about the social and ecological effects of a centralized, technological way of life. Not simply a question of population, Commoner claimed, ecological degradation was directly tied to the way society was organized and the manner in which its productive capacities were designed. Profits were being made by polluters. Americans could no longer naively revel in the promise of technology. They had to be concerned about its mischief.

Between 1949 and 1968, Commoner argued, in vehement opposition to Ehrlich, that technology not population caused most pollution and problems of distribution not global shortages of food caused most famines. The methods of industrial production and the synthetic and disposable things that were being produced by modern society were largely to blame for its environmental woes. Most radioactive materials, pesticides, detergents, and plastics were first produced in sizable quantities only after the Second World War. They began to wreak ecological havoc shortly thereafter.[22] Commoner also argued that a "root cause" of the population crisis was poverty and that poverty was largely caused by the exploitation of poor countries by rich countries. A redistribution of the world's wealth was in order.[23] Leaders from developing countries heartily agreed. As the Indian delegate at the United Nation's first conference on population in Bucharest said in 1974, "Development is the best contraceptive."[24]

Commoner and Ehrlich publicized two of the three most important agents of environmental degradation. The third agent, overconsumption, was explicitly brought to fore in 1972, with the publication of *The Limits to Growth*. Employing complex computer-aided analyses, its authors detailed the accelerating rate of natural resource depletion that modern technology, increasing human numbers, and rapid consumption had produced. If the growth trends in world population, industrialization, food production, pollution, and resource depletion continued unabated, the authors concluded, the planetary limits to growth would be reached within the next one hundred years. At that point, human beings would "overshoot the carrying capacity" of the planet and experience a "sudden and uncontrollable decline" in both population and industrial capacity. Within ecological circles, this predicted collapse was known as a "dieback."[25] The authors of *The Limits to Growth* explicitly pitted their doomsaying against the "technological optimism" that was then prevalent. Such optimism, they argued, promoted the folly of exceeding limits rather than the wisdom of living within them. A "Copernican revolution of the mind" would be necessary if society were to accept the notion of economic and ecological equilibrium.[26] The message struck a chord. Millions of copies of *The Limits to Growth* were sold. For many, the book demonstrated with statistical certainty the reality of a global environmental crisis.

The concerns that produced Earth Day and popularized a distinguished series of pessimistic publications were new to the American public. For the first time, the ever-increasing *quantity* of American economic production was no longer wholly reassuring. People became anxious about the deteriorating *quality* of their lives. The environmental effects of waste and wanton consumption mitigated the celebration of industrial power and commercial development. In turn, many began to worry about the consequences of population growth rather than naively endorse large economies of scale. In short, Americans realized in the late 1960s and early 1970s that growth in per capita income and gross national product was not the only or even the best measure of personal and national well-being. The social and ecological costs of industrial productivity, despite the innovations of technology and the interventions of a welfare state, were becoming difficult to ignore. The "affluent society," to employ John Kenneth Galbraith's term for Americans who grew up in the postwar boom, began to worry about the price of affluence.

During second-wave environmentalism, national organizations such as the Sierra Club, the Audubon Society, the National Wildlife Federation, and the Wilderness Society saw their memberships and revenues increase significantly. These well-established groups, however, neither had anticipated the rapid increase in popular support for environmentalism nor were primarily responsible for its mobilization. Most of these organizations were only peripherally involved in Earth Day. By the late 1960s, they had yet to integrate the more popular issues of pollution and population growth into their conservation programs. Only late in 1969, for instance, did the Sierra Club board of directors first debate whether issues such as air pollution were an "appropriate concern" for the Club. Given that membership was soaring and demands for broader involvement were multiplying, the board decided to expand the Club's agenda. That same year the board adopted its first policy on population growth.[27] For the next few years, however, the Sierra Club and other national conservation organizations barely kept up with the tide of second-wave environmental concern.

The growth of support for environmentalism at this time was reflected in the formation of a plethora of new activist groups. Between 1901 and 1960, an average of three conservation organizations formed each year. Between 1961 and 1980, an average of eighteen new groups were forming each year.[28] The Environmental Defense Fund, Friends of the Earth, National Resources Defense Council, League of Conservation Voters, Greenpeace, Sierra Club Legal Defense Fund, and Earth First! all set up shop at this time. Many of these groups were spinoffs of older, more traditional organizations. Unsatisfied with the programs and methods of the established conservation organizations, members joined together to form new groups with new mandates. These new groups expanded the environmental agenda and radicalized its operations.

The second wave of environmentalism was pushed to a crest by two calamitous events. In 1978, residents of a working class suburb of Niagara Falls called Love Canal drew nationwide attention to their plight of living atop 21,000 tons of toxic chemical wastes. The waste had been landfilled in the 1950s. Its burial marked the beginning not the end of the mischief. Having long complained of odors, dying lawns and shrubbery, skin irritations, respiratory illness, miscarriages, and birth defects, residents of Love Canal banded together under the leadership of a housewife named Lois Gibbs. Tired of being ignored by the powers that be, they marched on city hall and held Environmental Protection Agency officials "hostage." An environmental Rubicon had been crossed. New York State health officials eventually declared the leaching wastes at Love Canal a "grave and imminent peril." Initially 240 families were evacuated and later another 500 homes were boarded up. President Carter became involved in 1980, declaring Love Canal a federal disaster area, the first community so designated because of toxic contamination. The relocation of residents set the government back $30 million. The cleanup effort took the better part of a decade, with the cost mounting to $275 million.[29]

These grave events in upstate New York had not yet been put to rest when the nation's attention was diverted elsewhere. In March of 1979, Unit 2 of the Three Mile Island nuclear power station near Harrisburg, Pennsylvania, released unknown amounts of radioactive steam and came within 30 to 60 minutes of a core meltdown. Had the meltdown occurred, the cost in human lives and environmental damage would have been immense. The public was shocked and outraged. A "No More Nukes" rally announced shortly thereafter drew 120,000 people into the streets of Washington, D.C. As if to verify that the values and commitments that had launched the second wave were not mistaken, Three Mile Island brought the first environmental decade to a close with something perilously close to a bang.

Third Wave: The Danger and Merit of Co-optation

The day after Ronald Reagan was inaugurated as president of the United States in January of 1981, the executive directors of the major, national environmental groups met near the White House to discuss the difficulties of dealing with the new, openly anti-environmental administration. The participants of this meeting, the first of its kind, came to be known as the Group of Ten.

The executive rendezvous of the environmental elite was followed later that same year by a very different sort of event. Lois Gibbs, the Love Canal housewife turned activist, moved to Washington, D.C., to set up a clearinghouse for citizens concerned about toxic contamination of their neighborhoods. For two years after its low-key founding, the Citizens Clearinghouse for Hazardous Waste (CCHW) was located in Gibbs's basement. At this time, its subterranean headquarters consisted of little more than a telephone and a few fileboxes of information.

The meeting of the Group of Ten and the founding of CCHW symbolically marked the onset of the third wave of environmentalism. Both illustrated, in markedly different ways, the mainstreaming of American environmental thought and practice.

The Group of Ten (later known as the Green Group) included the executive officers of the most prominent environmental organizations of the time: the Environmental Defense Fund, Environmental Policy Center, Friends of the Earth (which merged with the Oceanic Society and the Environmental Policy Institute in 1989), Izaak Walton League, Natural Resources Defense Council, National Audubon Society, National Wildlife Federation, Sierra Club, the Wilderness Society, and the National Parks and Conservation Association. The umbrella group later expanded its membership to almost two dozen organizations. Initially, the Group of Ten distinguished itself by its dramatic clash with the Reagan administration.[30] Still, the orientation of its members was decidedly "by the book." Direct action and controversial tactics were not part of their repertoire. Thus, Greenpeace, a larger, more visible, more powerful but also more radical group than most of the founding organizations, did not become party to the original Group of Ten.

The January meeting of the country's environmental executives initiated a new program of action for the national environmental organizations. It also signaled a new method of operation. Environmentalism was becoming professionalized. With memberships in the millions and annual budgets ranging over $100 million, national environmental organizations were now run by administrators who controlled entire departments of scientists, lobbyists, lawyers, public relations personnel, communications and media consultants, fund-raisers, and membership recruiters. Predictably, the national organizations focused on reaching an ever-wider base of support. They were largely successful in their efforts. Many of these environmental organizations were critical of the political and business establishment. Nonetheless, they started looking and acting a good deal more like it.

In contrast, the grassroots toxics movement was less willing to bargain and more strident in making demands. As Barry Commoner observed, "[t]he older national environmental organizations in their Washington offices have taken the soft political road of negotiation, compromising with the corporations on the amount of pollution that is acceptable. The people living in the polluted communities have taken the hard political road of confrontation, demanding not that the dumping of hazardous waste be slowed down but that it be stopped."[31] Lois Gibbs insisted that the grassroots movement "was populist and would not allow the issue [of environmental degradation] to be compromised, bureaucratized, or intellectualized out of existence."[32] While mainstream groups found cooperation with government and industry expedient, the grassroots found confrontation unavoidable.

Mainstreaming Environmentalism

The environmental movement as a whole, despite ever-present infighting, enjoyed unprecedented popularity and growth in the 1980s. Membership and revenues increased markedly, reaching new heights. The number of new national groups also grew steadily and the number of local groups skyrocketed. Third-wave environmentalism effectively popularized—with the bane and benefit this term implies—the resource conservationism and nature preservationism of first-wave conservationists along with the pollution and population containment concerns of second-wave activists. Conservation and containment became part of the warp and woof of American life in the 1980s.

Environmentalism was now part of the mainstream. Because the mainstream was not particularly harmonious, however, divisions arose within the environmental movement. Effectively, the movement incorporated the discord inherent to American life. Widespread conflicts in society between business and consumers, corporations and residents, politicians and constituents, and classes and races were introduced into the environmental movement. As the environmental movement became increasingly popular, it reflected the social and political antagonisms that it was forced to navigate.

The growing realization that toxic contaminants might be found in anyone's backyard brought the environmental struggle into the daily lives of average Americans. In particular, it brought environmental awareness into the lives of minorities and low-income classes whose neighborhoods were disproportionately burdened with pollution and toxic contamination. Environmentalism was not only becoming popular, it was also becoming populist. While professional national organizers brought environmental concerns into the courts, into the halls of Congress, and onto the coffee tables of middle America, volunteer local organizers were bringing environmental concerns into the church basements, streets, and front lawns of urban and suburban neighborhoods.

The third wave of environmentalism expanded the roster of players in the game of environmental protection. This increase in the number and diversity of players inevitably posed a challenge to the way the game was being played. Many of the new players had families and neighborhoods hanging in the balance, and they were in no mood and no position to compromise. Organized locally yet linked nationally through clearinghouses, the toxics movement of the 1980s quickly grew in size and power. It boasted a steady stream of victories against corporate polluters. In contrast to the centralization and professionalization of the large national groups, the toxics movement was based on volunteer efforts. Citizens Clearinghouse for Hazardous Waste, for example, largely limited itself to providing support services to community groups. Its growth and success were measured by the number of groups that it assisted. In 1984, CCHW worked with 600 community groups. Four years later, it assisted over 5000 community groups, and the figure would rise to 8000 groups by the mid-1990s.[33]

The toxics movement was primarily organized to oppose the siting of new hazardous waste landfills and incinerators. Its overarching concern was the struggle for "environmental justice." Environmental justice arose as an issue in the late 1970s. It was defined as a "traditional" concern for justice under the law coupled with the knowledge that environmental hazards are not equitably distributed in society.[34] Isolated struggles for environmental justice originated as efforts to combat life- and health-threatening forms of waste and pollution that were dumped on people without their consent and often without their knowledge. The environmental justice movement truly came to life, however, in response to empirical studies carried out in the mid-1980s that demonstrated that environmental hazards were disproportionately borne by particular demographic groups—by the poor and even more noticeably by racial minorities. Deep ecologist Arne Naess characterized the "shallow ecology movement" of the second wave as a "fight against pollution and resource depletion" with a central objective of securing "the health and affluence of people in the developed countries."[35] Third-wave activists pointed out that environmental concern within developed nations was just as restrictively focused. The environmental health and welfare of the underprivileged was being systematically ignored.

Grassroots organizers framed the struggle against this inequity in the context of earlier and ongoing efforts to win democratic rights and social justice for oppressed groups. The demand for environmental justice, activists maintained, was being made "in the same way as women demanded the vote, workers demanded fair pay and humane working conditions, and African Americans demanded civil rights."[36] Civil liberties, minimum wages, safe workplaces, and the right to vote have, over the last century, become part of the basic sociopolitical infrastructure of American life. They remain the unchallenged, if still insufficiently observed, rights of every individual. In adding the right to a healthy environment to this list, grassroots activists were pushing environmentalism farther into the mainstream.

Initially, the environmental justice movement tapped into a specific current within the mainstream. While the national environmental organizations were primarily supported by middle-class and upper-middle-class America, the environmental justice movement primarily appealed to the working class, poor, and minorities. At the tenth anniversary conference of CCHW, organizers observed, 40 percent of the attendees were African American, Hispanic, Native American, or Asian.[37] These attendees came from low-income and minority neighborhoods that were disproportionally burdened with degraded and contaminated environments. In rallying formerly marginalized people to the environmental cause, the environmental justice movement expanded the demographic scope of and support for environmentalism in America.

The two-pronged mainstreaming of the environmental movement in the 1980s—from the top down through the large national organizations and from the bottom up through dispersed, grassroots efforts—may be best described as a kind of co-optation. The word *co-optation* has acquired a negative connotation, often signifying a caving in, a selling out, or a perverting of ideals. But co-optation has a more neutral meaning, denoting an appropriation, assimilation, and incorporation into an established order. To be co-opted, in other words, simply means to become part of the mainstream. This is the sense in which the term is employed here. The co-optation of environmentalism in the 1980s was a mainstreaming or popularization of environmental concerns and values.

The entry of the environmental movement into the mainstream of American life in the 1980s was carried out on a two-way street. The movement modified its ways in order to reach a broader base of support. But society also adapted. This mutual adjustment was principally displayed in three arenas: in the movement's relationship with government officials and agencies, with the business community, and with the public.

As national organizations reached out to secure broader public support, they found it necessary to tailor themselves to the values, attitudes, and (consumer) preferences of this expanded clientele. If in its infancy and adolescence the environmental movement sought to lead the public, in its adulthood it had learned to follow. Yet the movement was now following a public that had itself increasingly embraced environmental values. In turn, a groundswell of concern at the local level brought substantial numbers of low-income workers and minorities into the movement, changing its complexion and adjusting its priorities.

Environmentalists' co-optation of and by the business community accompanied its co-optation of and by the general public. Environmental groups began making things rough in the courts and at the cash register for those businesses that openly scoffed at environmental responsibility. These efforts pressured the private sector to incorporate environmental oversight into operations and public relations. By the 1980s, much of the business community realized that it would have to negotiate with environmentalists and appeal to the growing environmental concerns of the public. In getting corporate America to play ball, however, environmental organizations frequently found themselves involved in a new game. The stakes were higher and the rules had changed. As a result, market-oriented environmental protection gained adherents within the national organizations. In their eagerness to work with industry, these organizations often placed themselves in compromising positions. The purist, outsider status of environmentalism was coming to an end.

A similar co-optation was evident in the political arena. Owing to sophisticated lobbying and popular protest, environmentalists persuaded increasing numbers of politicians to endorse the environmental cause. In the 1980s, they gained access to the halls of Congress as never before. But here, as in the corporate board room, bargaining and compromise were the order of the day. The necessity of negotiation often put environmentalists in the awkward situation, ecologically speaking, of having to rob Peter to pay Paul. A senior attorney at the National Resources Defense Council said in 1986 that "[t]he environmental movement used to be about stopping things. Increasingly, it's about doing things."[38] In their effort to get things done, however, the national organizations often found themselves pressured into bargains that left certain constituents short-changed.

The assimilation of environmentalism into the mainstream widened the movement's appeal and increased its power, but also posed a serious threat to its integrity. That is the danger and merit of co-optation. There is a price to be paid for success.

Critics frequently wax romantic about the movement's past, when purer thoughts and higher ideals supposedly reigned. Whereas Earth Day 1995 came close to being "almost entirely bleached of critical political content," Tom Athanasiou remarked, "Earth Day 1970 was a child of braver times, and was intolerant of both commercialism and loose, forgiving rhetoric."[39] Brian Tokar likewise has insisted that "[t]he declining effectiveness of the mainstream environmental groups reflects, in part, the dominant political culture, which has become ever more subservient to the dictates of corporate America, with its reckless pursuit of unlimited financial gain."[40] These criticisms do not wholly miss the mark. Certainly the movement's growing corporate connections reflect its increasing professionalism and commercialization, which reached new heights in the 1980s and continued to grow thereafter. At the same time, the influence of moneyed interests on environmentalists is nothing new.

Stephen Fox, a renowned historian of the American conservation movement, observed that "wildlife conservation up to 1935 had suffered most from its commercial connections."[41] These connections were primarily limited to the gun companies that funded groups such as The American Game Protective Association and the American Wildlife Institute, which later merged into the National Wildlife Federation. The first Earth Day in 1970, in turn, was hardly free from commercial connections. Some concluded, having assessed the event's sources of financial support, that the intention of Earth Day was less to preserve nature than to foster "a more efficient rape of resources."[42] *Ramparts* magazine took aim at the "eco-establishment" at this time, charging environmentalists of the early 1970s with employing conservation as a device to perpetuate capitalistic profit.[43]

Although environmentalism was never a stranger to compromise or the allure of financial support, its co-optation undeniably reached new heights in the 1980s. This co-optation, to be sure, has a seamy side. No doubt many ideals were stretched, if not abandoned, to accommodate the movement's broader base of support. It is doubtful, however, that any social movement could become popular without also becoming subject to the pitfalls of popularity.

The Grassroots and the Nationals

In the 1980s, as professional "envirocrats" rotated through the "revolving doors" of national environmental organizations, government agencies, and business, antagonism from the grassroots grew. Many grassroots activists were critical of the nationals' conservative methods, restricted agendas, and tendency to seek political compromises. Local organizers claimed that the national organizations, staffed by professionals with Washington insider status, tended "to see grassroots people as potential donors or postcard signers, not as essential players in the creation of national strategies."[44] Some perceived the nationals' professionalism as striking "at the heart of what it meant to be an environmentalist."[45] One critic, voicing a growing concern among the grassroots, observed that environmentalists were increasingly focused on developing new executive skills rather than new public virtues.[46] Environmentalism was becoming a business.

Executives and staff of large environmental organizations were perceived to be fighting the wrong battles with the wrong weapons. They were spending increasing amounts of time in congressional offices lobbying for wildlife and wildland. Meanwhile, working-class people in urban and rural areas were suffering the effects of toxic contamination in their own backyards. For many grassroots activists—who spent their time protesting life-threatening pollution in their neighborhood streets—the political deals made by mainstream groups in Washington to protect charismatic species in faraway forests seemed irrelevant and elitist. That perception was reinforced by the fact that the leadership of the national organizations largely remained a "white-male island." The community-based movements against toxic waste and pollution, in contrast, were well represented in both membership and leadership by women and minorities.[47] This was cause for further estrangement and conflict.

Whereas grassroots activists fighting the dumping of toxic wastes in low-income neighborhoods derided the national groups' concern for wilderness and species preservation as elitist, other grassroots activists, adopting "deep ecology" as their philosophical nomenclature, derided the nationals' concern for human health and welfare as a "shallow" form of environmentalism. Earth First! founder Dave Foreman maintained that professionalized environmental reformers were chiefly concerned with the viability of their own organizations and while catering to human interests were achieving little in the way of wilderness preservation. Foreman graphically suggested that these professional reformers were simply rearranging deck chairs on the *Titanic*, oblivious to imminent ecological disaster. In adopting the organizational structure of the "corporate state," Foreman insisted, large environmental groups were also taking on its ideology.[48]

Many national organizations maintained broad, pragmatic agendas that tried to balance the goal of preserving nature with efforts to mitigate threats to human health and welfare. These organizations found themselves sniped at from both sides of the grassroots, from the toxics movement as well as the radical wilderness preservationists.

The national groups of this period did indeed isolate themselves from the populist pulse of environmental concern. Focused on political horse trading, a way of life in the nation's capital, the nationals often gave up "activism for access." Rather than sustaining the environment, grassroots activists charged, the national groups were most intent on sustaining their own revenue base. At best, the nationals were turning potentially active citizens into passive, check-writing clients.[49] At worst, they were selling out the grassroots to secure their own self-perpetuation.

Environmental scholars and journalists have speculated about the development of a new environmentalism linked to the grassroots movement. One thesis is that the professionalization, commercialization, and market-oriented approach of the nationals caused the mainstream environmental movement to fall out of favor with the American public and lose ground in the battle to preserve wilderness and contain pollution.[50] As the nationals toadied up to their corporate and foundation sponsors, compromise became a way of life and a means of financial life support. In such circumstances, we are told, the only hope lies with the rebirth of the environmental movement through a new form of activism, chiefly organized around the grassroots efforts of volunteers in local communities.

The assertion that the fate of a particular social movement lies in the hands of the grassroots has been frequently voiced by academics, the media, and spokespeople from labor, feminist, peace, and civil rights movements. It remains an inspiring, albeit romantic hope. History demonstrates that the grassroots seldom rises to claim a hegemonic role in fomenting social change and certainly never retains one. The grassroots has an indispensable role to play, but it is carried out in tandem with professional organizations. The grassroots stimulates the professionals through criticism and productive interaction, keeping them honest and egging them on.

Kirkpatrick Sale has maintained that during the third wave, "the movement divided into an increasingly professional mainstream and an increasingly radical grass roots."[51] While antagonisms were clearly growing, Sale's notion of a hard and fast split between mainstream professionals and grassroots radicals is untenable. The environmental movement has always been fractured along numerous fault lines. The nationals themselves have never been unified in their goals or strategies. As Rodger Schlickeisen of Defenders of Wildlife observed, the nationals have been "crossways more often with other nationals than with local and regional groups."[52] Likewise, the grassroots movement is not particularly unified. Dave Foreman suggested that the term "environmentalist" did not even apply to radical wilderness advocates who do not share the anthropocentric concerns of either the nationals or the grassroots toxics movement.[53]

Despite obvious divisions, there has always been and remains today much mutual support between and among grassroots activists and national organizations. A tensioned yet productive relationship will likely continue for the foreseeable future. This fertile tension is demonstrated within organizations as well as between them. Sierra Club members, for instance, periodically form dissident associations or stage internal campaigns to push through more radical agendas.

The environmental movement has certainly undergone a great deal of professionalization in the last three decades. Critics are rightfully wary of the trend. Compared to other social and philanthropic causes or interest groups, however, the environmental movement is not particularly overprofessionalized. The dozen most prominent national groups employ a mean of under 200 staff members.[54] Three quarters of the environmental organizations in America have three or fewer full-time paid staff. More than half of the environmental organizations have no paid staff at all, relying entirely on volunteers.[55] Even in highly professionalized organizations, such as The Nature Conservancy (TNC), volunteers far outnumber staff members. While TNC staff number a few hundred, about 25,000 of its members are actively involved in the organization, contributing 300,000 hours yearly as volunteers.[56]

The professionalized national organizations offer financial support, funding information, leadership training, scientific and legal expertise, and strategic coordination to grassroots groups, increasing these groups' effectiveness and stimulating their growth and proliferation. In 1984, for instance, the Sierra Club organized the National Toxics Campaign, which became an independent organization working with minority and other local groups to fight toxic contamination. In a five-year period, the National Resources Defense Council conducted more than seventy workshops nationwide, training over 1200 local activists to participate in reviews of industrial permit applications. In 1997, the Sierra Club trained 325 new volunteer grassroots organizers.[57] The Sierra Club currently publishes a bi-monthly newsletter oriented to volunteer activists and local organizing.[58] The Environmental Defense Fund, in turn, regularly publishes action-oriented environmental guides for local leaders to promote grassroots organization, such as the *Citizen's Guide to the Superfund Program* and the *Environmental Sustainability Kit*. Thus, grassroots activists and groups often depend on the nationals for basic infrastructure and office support, critical survey data, and sophisticated (and often costly) scientific and legal research. These resources are necessary to devise workable strategies to conserve resources, preserve wildlife and their habitats, and mitigate pollution. Even the more vocal grassroots critics make good use of the legal studies, scientific findings, and social and political surveys that their professional counterparts provide.[59]

Affiliation with the nationals generally stabilizes and perpetuates local grassroots efforts. Targeting responsibility for environmental injustices on specific corporations or their executive officers galvanizes grassroots protest. Yet such narrowly focused struggles often founder. Unless isolated incidents of environmental injustice, such as the toxic contamination of a neighborhood, are placed within a more encompassing historical and social framework, protest groups tend to collapse or disperse.[60] Affiliation with national movement organizations sustains grassroots efforts by supplying local activists with a broader context for their protest and advocacy.

When particular grassroots protests prove successful, they frequently take on a more institutionalized form. Their participants form more stable groups with expanded agendas, or they merge with larger, established organizations. It is unfortunate that grassroots groups that do not

grow, forge an institutional identity, or merge with other organizations
tend to disintegrate. Yet that is the norm. As one investigation of local
groups revealed, "there are no stable nonprofit equivalents of the small
family business that passes on from generation to generation. . . . Nonprofits that fail to grow become extinct. . . . There are no 'old,' small
nonprofits."[61] Such statements are needlessly categorical. Yet they underline an undeniable fact. Local environmental groups are quick to expire. Directories of environmental organizations usually become out-
dated on publication. The National Wildlife Federation's effort to
provide an up-to-date listing of these groups in its *Conservation Direc-
tory* has been described as an attempt to "catalogue a meteor shower."[62]
Of course, new groups are born just as frequently as others disperse.
The environmental movement owes much of its vibrancy to the frequent
birth of these local groups. To a great extent, however, the stability and
endurance of the movement lie in the hands of regional and national
organizations. These larger, more enduring organizations do not seem
to hinder the birth of new grassroots groups. Indeed, they often serve as
midwives.

In an attempt to reclaim the moral high ground, Mark Dowie suggests by way of anecdotal evidence that grassroots activists are largely
united in their irritation at if not enmity toward the nationals and remain
sharply critical of the current direction of the movement. In fact, non-
professional environmental activists are mostly positive in their attitudes
toward the environmental movement as a whole. In turn, their attitudes,
values, and strategies are often quite similar to those of the professionals.
Certainly that is the case regarding local volunteer members of the national organizations. These volunteers, for example, rate lobbying lawmakers as their second highest priority, following educating through the
media.[63] When asked what the most important activity of their organization should be, 45 percent of polled Sierra Club members answered
"lobbying and influencing legislation," with 31 percent choosing "educating people about the environment." Only 8 percent thought "local
grassroots activities" should be their group's most important activity.[64]
Likewise, a Friends of the Earth survey of its membership demonstrated
that public education, research and analysis, and political advocacy were
all ranked above grassroots mobilization as the methods most likely to
effect change.[65]

Support for the national organizations remains quite strong among
local activists. Indeed, professionalization within the major organizations has been widely applauded by rank-and-file environmentalists.

Many see professionalization as the natural path of a maturing move-ment and the necessary means to fight more sophisticated battles against more powerful opponents in government and industry. A large propor-tion of both the membership and staff of environmental movement or-ganizations endorses further professionalization.[66] In sum, the antago-nism between the grassroots and the nationals is much more ambiguous and sporadic than is often suggested. The vast majority of grassroots activists both support the national organizations and utilize their re-sources whenever possible. In any case, grassroots groups, like the na-tional organizations, are as varied as they are numerous. The picture we paint of the grassroots movement should be no more homogenized or sanitized than that offered of the mainstream nationals.

Mark Dowie attempts to support his thesis that the rise of the grass-roots and the demise of the nationals is imminent (owing to the latter's abdication of radical activism and its overprofessionalization) by noting that contributions to and memberships of national organizations de-clined in the early 1990s while increasing for other environmental groups.[67] There are better explanations of these fluctuations. Decline in financial support to the national groups at this time was precipitated by the lingering economic recession and much public interest diverted to the Gulf War.[68] In turn, the political success of certain Washington organizations, groups largely identified with the national political scene, came back to haunt them after the 1992 presidential election. Environ-mental organizations that took credit for helping elect environmentalist Al Gore to the vice presidency subsequently suffered significant losses in membership and revenues owing to the public's perception that large-scale environmental problems would be adequately addressed by the new administration in Washington.[69]

In any case, periodic rises and declines in memberships and revenues are endemic to all social movement organizations. While certain national environmental groups did lose ground in the early 1990s, most have rebounded. Many have gone on to attain record levels of support. Others never suffered net losses, experiencing steady growth in memberships and revenues. Indeed, between 1990 and 1995, nine of fourteen of the major national groups increased their memberships. Net membership in all fourteen groups grew by almost 1.5 million.[70] Moreover, national organizations that made rapprochements with business and government have gained increasing public support and show little if any indication of impending collapse. This directly contradicts Dowie's contention that public support for environmental professionalism and market-oriented strategies declined. Indeed, some of the biggest losses in memberships and revenues in the early 1990s were suffered by groups that traditionally maintained the greatest grassroots structure and orientation. The Sierra Club, with sixty-five chapters and almost 400 local groups, precipitously declined in membership and revenues after a 1990 peak. An even more dramatic loss of 800,000 members and a 40 percent reduction in revenues was suffered by Greenpeace, an organization that has succumbed less than most of the nationals to commercialization, consistently opposes market-oriented approaches to environmental protection, refuses corporate support, and maintains some of the closest links to the grassroots movement.

The professionalization and commercialization of the national groups should remain an object of critical assessment. It is simply not the case, however, that these developments portend the imminent demise of the nationals, undermine the growth of the grassroots, or signal an unbridgeable gap between grassroots environmentalism and large movement organizations. A continuum of groups span the spectrum from the most conservative mainstream and institutionalized to the most radical, progressive, and spontaneous. There is no single divide and thus no one bridge to be constructed. Multiple linkages exist and more should be encouraged.

Stephen Fox has observed that from its earliest days the conservation movement was characterized by a fertile tension between professionals and amateurs, with the amateurs, in Fox's opinion, supplying the "driving force."[71] The local efforts of citizens have always been crucial to the environmental movement. Grassroots activism is the seedbed of more organized and enduring efforts and institutions. Yet grassroots activism is not a substitute for the type of work carried out by large national organizations, such as expertly researched and prepared environmental litigation, scientific research, and high-level lobbying. Local volunteers cannot muster the resources and do not have the expertise to accomplish these tasks. In the same vein, the nationals remain dependent on the grassroots, which carries out the local protest and hands-on environmental protection that more professionalized organizations are ill-equipped to initiate or sustain.

Although cooperation and mutual support between and among national organizations and grassroots activists was pronounced, the third wave of environmentalism undeniably was marked by sharp disagreements as to what the fundamental goals of the movement should be and what strategies should be employed to attain them. The national organizations as a whole, despite the diversity of their agendas, drifted into middle-class respectability. Grassroots groups meanwhile, vaulted into community organizing, mass protest and, in rare cases, ecological sabotage. At the same time, conflict within the grassroots—biocentric Earth First!ers championing wilderness preservation and anthropocentric community organizers fighting toxic threats to urban and suburban neighborhoods—was often as marked as that between the grassroots and the professionals.

Third-wave environmentalism was a mainstreaming. Environmental values were appropriated by politicians, by business interests, and by the general public, including its minorities and low-income classes as well as the middle class. The environmental movement assimilated these constituencies and was significantly transformed by this assimilation in both beneficial and dangerous ways. Generously interpreted, the co-optation of environmentalism that began in the 1980s produced increased support from a wider cross-section of the general public and initiated a constructive engagement with former antagonists in business and government. Critically interpreted, it signaled a crass commercialization, a self-serving professionalization, and a caving in to the powers that be. In either case, one thing is clear: throughout the 1980s environmentalism was making more inroads to a broader section of society and gaining greater access to its economic and political institutions than ever before.

The allure of wider public support, corporate contributions, and political horse trades undoubtedly colored the decisions and shifted the principles of many environmental organizations. Now there were many shades of green. Environmentalism had come of age, and with its new-found maturity, some of its youthful innocence and vigor was inevitably lost. Yet the incorporation of the struggle for environmental justice into the environmental agenda, initiated and sustained by grassroots activism, ensured that the movement would retain its role as watchdog and critic of economic privilege and undemocratic power. It is appropriate, albeit ironic, that the mainstream environmental movement itself became the object of much of this surveillance and criticism.

The Challenge
of Coevolution

Interdependence and
Sustainable Development

The first wave of environmentalism, the period of conservation (from the mid-1800s to the early 1960s), focused on the protection of natural resources, scenic pockets of nature, and isolated species from degradation and destruction by humans. The second wave of environmentalism, the period of containment (from the mid-1960s through the 1970s), introduced the need to protect humans from the effects of an industrially abused and overpopulated world. The third wave of environmentalism, the period of co-optation (the 1980s and early 1990s) popularized and mainstreamed these efforts, revealing in the process that certain demographic groups disproportionately suffered from the effects of a degraded environment.

The fourth wave of environmentalism first arose in the late 1980s. By the mid-1990s, it was cresting above the third wave. Like its immediate predecessor, which achieved its own breadth and height from the combined force of the first two waves, the fourth wave of environmentalism is a synthesis of earlier efforts. It represents not so much a new environmental movement as an expanding horizon of sensibilities and a rechanneling of efforts. Fourth-wave environmentalism amplifies certain features of earlier waves, mitigates some of their conflicts, and integrates many of their cross-currents. At the same time, fourth-wave environmentalism bears its own distinct features.

Environmentalism has been and for the forseeable future will continue to be defined by many different voices. Each cause has its own

chorus of supporters and detractors. When advancing their own particular interests and posturing for their own political ends, environmentalists may seem less in the business of galvanizing public commitment than
dispersing it. The fourth wave of environmentalism is no less diverse
than its predecessors. If anything, its broader reach, largely a product of
the third wave's co-optation of and by the public, the political establishment, and the business community, makes it even less homogeneous.
A certain level of infighting is ever present. People for the Ethical Treatment of Animals (PETA) for example, is harshly critical of many mainstream groups for their tacit support of game hunting. The animal rights
group counsels its members to tie up the toll-free telephone lines of the
mainstream organizations with costly calls. The Citizens Clearinghouse
for Hazardous Waste (CCHW) continues to attack the National Wildlife
Federation (NWF) for its corporate connections to the waste trade.
Greens accuse population organizations that advocate immigration restriction, such as Negative Population Growth, with racism and elitism.[1]
The list could go on. Brent Blackwelder, president of Friends of the
Earth, cautions that "[i]t's a mistake to think there is one direction everyone [within the environmental movement] is headed."[2]

Environmentalism may not be heading in one direction, yet a convergence is evident. Convergence does not connote unity of purpose. It
does not signify identical paths or goals. A convergence is a bending or
turning together, a rapprochement, a drawing nearer. Fourth-wave environmentalism is converging toward a coevolutionary perspective.

Activist and poet e b bortz has stated that the integration of "human
beings into the fabric of an interdependent universe" is the goal to which
Greens aspire. "Whether or not human beings flourish into the next
millennium," bortz insists—indeed, the "very continuation of the human
species"—depends on whether we can achieve this "living co-
existence."[3] Fourth-wave environmentalism is grounded in the effort to
integrate humankind into natural systems. The ultimate goal is a dynamic, shared existence—a coevolution. This coevolutionary integration
has been described by environmental philosophers in terms of a "partnership ethic" between human and natural communities.[4] The coevolutionary perspective is grounded on an appreciation of the dynamic
and complex interdependencies within social life, within nature, and between society and nature. John Sawhill, president of The Nature Conservancy (TNC), speaks of the "larger and more important truths" that
will define the "new conservation ethic" of the foreseeable future. "In
the long term," Sawhill maintains, "conservation success will hinge upon

the ability of people to embrace that interdependence of people and nature."[5] The effort to put the ethic of interdependence into practice is chiefly concretized in the struggle for sustainable development. It is an inclusive, future-focused struggle that is both local and global in scope.

Dynamic interdependence has always been the watchword of ecological thought. In the fourth wave of environmentalism, however, the concept of ecology has broadened. Its domain is now an intricate weave of social and natural relations. The environmental movement, despite its endemic diversity, has transformed formerly isolated concerns with the conservation of resources, the protection of wilderness and biodiversity, and the containment of pollution and population, into an increasingly integrated effort to sustain the adaptive, interdependent social and ecological relationships that define earthly life.

Community in Diversity

Conflict is intrinsic to life. Conflict is also an inevitable feature of the life of a social movement. The third wave of environmentalism was marked by sustained and often quite intense conflict between professionals and the grassroots, radicals and reformists, anthropocentrists and biocentrists, compromisers and noncompromisers. None of these conflicts has disappeared. But most have been transformed. In large part, this transformation has not occurred simply because the differences in orientations and opinions that produced the original conflicts have disappeared. Rather, the benefit of maintaining differences of orientation and opinion has become evident.

For the most part, the multiple antagonisms of earlier waves of environmentalism have proved productive even though many remain unresolved. The diversity of the movement is increasingly valued as a creative force, despite the inevitable problems it fosters. The conflicts among biocentric and anthropocentric grassroots organizations and between the grassroots and the nationals that characterized the third wave of environmentalism, for example, paved the way for the community-based environmentalism of the fourth wave.

Fourth-wave convergence does not signal a homogenization of the environmental movement. It is an affirmation of interdependence. Interdependence presumes an elaborate and dynamic diversity of life—multilayered relations grounded in mutual but not necessarily equal dependence. To affirm interdependent change is to gain a broader and deeper appreciation of the distinct, often conflicting parts necessary for the healthy functioning of the social and ecological whole.

Environmentalists are increasingly appreciative of the dynamic diversity of nature, of society, and of the environmental movement itself. A Wilderness Society promotion applauds interdependence as a means of heightening strength and viability by way of "symbiotic" connections: "It is one of the axioms of ecological science that any viable biological system is replete with interdependent relationships whose needs complement one another and nurture the health of the whole system. As with biological systems, so with human institutions. The conservation community at large is no exception to this rule. Neither is the Wilderness Society."[6] The coevolutionary perspective celebrates the interdependencies within a diverse social movement no less than those within a diverse society and a diverse ecosystem.

Ecology is the science of the relations of interdependence that develop among organisms. One of its principles is that the strength and beauty of an ecosystem lie chiefly in its complex diversity, that is, in its multilayered relations of interdependence, including relations of competition and conflict. Fourth-wave environmentalists have effectively internalized this ecological understanding. They have applied it to the environmental movement itself. Differences of orientation and approach, and the competition and conflicts these differences give rise to, are increasingly approached from a holistic perspective that testifies to their ecological merit.

Donald Snow has observed a "hot competition among the empire builders of the national [environmental] movement."[7] Anyone subjected to the onslaught of direct-mail or telephone solicitations from these groups will agree that this heated competition exists. As Michael Mc-Closkey of the Sierra Club has observed, almost all environmental organizations face a "competitor trying to occupy the same market niche and competing for visibility, leadership, membership, and funds."[8] Yet the competition for support, real as it is, is not perceived by movement leaders and members as a zero-sum game wherein one organization's gain is always another's loss.

The gain of a member or financial supporter by one group does not necessarily mean the loss of a member or financial supporter by another group. Most "checkbook" environmentalists as well as activist supporters belong to more than one group at a time. They often choose to join multiple groups based on their desire for varied sources of information and their endorsement of varied tactics and strategies. A staff member of the Cousteau Society explained that he belongs to groups that "represent a cross-section of the environmental movement: [The] Cousteau [Society] because of their emphasis on education; Greenpeace because I think a certain amount of [direct] activism is needed; Sierra [Club] because they're so big and get involved in lobbying and litigation." His multiple memberships were grounded in an appreciation of the specific virtues and strengths of each organization. Membership in one group often stimulates the joining of other groups as increased knowledge fosters broader concerns. Indeed, that is the theory behind the lucrative business engaged in by most large environmental organizations of selling direct-mail address lists to each other.

The large national organizations compete for members and financial support. As McCloskey has explained, however, "[t]he competition between mainstream environmental groups has always been muted and subtle. It does not get in the way of practical coordination on issues, nor cooperation on such matters as employee salary surveys and dealing with regulatory agencies."[9] The Group of Ten, for example, formed in the early 1980s to coordinate their activities and agendas. In an effort to reduce exclusivity, it changed its name to the Green Group in the late 1980s and broadened its membership by reaching out to the grassroots and radicals. The Green Group now meets quarterly with about thirty organizations involved as members or guests, including more radical groups such as Greenpeace.

Movement organizations frequently coordinate their efforts on an ad hoc basis as well. Congressional efforts to undermine environmental protection during the early 1990s, for instance, facilitated the creation of a united front of environmental groups. Fifteen of the largest nationals joined together in 1994 to coordinate a response to the anti-environmentalist lobbying and legislation of the Wise Use groups and the newly elected Republicans. These groups asked their own members to support "competing" environmental organizations. Jane Perkins wrote to her membership that "it breaks all the fundraising rules for me to suggest that you not only send a contribution to *Friends of the Earth*, but to the other environmental groups you support as well. . . . But, this is no time for business as usual. . . . because we're all in this together."[10] Such formal, public alliances are usually issue specific and do not endure. Nonetheless, they demonstrate the potential within the movement for concerted action.

When coalitions do form, and victories are won as a result, each environmental group typically lays claim to the leadership and initiative that proved key to success. This is a relatively benign form of rivalry. As McCloskey observes, such "healthy competition" keeps the organizations on their toes without stifling beneficial coordination. The problem, he has suggested, lies not in the friendly rivalry between nationals, but "in the absence of healthy interaction between the more radical groups and the mainstream groups."[11]

A Sierra Club lobbyist found the most unsatisfying aspect of her work to be negotiating with "people who consider themselves pure environmentalists who think that the whole system needs to be immediately and dramatically changed, and who consider people like me not pure enough." Likewise, Andy Stahl of NWF takes issue with Earth First!'s efforts to protect old-growth forests through direct action. "They're incredibly sincere," Stahl said of the radicals, "but they made all of us who were trying to protect old growth seem marginalized, not a cross section of middle America."[12] Stahl worries that it might take decades to rehabilitate the reputation of environmentalists owing to the radicals' resort to ecological sabotage or ecotage, also known as monkeywrenching. Monkeywrenchers have engaged in activities such as disabling earthmoving and timbering equipment, spiking trees to prevent their harvesting, cutting down billboards, and pulling up survey stakes. Meanwhile, Earth First!ers deride the mainstream national groups at "Green Central in DC." They maintain that "[a]ll positive change has come and will only come through the efforts of impoverished, dedicated bands of idealists. That's what most environmental groups started out as—it's what EF! always has been and must remain. . . . Compromise has never worked. It never will."[13] The conflict between radicals and reformists remains, in many instances, sharp and bitter.

Despite marked differences in philosophical perspective and tactical approach, radicals and reformists are beginning to appreciate their differences, even their protracted conflicts, as indications of ecological strength. "This is an area where you let 1000 flowers bloom. Even when you are at odds with their views, everybody makes a contribution," Rodger Schlickeisen of Defenders of Wildlife observed about the conflicts between radical and mainstream environmentalists.[14] For the most part, reformers today appreciate the efforts of radicals. David Brower of Earth Island Institute has praised Earth First! for giving CPR to a "drowsy" environmental movement.[15] Brent Blackwelder of Friends of the Earth likewise acknowledges the beneficial role of grassroots critics, admitting that "[t]he nationals have taken lots of compromising positions when we should have been leading and setting the way."[16]

Politicians tend to adopt the path of least resistance. For this reason, environmental reformers can often succeed best when more extreme elements within the movement channel support to the center. Asked about the controversial tactics and demands of the radicals, Robert Hattoy, Sierra Club's Southern California representative, responded: "Frankly, it makes us look moderate. When Earth First! is out there demanding a hundred million acres of wilderness and we know we can only get ten million, I can turn to a congressman and say, 'Look, we're the voice of reason.' "[17] Expressing a similar point of view, a prominent member of the EDF board admitted that he supports Earth First! with larger donations than he gives to his own organization. There are distinct advantages, he suggested, to fighting a war on two fronts.

For this reason, radicals and reformers of the fourth wave are significantly increasing their "healthy interactions." Mary Hanley, vice president of communications for the Wilderness Society, observed that "[r]adical groups serve a purpose. We don't approve of violence. But these groups often highlight important issues, even if they can't follow up and solve them. We understand the dynamic when they have to keep an arm's length from us. We have dialogues off the record. There's room for all of us. There has to be a spectrum. But laws are still made in [Washington] D.C. and someone has to be here. We should not be ashamed of being lobbyists for the environment."[18] From the perspective of the reformist national organizations, radical environmentalists do the "dirty work" that needs to be done. Even when this work becomes too hot for the mainstream groups to handle, they remain appreciative of those who get burned on their behalf.

The radicals, for their part, are well aware of the indirect support they lend to the mainstream organizations. They understand their job to be threefold. First, they attempt to contribute directly to environmental protection. They also aim, through their demands and actions, to "keep established groups honest." Finally, they provide a "productive fringe" to the movement as a whole. As a productive fringe, radical groups effectively take the "political heat" so that mainstreamers can do their jobs more effectively. Demonstrating the reasonableness of the mainstream by way of their own militancy, radicals acknowledge, is "a major accomplishment."[19] Howie Wolke, co-founder of Earth First!, considers his organization the "sacrificial lamb" of the environmental movement.

Earth First! makes the demands of mainstream groups seem more acceptable "by taking positions that most people would consider ridiculous."[20] Dave Foreman explained that "[b]ack at the beginning of the Reagan administration, the Sierra Club was being called a bunch of environmental extremists. Well, we in Earth First! put an end to all that. . . . [We] tried to create some space on the far end of the spectrum for a radical environmentalist perspective. And, as a result of our staking out the position of unapologetic, uncompromising wilderness lovers with a bent for monkeywrenching and direct action, I think we have allowed the Sierra Club and other groups to actually take stronger positions than they would have before and yet appear to be more moderate than ever. What's different now is that they are compared to us."[21] Shifting the eco-ideological spectrum entails conflict, competition, and critical exchanges between radicals and mainstream reformers.

Radicals often acknowledge the merit of the more conservative methods of environmental protection. Earth First!ers Gary McFarlane and Darryl Echt have observed that "[t]here has been no movement, struggle or revolution throughout history (that we know of) that has succeeded via a single strategy or tactic. . . . We are in no position to limit our options—we simply must diversify our strategy, our tactics and our movement by empowering others to carry on the fight with whatever skills they possess."[22] Dave Foreman likewise has observed that ecological sabotage is not the sole means of achieving sought-after ends. "It's one tool," he observes. "Sometimes you lobby; sometimes you write letters; sometimes you file lawsuits. And sometimes you monkey-wrench."[23] Those who write letters, lobby, and litigate typically belong to different groups than those who monkeywrench. The important point, from Foreman's perspective, is that all the needed tasks are getting done. Moreover, lobbying, litigating, and making political compromises are made easier if strong positions defended on the ground by radical groups provide a point of reference. Foreman himself joined the Sierra Club board of directors from 1995 to 1997, maintaining that his effort to protect wilderness was at that time better served within a large national group. His aim was to balance the mainstream group's focus on Washington politics with the "old organizing skills" of grassroots mobilization.[24] In his role as a Sierran, Foreman explicitly rejected the "idealistic notion of no compromise." In this vein, he argued unsuccessfully against the club's decision to advocate a total ban of commercial logging on public lands. A zero-cut policy, he feared, would allow Congressional conservatives to "label us as extremists" when the role of extremist would be better played by other groups.[25] The strategy is consistent, given Foreman's new niche in the political ecosystem. Not surprisingly, certain former comrades in Earth First! interpreted the change of tactics as a betrayal of values.[26]

There are those who maintain that the "eclectic" reformist voices of mainstream environmentalism are being "drowned out" by radicals and that "extreme positions usually provoke fervent opposition among non-believers, not partial, lukewarm conversion."[27] Extremism does on occasion alienate the general public. This alienation generally becomes sustained and widespread, however, only if radical ideological positions give way to violent practices. Civil disobedience is embraced by most activists as a legitimate and even dutiful response to certain forms of environmental destruction. Violence is rejected. Ecotage is viewed by many as a form of civil disobedience.[28] That may be too generous a claim. As advocated by Earth First!, however, ecotage is neither violent nor terroristic. It is limited to the destruction of inanimate property that is itself employed in the destruction of nature. The destruction of human life is explicitly repudiated, and the destruction of innocent human life (terrorism properly speaking) is not even contemplated. Clearly, environmental sabotage is a dangerous business. Like any other illegal activity designed, in part, to win public support, it can easily backfire. All forms of extremism risk alienating the public.

Whatever public alienation radical environmentalists suffer, they have not drowned out the voices of moderation. While radical groups have grown in numbers over the years, even the largest of them remain dwarfed by the smallest of the mainstream nationals. (Earth First! and the Sea Shepherds claim between 10,000 and 15,000 members each. That is about one tenth of the membership held by the smallest of the mainstream nationals.) Nor is this imbalance changing over time. Radical environmentalism, like most radical social endeavors, remains a "fringe" movement that will achieve its greatest productivity as a fringe: by marginally shifting the ideological spectrum and by being incorporated, piecemeal and in compromised form, in the agendas of more moderate actors and organizations.

One-sidedness is generally viewed as a fault in individuals. It need not be a vice in organizations, at least not in those organizations that operate within a larger, more encompassing system. Here one-sidedness often represents little more than an efficient, and politically necessary, division of labor. John Sawhill of TNC observed that his organization is frequently criticized for its "nonconfrontational approach to politics."

The Nature Conservancy does not make political endorsements. Sawhill explained that "[t]he Conservancy does take sides. We are fiercely partisan about our biodiversity protection mission—and about nothing else."[29] By remaining politically nonpartisan, TNC receives financial support to buy and manage ecologically valuable tracts of land from a broad ideological constituency. When members inquire why TNC does not take stands on issues such as overpopulation, the answer is predictable: "Different organizations fill different needs—all very important—under the environmental umbrella. We defer to organizations with the expertise and means to handle other key issues. . . . We need diversity in our political styles and conservation missions just as we need diversity in nature."[30] An ecological overview frames the response.

Were every environmental organization to take a restricted, nonpartisan approach, the environmental cause would be poorly served. Yet TNC does well from its nonpartisan and narrowly focused position—in no small part because grassroots groups and other large organizations, such as the Sierra Club and Greenpeace, are doing the openly political work that *they* do best and because radical groups such as Earth First! have helped to make TNC's brand of environmental advocacy uncontroversial, an accepted part of the mainstream. Some observers have claimed that the "schism" between the national "corporate-minded" environmental organizations and the grassroots groups is widening. In fact, a formerly wide breach is systematically being bridged on many fronts. Even those with a less sanguine prognosis admit that organizational differences among environmental groups generally prove "mutually reinforcing." In turn, the differences that exist create a movement that "offers something for practically everyone."[31] Diversity fosters resilience.

A healthy ecology is not composed of identical organisms engaged in identical life processes. Individual organisms and species do what they do best, and the whole gains in strength and beauty from a range of cooperative, competitive, and conflictual interaction. With this in mind, Sierra Club executive director Carl Pope has insisted that environmentalists entertain if not exploit all potential alliances. He has observed that the publications of today's sporting community, such as *BASS Times, Sports Afield*, and *Field and Stream*, "read like environmental-activist newsletters, alerting readers to the assault on wetlands, water quality, and wildlife habitat." Anticipating sharp criticism by animal rights activists within the club for its alliances with sportsmen's groups, Pope reminds Sierrans that "[a]lliances, like marriages, require work on both sides. In order to build the broadest possible coalition for wilderness and wildlife, both sides will have to agree to disagree on some issues."[32] Animal rights groups fiercely object to the slaughter of wild animals. Efforts to protect the availability of game for hunters, however, typically enhances overall biodiversity. In its efforts to provide duck hunters with an abundance of prey, for instance, Ducks Unlimited has contributed $900 million since its founding in 1937 to the conservation of over seven million acres of wildlife habitat in North America. The efforts of Ducks Unlimited's 500,000 members to preserve and restore wetlands benefits not only waterfowl, but about 600 other species.[33] The interdependent relationships that compose the web of life mean that efforts to protect any particular strand usually strengthen the whole.

When environmentalists and sportsmen join forces, alliance advocates observe, they represent an "irresistible coalition" comprising an estimated 60 to 70 percent of the population as a whole.[34] Such power in numbers is not to be discounted, even though the dilution of principles that accompanies coalition building remains a danger. What facilitates bridge building among diverse environmental organizations in the face of inevitable conflicts is an ecological sensibility. Interdependence grounded in diversity is intrinsic to healthy social and natural communities. For fourth-wave environmentalists, the cost of maintaining any particular alliance may be too high, but an openness to alliance building is indispensable.

In their overview of the environmental movement since 1970, Dunlap and Mertig wrote: "Compared with 20 years ago, the movement has a much more diverse organizational base (from local to national and international), draws hard-core activists and volunteers from a wider range of social strata (especially more from the working class, minorities, and women), and addresses a far wider range of problems (from local health hazards to wildlife habitat to global ecosystem protection) with a greater range of tactics (from lawsuits to green consumerism to tree-spiking)." The authors concluded that "increased diversity may lead to greater resiliency in social movements. Nevertheless, we cannot ignore the potentially fragmenting effect of such diversity."[35] The point is well taken. The environmental movement is more diverse today than ever. The threat of a debilitating fragmentation is real. At the same time, the proliferation of issues is matched by a growing appreciation of their interdependence.

Involvement in one area of environmental advocacy often stimulates interest in other areas.[36] On the one hand, this issue linkage may cause a diffusion of energies. Certainly it does not produce the most efficient division of labor. That is a significant cost. On the other hand, the perception of issue interdependence promotes the confrontation of problems from novel perspectives and broadens the base of support for related concerns. That is an overwhelming benefit.

With an eye to issue interdependence, most large environmental organizations significantly expanded their agendas and refined their strategies in the late 1980s and early 1990s.[37] Even organizations once known for their single-cause explanations of environmental degradation have moved beyond these positions. Zero Population Growth, for instance, originally focused on the use of birth control technology as the singular means to address its central concern with overpopulation. The organization now proposes a much broader and more nuanced approach. Its Campaign for a Quality Future, launched in late 1994, "addresses a sweep of interrelated issues—population growth, poverty, environmental degradation, the low status of women, and wasteful use of resources. But it also recognizes that there is no magic silver bullet, that *all* these issues need to be addressed to reduce population pressures and ensure a high quality of life for future generations."[38] Likewise, the National Audubon Society is acutely aware that the decline of bird species is mostly a product of habitat lost to population growth. Consequently, it has established over half a dozen field offices to work on population issues. Environmental justice organizations are particularly aware of environmental issue linkage. Oppressed people usually do not have compartmentalized problems. Consequently, the environmental organizations that address these problems typically endorse holistic solutions that blend social, economic, political, and ecological strategies.[39]

Across the board, fourth-wave environmentalists are more sensitive to issue interdependence. Resource conservationists are aware of the need for wilderness protection, preservationists acknowledge the importance of human health and welfare, local pollution fighters understand that safeguarding human health depends on worldwide ecological care, and globalists admit that planetary problems often stem from local dysfunctions that must be combated one village, region, or ecosystem at a time.

John Nichols, author of *The Milagro Beanfield War*, observed that the failure to understand the interdependence of environmental issues leads to an impoverished movement. He wrote:

> The tragedy is "environmental action" is often seen only as an effort
> to save the spotted owl . . . without giving a hoot about the ghettos

of Houston, Cleveland, North Philadelphia and the situation of the *people* on this globe, which is atrocious. Until environment is seen as the entire picture, both natural and human, we're going to have a real problem. . . . If you're a member of the Sierra Club you should also be against death squads in El Salvador, if you're a member of Audubon Society you should also have been out there trying to stop the Gulf War; if you're a member of National Wildlife Federation you should also be worried about the total destruction of our inner cities and the fact that unemployment among black children between 15 and 24 is 80 to 90 percent. Otherwise, it becomes irrelevant to worry about sea turtles. The clue to any kind of survival of the planet is training people to the macroscopic overview, to understand how their lives are connected to everything else, to understand that if you kill one species you're endangering all others.[40]

To affirm issue interdependence is not to say that every organization must actively oppose death squads or expend its resources saving spotted owls. An efficient division of labor between organizations within an increasingly integrated movement is often preferable. As is the case with ideological and tactical differences, there is no need for homogeneity of ends or means. Complementarity will do just fine. Again, nature provides that model: diversity is a strength.

To say that the concerns of environmentalists have recently converged, then, is not to say that the number or variability of environmental concerns has decreased. Today, more environmental groups with more varied agendas are fighting more battles in more arenas than ever before. The interdependence of these battles, however, is increasingly appreciated. Coalitions and alliances, tacit or explicit, are more common. It is unlikely that environmentalists will ever achieve a "consensus" about their values or strategies.[41] Yet individuals and organizations are now conceptually and concretely integrating formerly disparate concerns of resource conservation, nature preservation, economic development, and social justice. The result, for the most part, is a holistic orientation and practice.

Fourth-wave environmentalists do not all have the same goals or endorse the same strategies for achieving their goals. Yet they agree that viable solutions to the multiplicity of problems they face will have to be grounded in ecological processes of interdependent change. For this reason, it is best to describe fourth-wave environmentalism not as the singular endpoint of a long and fractious history of environmental struggle but as the multiple starting points of a converging, coevolutionary project. Environmentalists are well aware of the differences among them, differences that often give rise to conflict. Yet an overarching ecological perspective, grounded in the affirmation of coevolutionary interdependence, allows them both to maintain their particular approaches and agendas and to appreciate the benefits of diversity within their ranks.

Stability in Change

The coevolutionary perspective is grounded in the belief that, viewed globally and in the long term, the protection of human welfare and the preservation of the natural environment are mutually reinforcing. Coevolutionary biology emerged as a field of study in the 1960s. In its original context, coevolution describes the processes whereby the evolutionary paths of two or more species that maintain a close ecological relationship largely depend on the patterns of their interactions.[42] Coevolution thus pertains to the adaptational changes in relationships that develop among interdependent organisms. In time, coevolutionary perspectives developed in economics and other social disciplines. The domain of these coevolutionary studies was generally the complex, mutually adaptive changes in relationships that occur between human beings and the natural environment. While only a handful of scholars explicitly employ the terminology of coevolution, the coevolutionary perspective is implicitly invoked and concretely manifested in the words and actions of growing numbers of environmentalists today. Environmental organizations increasingly describe their work in terms of a dynamic, complex, interdependent world. Their celebration of this interdependence bespeaks a holistic socioecological perspective that affirms the integrated webs of natural and social life.

Fritjof Capra has observed the emergence of a new "social paradigm" grounded in the recognition of a dynamic connectedness that spans diverse fields of inquiry and action. This paradigm is defined as "a constellation of concepts, values, perceptions, and practices shared by a community, which form a particular vision of reality that is the basis of the way the community organizes itself." The paradigm is grounded in a vision of "the fundamental interdependence of all phenomena and the embeddedness of individuals and societies in the cyclical processes of nature."[43] Capra wrote that "[w]e live today in a globally interconnected world, in which biological, psychological, social, and environmental phenomenon are all interdependent. To describe this world appropriately we need an ecological perspective."[44] For Capra,

> [a]ll members of an ecological community are interconnected in a
> vast and intricate network of relationships, the web of life. . . . In-
> terdependence—the mutual dependence of all life processes on one
> another—is the nature of all ecological relationships. . . . Under-
> standing ecological interdependence means understanding relation-
> ships. It requires shifts of perception that are characteristic of sys-
> tems thinking—from the parts to the whole, from objects to
> relationships, from contents to patterns. A sustainable human com-
> munity is aware of the multiple relationships among its members.
> Nourishing the community means nourishing those relationships.[45]

Capra explicitly characterizes the dynamic nature of these interdependent relationships in nature and society in terms of processes of coevolution.[46] Here Capra adopts the worldview developed by Gregory
Bateson. Bateson argued for an ecological perspective that focuses not
on isolated, static individuals but on the "organism-in-its-environment"
and the dynamic, interdependent relations composing this environment.[47]

Nature never stands still. Humans, likewise, are enterprising, progressive creatures. The relationships between humankind and nature are
constantly changing, therefore, because the constituents of these relationships are constantly changing. The means of protecting and enhanc-
ing these relationships, it follows, must also be amenable to change. The
coevolutionary challenge issued by fourth-wave environmentalists calls
for the safeguarding of nature's and society's capacities for adaptive
change.

In their affirmation of adaptive change, fourth-wave environmentalists employ the natural world as a model. Natural systems are always
in some state of flux. Efforts to preserve a natural system in a static
state—given our ignorance of its complex interrelationships and our inevitably abbreviated time frames—generally result in more disruption
than preservation. Hence, the notion of an unchanging "balance of nature" has been replaced with a focus on periodic disturbance. Nature is
not a homeostatic system resting in equilibrium, at least not on a spa-
tiotemporal scale relevant to the preservation of particular species or
habitats. Ecosystems are "inherently erratic, infinitely malleable creatures" whose characteristics shift with the tides of time. Incessant small-
scale change and evolutionary adaptation are what allow for large-scale
stability in nature's laboratory. This is known as resilience.[48]

The dynamic nature of earthly life is by no means wholly a product
of evolutionary adaptation. Ecosystems are dynamic entities in them-

selves, beyond whatever evolutionary adaptations their constituent species undergo. Indeed, drastic ecosystemic change often stimulates evolutionary adaptation. Certain forests, for instance, depend on periodic fires to rejuvenate. The cones of Florida's longleaf pines will open only after a good charring. In this way, they disperse their seeds after fire has swept their habitat and cleared out undergrowth that would hinder the pines' germination and growth. Likewise, Douglas fir will not regenerate in its own shade. Periodic burns are required to maintain Douglas fir forests. Similarly, in the absence of light ground fires, oaks forests will be overtaken by invading maples.

Such ecological paradigms provide environmentalists with the means of understanding social dynamics. Disturbance is considered to be "healthy" in both natural and sociopolitical systems. "Elections in democratic governments and revolutions in totalitarian systems demonstrate the importance and impact of disturbance in different time frames," a Cousteau Society publication suggests. "In the United States, we have a controlled governmental 'forest fire'—called an election—every four years that serves to revitalize all or part of the system. As in a forest, the turnover eliminates dead wood, maintains diversity and restores vigor. Suppressing such disturbance means that a change, when it does come, will be greater and more drastic—as in the former Soviet Union. A healthy government seems to require an appropriate level of disturbance."[49] In society no less than in the natural world, change, even disturbing change, is intrinsic to long-term health.

Efforts to preserve wildlife in the absence of a larger vision of ecosystemic dynamics have significant limitations. Environmentalists' formerly narrow focus on preserving individual species has been largely abandoned. Safeguarding the "interrelationships among species" is now the key ecological strategy.[50] Early attempts to preserve small parcels of wilderness in a static state were often counterproductive because they did not account for the natural fluctuations of ecosystems. If one is going to "build an ark to last," habitat managers discovered, "one had better build it big enough to roll with the storms."[51] Many of the smaller nature reserves were not up to the poundings of nature's storms or the incursions of humankind. Protecting nature, fourth-wave environmentalists learned, entails protecting nature's capacity to adapt. The environmental mandate to preserve biodiversity, therefore, has nothing to do with maintaining nature as a museum of species. Preserving biodiversity entails safeguarding "the variety of life and its processes." This includes "the variety of living organisms, the genetic differences among

them, the communities and ecosytems in which they occur, and the ecological and evolutionary processes that keep them functioning, yet ever
changing and adapting."[52] As Dave Foreman has observed, "species
can't be brought back from the brink of extinction one by one." Rather,
the dynamics of entire ecosystems must be protected. The task is to
maintain "the flow and dance of evolution."[53] The goal, in other words,
is not to eliminate change in nature. The goal is to facilitate (or at least
not impede) nature's resilience in the face of change.

Ultimately, protecting nature's resilience in the face of change entails
benignly integrating humankind into natural systems. T. H. Watkins
shares this orientation with members of the Wilderness Society:

> The idea that the ecological systems represented by wilderness are
> static natural enclaves that have remained pretty much the way they
> are for whole geological epochs has long since ceased to be part of
> the preservation argument. It may still linger among a few clots of
> ignorant New Age enthusiasts or particularly uninformed members
> of the "X" generation, but the science that underlies most of the
> wilderness preservation movement today has for decades recognized
> the fact that natural systems are anything but static. . . . Just like
> natural systems, whose interdependent parts function in a dynamic
> of change, the arguments for wilderness preservation also must
> evolve or die. It is no longer enough to identify a landscape, draw
> a line around it, add it to the National Wilderness Preservation Sys-
> tem, then rest on our laurels, satisfied that we have just saved one
> more piece of the natural world forever. The brutal fact is that wil-
> derness areas so conceived—even if we monitor their use and man-
> agement diligently—cannot in the long run survive the pressures of
> the world all around them. . . . Somehow, we must learn to dem-
> onstrate that wilderness areas can only be truly preserved if they are
> also made the core of larger bioregional systems of protected land
> and human communities whose overall function is the sustainability
> of all life contained within them—including human life.[54]

With this in mind, fourth-wave environmentalists embrace the coevolutionary development of society and nature.

John Muir once said that "[w]hen we try to pick out anything by
itself, we find it hitched to everything else in the universe."[55] Only quite
recently has the sensibility behind this assertion permeated the environmental movement and redirected its efforts. In large part, a changing
world has dictated this shift in perspective and practice. Environmentalists today are aware of the futility of efforts to preserve isolated bits

of scenic land in the absence of broader efforts to ensure the economic viability of local inhabitants, curb air and water pollution in the region, and stem the destruction of wildlife habitat beyond park borders. As a Sierra Club writer has observed, "not even Muir—apostle of interconnectedness of all living things—could foresee the dangers looming beyond these designated boundaries."[56] Fourth-wave environmentalists perceive social interdependence as part of the multiple systems of interdependence that define the biosphere. To ensure a rich diversity of human and nonhuman life on the planet, nature and society must coevolve.

Gregg Easterbrook charges environmentalists with harboring a "horror of change" that places them quite out of touch with a truly ecological perspective.[57] In fact, Easterbrook himself proves quite out of touch with contemporary environmental thought. Fourth-wave environmentalism is distinctly receptive to change. "Change," activists repeatedly affirm, "is the only thing that's guaranteed." Death is part of life, extinction is the prerequisite for the evolution of new species, and the only certainty for human beings, as for all species, is that nothing stays the same over the long term. Environmentalists ground their work on the understanding that the inhabitants of the biosphere, including the human species, are slated for interdependent, coevolutionary transformation. As one environmental advocate stated, "we are evolving, and our responsibility to the planet is evolving as we change."

Observing that the World Wildlife Fund's traditional goal of securing "a healthier, more biologically abundant planet for future generations" remains intact, president Kathryn Fuller insists that her organization, precisely for that reason, cannot remain static. The tools and approaches of WWF will necessarily transform with time, and the organization itself "must remain ever adaptable, ever ready to change."[58] Likewise, TNC president John Sawhill affirms:

> The Conservancy learns and evolves, constantly adapts to changing conditions and new knowledge. This is as it should be, because in nature, the only real long-term constant is change. . . . Certainly, no rigid or static conservation approach will succeed over the long term. But an approach that is predicated on change—one that adapts and responds to shifting conditions—can succeed on significant time scales. . . . Effective conservation must be as dynamic as the natural areas and processes we seek to save. Little wonder, then, that the Conservancy places such a premium on innovation, creativity and enterprise. The Greek philosopher Heraclitus said, "You

never step in the same stream twice." I like to think of The Nature Conservancy as that stream: ever-moving, ever-learning, ever-adapting. And as a result, enduring—even in the long term.[59]

Developing a similar orientation, NWF's 1994 annual report states that "[e]volution, the adaptation to change, is the natural course of life. We . . . believe firmly that we must welcome the change, transformation and evolution ahead."[60] For all their talk of conservation, then, what environmentalists seek to conserve is not a static balance, but ever-transforming, interdependent social and ecological relationships. The goal is not to construct and manage a planetary museum, but to allow the biosphere various pathways to adaptive change. Ongoing human development is an important part of this process.

The problem, of course, is that much human (that is, industrial, commercial, and technological) "development" occurs at the expense of other species. Today human beings directly appropriate for their use or otherwise redirect or destroy up to 40 percent of the solar energy that is biologically fixed on the earth's landmasses.[61] That share may double to 80 percent within the next three decades, given current trends.[62] Humanity's growth in numbers and its increasing consumption of resources translate into the widespread expropriation, fragmentation, and spoliation of natural habitats. Human beings are driving countless species to extinction at a rate that rivals that caused by global geological disruptions of the past. Humans are also hindering, and in some cases terminating, the very processes of evolution. A precondition of biological diversity is large, intact, undeveloped habitat. That is also the precondition for the evolution of new species (speciation). With insufficient biological diversity and habitat to exploit, species are unable to adapt to environmental flux. Rather than slowly evolving, they simply perish. The increasing loss of biodiversity and, more significantly, the rapid diminishment of habitat mean that the opportunity for the evolution of new species is rapidly decreasing.

Highly adaptable species, such as certain insects, can reproduce and mutate quickly and subsist in very small and disrupted ecosystems. These species will likely continue to produce new (sub)species despite any human malfeasance. The same cannot be said for more complex and ecologically demanding forms of life. Regarding tropic populations, conservation biologist Michael Soulé wrote that "[f]or the first time in hundreds of millions of years significant evolutionary change in most higher organisms is coming to a screeching halt. . . . Even the largest

nature reserves are probably too small to guarantee the permanent survival of large herbivores and carnivores. This century will see the end of
significant evolution of . . . terrestrial vertebrates."[63] Environmentalists
today are acutely aware of the threat posed by humans to the processes
of evolutionary change, processes that have defined earthly life from the
start.

The goal of restoring the integrity of ecological systems, David
Brower observed, "is not an effort to stop the clock, but rather a chance
to keep the clock running." It is, Brower has insisted, "our best chance"
to safeguard evolutionary development.[64] Early conservationists often
assumed the role of custodians of the museum of nature. In many respects, they wanted to stop the clock. Fourth-wave environmentalists, in
contrast, increasingly take on the task of keeping the evolutionary clock
running. As the stated goal of the Wildlands Project demonstrates, the
intent is "to protect and restore evolutionary processes and biodiversity."[65] Fourth-wave environmentalists want to preserve the conditions
that facilitate adaptive change. They realize that safeguarding the processes of evolution is the best means of preserving biological diversity.

Environmentalists' affirmation of adaptive change would contradict
their struggle for sustainability if this struggle were understood as an
effort to maintain the biosphere in a static state. But that is far from
what sustainability means. As Richard Norgaard wrote,

> [s]ustainability does not imply that everything stays the same. It
> implies that the overall level of diversity and overall productivity of
> components and relations in systems are maintained or enhanced.
> . . . The shift towards sustainable development entails adopting pol-
> icies and strategies that sequentially reduce the likelihood that es-
> pecially valuable traits will disappear prematurely. It also entails the
> fostering of diversity *per se.* . . . This definition of sustainable de-
> velopment applies to belief systems, environmental systems, orga-
> nizational systems and knowledge systems equally well. And nec-
> essarily so, for the sustainability of components and relations in
> each subsystem depend on the interactions between them. . . . Sus-
> tainable development will entail a return to the coevolutionary de-
> velopment process with the diversity that remains and the deliberate
> fostering of further diversity to permit adaptation to future sur-
> prises.[6]

In the face of inevitable disturbances, diversity maximizes the chances
of adaptational success. For this reason, diversity has been called "nature's insurance policy against catastrophes."[67] What is being sustained

in sustainable development, therefore, are the social and ecological elements of diverse, interdependent life systems. These diverse elements and relationships facilitate successful evolutionary adaptation.

Because the interdependence that environmentalists embrace is dynamic, the ethic they aspire to is responsive to change. Aldo Leopold observed that ecological ethics are never really "written," and certainly never written in stone. They are "tentative" products of "social evolution." They are tentative, Leopold has stated, "because evolution never stops."[68] It follows that the convergence demonstrated by fourth-wave environmentalism does not signify any sort of permanent consolidation of interests or sensibilities. The environmental movement itself will certainly continue to change over time. The fifth wave will bear its own distinguishing features and produce its own distinct fruit. Fourth-wave environmentalism, therefore, is not a final state of affairs. Nonetheless, I believe that the key features of fourth wave convergence—the affirmation of interdependence, the struggle for sustainable development, and an overarching coevolutionary framework—will define the environmental movement well into the twenty-first century.

Chattanooga Councilman David Crockett, who has helped his city move from having the second worst air quality in the country to acquiring flagship status within the "sustainable cities" movement, stated:

> If you look at the industrial age in this country, and all the implications it had, it changed everything, from how we worked to family structure. Then along came cars, and they changed politics, agriculture, the environment, how cities grew. Now we have the age of computers, and you can see that it's changing things again. We believe that the next big shift will be the age of sustainability. You don't have to be a member of some environmental group—you don't even have to be real good at math—to see that things have to change. We can't just go on consuming resources and dumping waste into holes. We don't have enough resources; we don't have enough holes in the ground. We have started to recognize that we're at the doorstep of this great change and that we should try to think of things in a different way. In Chattanooga we believe it's to our advantage to be part of that.[68]

Fourth-wave environmentalists argue that it is to everyone's advantage to be part of this transformation. They have learned, and internalized, the key lessons of ecology. They are embracing the coevolutionary change inherent to the interdependent relations of the web of life.

The coevolution of humans and other species actually goes back tens of thousands, if not hundreds of thousands of years. The indigenous peoples of various parts of North America may be considered the "keystone species" of their ecosystems, for example, because they noticeably altered the landscape and its ecosystemic relationships through predation and fire. Today, however, the human impact on nature is quite overpowering. Environmentalists insist that this impact must be decreased. At the same time, they recognize that nature occasionally must be managed or restored if it is to be saved. The integrity of many wilderness preserves, for instance, cannot be maintained without human intervention. Limiting humanity's impact on the natural world and interacting with nature in more benign ways describes the coevolutionary challenge. In this sense, coevolution might cease to be a forgotten feature of our distant past and become a conscious goal for our future.

Coevolution, Growth and Sustainable Development

The challenge of coevolution links the improvement of human welfare to the safeguarding of nature's integrity. For this reason, it is primarily played out in the struggle for sustainable development.[70] The term *sustainable development* originated in the early 1980s. It was popularized in *Our Common Future*, the report of the United Nations' sponsored 1987 World Commission on Environment and Development. Chaired by Gro Harlem Brundtland, former Prime Minister of Norway, the Commission's publication became known as the Brundtland Report. Here sustainable development received its classic definition as "development that meets the needs of the present without compromising the ability of future generations to meet their own needs." The Brundtland Report went on to stipulate that "[t]he concept of sustainable development does imply limits—not absolute limits but limitations imposed by the present state of technology and social organization on environmental resources and by the ability of the biosphere to absorb the effects of human activities."[71] Sustainable development was not defined as a fixed state of affairs. It does not dictate an absence of change. Rather, it sponsors social and technological change within boundaries determined by the ecological resiliency of the earth as a whole. The 1992 Earth Summit in Rio de Janeiro endorsed sustainable development as the dominant environmental paradigm. Following the conference, the United Nations formed a Commission on Sustainable Development to implement the chief document derived from the Rio conference, *Agenda 21*.

The term *sustainable development* has been much misused and misapplied in the last decade, prompting certain environmentalists to label it a "thundering oxymoron," and a "gigantic exercise in self-deception."[72] The problem is that sustainable development has become a convenient phrase for rallying support in the absence of concrete efforts to implement change.

Critics rightfully argue that advocates of sustainable development often want to have their cake and eat it too. The President's Council on Sustainable Development, for instance, insists that sustainable development will yield both an increase in profits *and* an increase in wages, an increase in capital *and* a decrease in poverty, an increase in industrial productivity *and* a decrease in pollution and waste. Faced with these potentially incompatible goals, sustainable developers may shirk their ecological mandate in favor of economic growth. The environmentalists' worry, Donald Worster wrote, is that in the partnership of sustainable development, it "will be 'development' that makes most of the decisions, and 'sustainable' will come trotting along, smiling and genial, unable to assert any firm leadership, complaining only about the pace of travel."[73] Wolfgang Sachs likewise observed that sustainable development primarily "calls for the conservation of development, not for the conservation of nature."[74] The problem is that the developmental focus of sustainable development often serves as a guise for business as usual. As one critic has observed, sustainable development is "offered as the magic solution to the world's catastrophic ecological problems. Yet this resolution is shaped in such a way as to make environmental progress synonymous with the sustainable development of *capitalism*."[75] There is little reason to believe that social equity and ecological integrity will receive their due in an economic system driven by the unfettered pursuit of profit. What certain sustainable developers chiefly want to sustain, it appears, is unlimited economic growth.

In 1972, John Connolly, U.S. Secretary of the Treasury, said, "Never has growth been more important. You can never feed the poor or ease the lives of the wage-earning families, ameliorate the problems of race or solve the problems of pollution without real growth."[76] This view is still widely held.

The Brundtland Report deems "absolutely essential" a "new era of economic growth." While this growth is to be constrained by the principle of "intergenerational equity" and is supposed to heed the biospheric limits of the planet, these principles and parameters remain largely undefined. The Rio Declaration demonstrated a similar tendency in its unqualified declaration of nations' sovereign "right to development." Principle 12 of the Rio Declaration on Environment and Development specifically calls for continued economic growth. The limits that must be placed on this growth remain undisclosed.

The Clinton-Gore administration has consistently celebrated economic growth as an answer to almost every problem that ails America. President Clinton's 1993 Executive Order (No. 12852) setting up the President's Council on Sustainable Development explicitly defines sustainable development as a form of economic growth. The Council itself concluded that "[t]he issue is not whether the economy needs to grow but how and in what way."[77] William Ophuls and A. Stephen Boyan have observed that "Growth is the secular religion of American society, providing a social goal, a basis for political solidarity, and a source of individual motivation. . . . Especially in recent years, growth has become an all-purpose 'political solvent.' "[78] As T. H. Watkins suggested, the belief that "economic growth is always and limitlessly good" has become the "modern secular orthodoxy."[79]

Many environmentalists believe that economic growth and environmental protection generally go hand in hand. They hold that, other things being equal, economic growth improves everyone's standard of living and consequently enhances everyone's capacity to preserve the environment. The Group of Ten's 1985 *Environmental Agenda for the Future* maintains that "[c]ontinued economic growth is essential. Past environmental gains will be maintained and new ones made more easily in a healthy economy than in a stagnant one with continued high unemployment."[80] Poverty, it has been said by environmentalist proponents of growth, is the worst polluter and overpopulator. Decrease poverty through economic growth and the environment necessarily benefits.

By loudly celebrating the promise of economic growth while paying only lip service to environmental preservation, the language of sustainable development often gains political currency at a substantial ecological cost. Thus, while sustainable development offers "a broad, easy path where all kinds of folk can walk along together," environmentalists worry that many of its advocates are "going in the wrong direction."[81] As economist Denis Goulet observed at the World Bank's own 1995 conference on sustainable development, the drive for a globalized economy and unrestricted free trade, the high level of consumption in the developed countries, the widening gap between the world's rich and poor, and the growing power the world's economic elites all militate against sustainable development. Few if any of these trends show signs of reversal.[82]

The language of sustainable development has certainly been appropriated by the politicians, the business community, and to some extent the general public. Like democracy, sustainable development is widely endorsed and remains highly variable in its meanings and applications. Like democracy, sustainable development has its banners raised by many yet its demands satisfied by few. And like democracy, sustainable development is here to stay. For better or worse, the language of sustainable development has become the lingua franca of environmentalism. With this in mind, critics would do better struggling to enhance the term's ecological integrity and practical implementation than foregoing its use because of its potential for abuse.

Environmentalists bear the burden of reworking the meaning of development. Development is not synonymous with economic growth as it is most commonly understood and measured. Development refers to improvements in a wide array of the standards of human well-being. These standards may, but need not, include higher aggregate levels of material production and consumption. Factors such as longevity, access to education, and the protection of human rights are generally considered a part of social development yet are not automatically linked to a country's economic growth or increases in its per capita income.[83] This is especially true when income is very unevenly distributed within a society. By definition, sustainable development describes those improvements to human well-being that are ecologically sustainable. Economic growth may facilitate sustainable development. At present, it seldom does.

The distinction between development and growth grew out of efforts in the 1970s to promote a "steady-state" or "no-growth" economy. Herman Daly observed at this time that "[n]ot only is quality free to evolve [in a steady-state economy], but its development is positively encouraged in certain directions. If we use 'growth' to mean quantitative change, and 'development' to refer to qualitative change, then we may say that a steady-state economy develops but does not grow, just as the planet earth, of which the human economy is a subsystem, develops but does not grow."[84] Daly observed that human societies have always developed, but only in the last two centuries has economic growth been sufficiently rapid to be felt within the span of a single lifetime, and only in the last half-century has economic growth become an overarching sociopolitical priority.

For Daly, the near-universal pursuit of growth suggests a near-universal blindness to the present generation's theft of resources from future generations. It also masks an unwillingness to address present social and economic inequities. Daly wrote: "Our system is hooked on growth per se, and does *not* see growth as a temporary *means* of attaining some optimum level of stocks, but as an *end* in itself. Why? Perhaps because, as one prominent economist so bluntly put it in defending growth: Growth is a substitute for equality of income. So long as there is growth there is hope, and that makes large income differentials tolerable. We are addicted to growth because we are addicted to large inequalities in income and wealth. To paraphrase Marie Antoinette: Let them eat growth. Better yet, let the poor hope to eat growth in the future.' "[85] One suspects that many social activists concerned with the welfare of the disadvantaged advocate growth because of this promise. For this reason, environmentalists often find themselves isolated in their reluctance to join the popular celebration of economic growth.[86]

Sustainability is often best ensured by reducing consumption or restricting consumptive practices. In certain circumstances, however, sustainability is abetted by forms of economic development that increase the production of certain goods or services and raise the standard of living such that greater consumption of these goods or services is made possible. Zero-emission and low-emission automobiles are available, but their cost is prohibitive for most people. Well-constructed, well-insulated housing that conserves energy is also beyond the economic reach of many. As an activist with the Greens observed, "Poverty means you can't afford to live in an energy-efficient house." Likewise, kerosene lamps employed for lighting use fifty times more energy per watt than light bulbs. Yet the poor in developing countries often cannot afford electricity. Cooking with earthen pots over an open fire uses eight times as much energy as cooking with gas stoves and aluminum pans. Yet the latter remains unavailable to many. Solar cookers have been developed to cook food and pasteurize water, potentially saving vast amounts of fuelwood and preserving trees, which are the primary source of energy for cooking and heating for 80 percent of world's people. Yet many people cannot afford solar cookers. People in developing countries who lack access to adequate toilets or latrines, in turn, spend up to six times as much on medical bills as those in areas with better sanitation. In these and many other cases, the increased conservation of certain resources (e.g, clean air, fossil fuels, trees, and medical care,) would be made possible through economic and technological developments that increase the production and consumption of specific goods and services (e.g., energy efficient cars and homes and sanitation systems).

Economic and technological development is often a precondition of environmental preservation. That is why the struggle for sustainable development arises. Sustainable development broadens the concept of sustainability beyond the focus on individual responsibility. It encompasses concern for the socioeconomic structures that facilitate or hinder sustainable behavior within the population at large. It aims to foster the socioeconomic conditions that make sustainable practices, relations, or processes feasible to a wider, and ideally all-inclusive, population. Sustainability simply means that you don't eat your seedcorn. Sustainable development means that people are not forced to eat their seedcorn owing to poverty or powerlessness. Advocates of sustainable development argue that poverty and economic domination are environmentally destructive social relations in themselves and that the development of more equitable forms of economic production and exchange will facilitate greater environmental care.

There are good arguments to be made in favor of certain forms of economic growth. Fourth-wave environmentalists frequently make these arguments but refuse to give economic growth a blanket endorsement. Fourth-wave environmentalists reject the unqualified celebration of economic growth even when they find it necessary or useful to employ the term. When asked if and why he favors growth, David Brower replied, "If you want growth—I'll call it that because people love the word— we want growth based on [environmental] restoration rather than destruction. There are jobs in it. There's money to be made in it, and our children are going to like it."[87] Contemporary environmentalists define growth in terms of "quality" rather than "quantity" and identify good growth with environmental protection and restoration. As Bryan Norton wrote, "Growth through restoration, which evokes Leopold's idea that some forms of land use are harmonious with the contextual landscape, and protect and even enhance its 'integrity,' may well represent the banner behind which environmentalists rally. . . . Restorative growth is growth that protects and enhances the health of larger, contextual systems."[88] Growth is not a panacea. But certain kinds of growth, environmentalists argue, may be ecological boons.

To the extent that growth does not arise from an increase in the consumption of energy, raw materials, and habitat but from services and technologies that do not increasingly tax the environment, there is no ecological reason to condemn it. The production and use of "nonconsumptive goods" may grow indefinitely without exceeding the tolerance limits of the biosphere. Growth that serves to redistribute resources more equitably may also be environmentally benign. In turn, growth arising from increases in the value of goods may be ecologically beneficial if, say, more expensive biodegradables are substituted for less expensive materials that would create unmanageable waste streams. Finally, as Ophuls and Boyan have observed, sustainable social and economic life will never be realized without substantial intellectual, moral, scientific, and spiritual growth.[89] Just as the finite boundaries of the biosphere have allowed indefinite evolutionary development, so its limits do not preclude indefinite human development. Growth understood as qualitative development is not the problem. Unrestricted and unlimited economic or population growth, however, is socially and ecologically pathological.[90]

Rather than promoting increased per capita income and consumption, the standard indicators of growth, fourth-wave environmentalists insist on more specific criteria. Beneficial forms of growth must be distinguished from harmful ones. Rising gross national product or gross domestic product, the market value of all goods and services produced in a nation, is not an appropriate standard, for the absolute growth in the rate of use of raw materials and energy is patently unsustainable in a finite biosphere. Thus, environmentalists view their work in terms of preserving and enjoying things that really count, rather than stimulating the growth of those things (consumer goods and services) that economists traditionally count.[91] That is the gist of sustainable development.

The struggle for sustainable development is, by and large, a task of integration. During the first three waves of environmentalism, the advocacy of wildlife and wilderness preservation was largely carried out in opposition to or in ignorance of the demands of social justice. These were, for most environmentalists, mutually exclusive concerns. Fourth-wave environmentalists recognize the tension that remains between the preservation of nature and the demands of social justice but increasingly reject the antagonism. Because sustainable development focuses on improving the general quality of life while living within the carrying capacity of ecosystems, its advocates typically focus on ameliorating the conditions of those who suffer most from poverty, powerlessness, and environmental degradation. Sustainable development begins with the premise that poverty and economic domination are environmentally destructive and that more equitable forms of economic production, exchange, and distribution—apart from being morally and politically laudable—facilitate improved environmental stewardship. The presumption is not that social equity automatically ensures environmental sustainability or that sustainability in itself ensures social equity. The two ideals must be fit together on the ground. Theory must be matched with practice in specific contexts.

Meshing social equity and sustainability is the chief task at hand for environmentalists according to Dr. Ben Chavis, former executive director of the National Association for the Advancement of Colored People (NAACP). Chavis has observed a convergence toward this goal. He stated that "[t]here are several growing movements. One is called the environmental justice movement—that's primarily people of color. There's another activist environmental movement that's primarily white. Both of these movements are moving towards convergence, which I'd call a sustainable development movement." Reflecting on the future of environmentalism under the sustainable development paradigm, Chavis remarked: "I'm optimistic, because as I've traveled throughout the United States, throughout all regions, I see a larger movement, a much more credible movement that's really rooted where people live . . . [that] will take our society into the next century and that will really transform American society."[92] Chavis's optimism is grounded in the maturation of a social movement that understands how to capitalize on its own diversity.

The Threefold Nature of Interdependence

For fourth-wave environmentalists, interdependence has three distinct dimensions. Interdependence displays itself across time, across space, and across species. *Interdependence across time* refers to the rights and duties and accompanying relations of mutual risk and benefit that we share with future (and past) generations. These are the rights and duties to receive from ancestors and pass on to progeny the legacy of a biologically rich, life-supporting planet. The affirmation of this generational interdependence marks a concern for intergenerational justice. It is primarily oriented to the development of lives and livelihoods that can be environmentally sustained into the future. At base, it is grounded in our love for and obligations to ancestors and offspring.

Interdependence across space refers to the rights and duties and accompanying relations of mutual risk and benefit that we share with our local, regional, national, and global neighbors. These are the rights and duties to share equitably with human contemporaries the benefits of a biologically rich, life-supporting planet. The affirmation of this social interdependence is commonly referred to as a concern for environmental justice. It is primarily oriented to the development of lives and livelihoods that can be environmentally sustained across social and geographic space. At base, it is grounded in our love for and obligations to neighbors, wherever they may live on this planet and regardless of their nationality, class, or race.

Interdependence across species refers to the rights and duties and accompanying relations of mutual risk and benefit that we share with other species on Earth. These are the rights and duties to integrate ourselves harmoniously with other life forms in the natural processes of a biologically rich, life-supporting planet. The effort to safeguard this biological interdependence might be thought of as a concern for ecological justice. It is primarily oriented to developing lives and livelihoods that acknowledge and affirm our ecological connectedness in a diverse biosphere. At base, it is grounded in our love for and obligations to nature, often voiced in terms of a spiritual connection, and in our knowledge of the complex and dynamic web of ecological relationships that sustains us.

Bryan Norton wrote that the "linchpin" of the modern environmental movement is the belief that the study of nature, by underlining our interdependence, will "inspire a shift to a new perspective."[93] Greens identify this interdependence in terms of the basic "solidarities" that define their work, namely, solidarity with future generations, solidarity with people in need at home and across the globe, and solidarity with nature.[94] Their perspective encourages the embrace of our membership in a community that extends across time and space and, biologically speaking, across species. From this point of view, one's role and responsibility, as a citizen of the biosphere, is to participate rather than dominate. The affirmation of interdependence has this moral dimension and integrative capacity. It predisposes us to an ecological ethics.

Ecology is the science that equips us to understand and integrate ourselves into the vast web of interdependent life. Ernst Haeckel coined the term *oecologie* in 1866. He wrote that ecology was "the body of knowledge concerning the economy of nature—the investigation of the total relations of the animal both to its inorganic and to its organic environment; including, above all, its friendly and inimical relations with those animals and plants with which it comes directly or indirectly into contact."[95] Ecology, in other words, is about relations of interdependence. Ecology is also about sustainability. The two concerns are mutually reinforcing. Relations of interdependence only arise after the interaction of the parts of a system becomes sustained. In turn, these multiple interdependencies sustain the ecological whole. What makes an ecosystem an ecosystem, in other words, is the sustained interdependent relationships of its component parts.

Owing to its focus on sustainable interdependence, ecology proves amenable to ethical formulation. Ethics pertains to norms and obligations that arise out of and are designed to maintain relations of social interdependence. In a fashion reminiscent of Aristotle, Aldo Leopold observed that "[a]ll ethics so far evolved rest upon a single premise: that the individual is a member of a community of interdependent parts." The "land ethic," Leopold continued, "simply enlarges the boundaries of the community to include soils, waters, plants, and animals, or collectively: the land."[96] Most efforts to foster environmental ethics, following Leopold's lead, have sought to expand our sense of community. Ethical consideration is extended to the ecosystem and the biosphere. A broader sense of community is achieved by cultivating our natural sympathies with other forms of life, by fostering moral principles that extend their reach beyond the human, and by underlining the rationality of caring for the biological web that sustains our own lives. The effort to secure intergenerational and social justice combined with the impetus to safeguard ecological integrity produces the struggle for sustainable development, a struggle at once pragmatic and ethical.[97]

Ethical practice entails viewing others along with the self as integrated parts of a greater whole. Ethics pertains to the obligations we have to sustain the community that sustains us. Environmental ethics concerns our dutiful integration in the community of life. Ecology becomes an ethic when sustaining a natural and social community becomes a duty and living within a sustainable natural and social community becomes a right.

The affirmation of temporal, spatial, and biological interdependence begets the imperative to pursue three forms of justice: intergenerational justice, environmental or social justice, and ecological justice. Interdependence in itself is not a moral imperative. It is simply a fact of life. The manner in which we choose to exercise or ignore our interdependence, however, does have moral implications. We may choose to live in denial of our generational, social, and ecological interdependence, just as we may choose to live in denial of our own mortality. Alternatively, we can more fully and consistently embrace the interdependence of our earthly lives. Either way, we cannot escape its effects. The self-conscious embrace of interdependence gives rise to a concern for justice. Justice has been defined since ancient times as rendering to each what is due. For fourth-wave environmentalists, what is due to the diverse inhabitants of Earth—unique human beings of present and future generations and unique species of plants and animals—is the opportunity to flourish and evolve.

Geographer David Harvey observed that "the history of social change is in part captured by the history of the conceptions of space and time, and the ideological uses to which those conceptions might be put."[98] One may expand Harvey's observation to describe the moral change advocated by fourth-wave environmentalists who seek, conceptually and practically, to extend obligations of caretaking across time, across space, and across species. The struggles for environmental sustainability, environmental or social justice, and environmental integrity display this threefold extension of ethical concern.

The following chapters describe how the seeds of fourth-wave environmentalism are being cultivated to meet the challenge of coevolution. What follows is not a catalog of novel orientations and actions. Rather, it is a description and analysis of certain concepts and practices that have become driving forces within the contemporary environmental movement. The subsequent chapters offer an account of ideas whose times have come.

The Quest for
Environmental Sustainability

*Generational Interdependence
across Time*

"The fate of mankind," Teilhard de Chardin once said, "depends upon the emergence of a new faith in the future."[1] Environmentalists might well agree. Yet their orientation is less that of the religious faithful than that of the morally forward-looking. What is intrinsic to environmentalism is a sense of obligation to those who will come after us.

This "future focus" constitutes the most widespread and enduring feature of contemporary environmentalism.[2] It is concisely represented by a popular environmental slogan: "We do not inherit the Earth from our parents, we borrow it from our children."[3] A sense of obligation to progeny grounds environmental ethics. Environmental organizations make this commitment explicit. "If there is anything that has distinguished the environmental movement during the past 100 years," Sierra Club executive director Carl Pope wrote, "it has been our insistence that we not plan for a one-generation society, that the future matters."[4] The founding of the Natural Resources Defense Council (NRDC), representatives observe, was impelled by the commitment not to "steal the future from our children and our children's children—the clean air and water they will need, the natural resources that should sustain their economy and their jobs, and their chance to experience wilderness and wild creatures and to find spiritual renewal in nature."[5] John Sawhill, president of The Nature Conservancy (TNC) identifies the central task of his organization as "ensuring that our children will inherit a rich and diverse natural world."[6] John Flicker, president of the National Audubon Society, has stated succinctly that "[o]ur agenda is to safeguard our children's natural inheritance."[7] The World Wildlife Fund (WWF) operates under the banner of "saving wildlife for future generations." The concern for future generations is common to all environmental groups.

While it is appropriate to highlight the future focus of environmentalists, their temporal orientation is actually Janus-faced. Environmentalists operate with a view to the past, and a rather extended past at that. The aforementioned slogan is misleading in this regard. Environmentalists observe that we *borrow* the earth from our children and grandchildren because we have inherited the earth *as a trust* from parents and grandparents—and from a tree of life billions of years old. Eyes are turned toward the future in large part owing to a profound appreciation of the past. What environmentalists feel obligated to preserve for future generations is a vast ecological legacy.

Aldo Leopold wrote that Darwin's discoveries demonstrated that human beings "are only fellow-voyagers with other creatures in the odyssey of evolution. This new knowledge should have given us, by this time, a sense of kinship with fellow-creatures; a wish to live and let live; a sense of wonder over the magnitude and duration of the biotic enterprise."[8] For fourth-wave environmentalists, it decidedly has. As one Greenpeace activist acknowledged, Darwinian evolution "puts in perspective what little blips of dust we are in the huge and infinite universe, and how aware we have to be about destroying things that have taken millions of years to evolve." Future focus begets an obligation to safeguard an evolutionary legacy grounded in complex biological diversity. The environmental community, in this sense, is perforce what Edmund Burke considered human society to be, namely, "a partnership not only between those who are living, but between those who are living, those who are dead, and those who are to be born."[9] Even religiously oriented environmentalists who reject the evolutionary aspects of life and subscribe instead to some form of creationism often look to the past to justify their obligations to the future. They invoke the story of Noah's Ark or other scriptural lessons to demonstrate that humanity's ongoing responsibility of stewardship is grounded in the obligation to preserve a legacy of life.[10]

In his assessment of the environmental movement, Robert Paehlke observed that "[t]ime horizon may be the single most important distinction between environmentalists and others."[11] To the extent that environmentalists have cultivated a sense of community, it largely arises owing to their sharing an expanded time horizon. They invoke the past because the future is held at a premium. Whether their chief concern is with wildlife preservation, population control, or pollution abatement, environmentalists think across generations and act with this temporally extended community in mind. The Wildlands Project, for example, has the aim of creating and maintaining viable populations of the indigenous flora and fauna of the American continent. To achieve this feat, the creation of a pattern of wilderness areas and corridors is proposed within which populations of wildlife could be sustained. It is foreseen that 25 percent of the land area of North America, including a substantial portion of developed and partially developed land, would need to be preserved or restored as wilderness to achieve these goals. Such an extensive reclamation, project coordinators admit, could be achieved only slowly and over the long term. The strategic program necessarily extends through the next century, with 200-to 500-year projections for recovery in some areas.[12] These achievements, it is hoped, will endure for millennia. As Michael Soulé, a co-founder of the project, has stated, "The Wildlands Project does not accept the limitations of time and space that we so often [find] constraining."[13] Environmentalists primarily concerned with overpopulation estimate the growth of national and world populations and argue for their reduction to acceptable levels in time scales that extend at least to the end of the twenty-second century.[14] Likewise, environmentalists fighting the toxic contamination of their neighborhoods must reckon with the multigenerational effects of carcinogens and other toxic materials. When the issue at hand is radioactive contamination, which may persist for tens of thousands of years, the time horizon must be expanded even further. The nature of their struggles demands that environmentalists radically extend their temporal vision.

Managing Time Horizons

The expanded time horizon of environmentalists is well illustrated by a thought experiment that environmental thinkers often employ.[15] We are asked to imagine the life of the planet condensed into a single year. Each day of that year, beginning on midnight of January 1, represents 12,602,740 years of the earth's 4.6 billion year existence. A month represents about 400 million years.

During the first few months, volatile conditions likely forbid the development of any life forms whatsoever. Single-celled organisms—bacteria and blue-green algae—form in the planet's cooling seas sometime in March. It will be summertime before a protective layer of ozone develops in the stratosphere and multicelled organisms appear in the oceans. Only in late November does the sea give up its monopoly on life. Land plants germinate. They are followed by insects and spiders. It takes until the end of the first week in December before reptiles crawl around on dry land. Dinosaurs inaugurate their extensive reign a week later. In another few days, the central land mass of Pangaea rifts apart, forming the earth's continents. At this time, mammals and birds originate. Before the fourth week of December is over, the dinosaurs have become extinct, giving way to an ever more abundant speciation of mammals.

It takes until the early morning hours of December 31 for the first hominids, *Australopithecus afarensis*, to emerge. By about 11 P.M. of that final day, Stone Age humans gain mastery over fire. Fourteen minutes before midnight, Neanderthals make their brief appearance. Six minutes later, with Neanderthals on their way to extinction, the contemporary human race, *Homo sapiens*, begins its brief odyssey. Less than two minutes before the year expires, at the end of the Ice Age, human beings learn to domesticate plants and animals. Another minute and a half goes by before the great pyramids of Egypt are built. At 4 seconds to midnight, Columbus accidentally bumps into the New World. The human population now stands at 500 million. Before the day is done, it will increase twelvefold.

Industrialization begins to change the face of Europe and spreads to America at two seconds to midnight. With less than one second remaining, the first national parks and wilderness refuges are created. These initiatives conserve a small part of the earth's natural legacy, now placed in great jeopardy by a growing human population. At this time, the effects of human civilization begin to be felt on a global scale as habitat loss, pollution, stratospheric ozone depletion, the overconsumption of resources, and global climate change threaten countless ecosystems and their inhabitants.

Looking ahead, one might doubt whether *Homo sapiens* will see daybreak. If humanity does survive the night, then it is even more unlikely that millions of currently existing species of birds, fishes, mammals, and other life forms will experience the dawn. The "progress" of the human race, particularly in the last milliseconds of its evolutionary odyssey, has not been conducive to the continued existence, let alone flourishing, of earthly cohabitants. Even those species that have propagated widely owing to their domestication or cultivation by humans, animals such as dogs, cats, chickens, pigs, and cows and crops such as wheat, rice, corn, and potatoes, have flourished only in terms of their absolute numbers. The diversity of their gene pools continues to shrink. Meanwhile, tens of thousands if not millions of other species have been pushed into extinction or to its brink by human hands. That is a sobering indictment for creatures that have been around such a short time.

It is estimated that up to one fifth of all known living species are currently threatened with extinction. The threatened species yet undiscovered may number even higher. Environmentalists juxtapose the rapidity of ecological destruction today with the vastness of our evolutionary legacy. "If you look at it from a historical perspective," a Sierra Club activist remarked, "each generation has passed the planet on, so to speak, without harm to the next. This generation and our parents' generation are the first to pass it on in seriously depleted condition. We don't have a right to do this . . . No other species does that." A backward glance produces a future focus. An extended historical and evolutionary perspective motivates environmentalists to look far ahead.

Christopher Manes observed that "this single human generation now living will probably witness the disappearance of one third to one half of the Earth's rich and subtle forms of life, which have been evolving and blossoming for billions of years."[16] The scale of obliteration is shocking. Yet the destruction of species in the past was occasionally both swifter and greater. Since life on the planet originated, over 99 percent of all species have gone extinct. That might amount to as many as four billion extinct species. This figure is one hundred times larger than the number of species that currently inhabit the earth. (About 1.75 million species have been scientifically identified to date, with about one fourth of these being plants, one half insects, and one fourth other animals. The total number of species on Earth is commonly estimated at 5 to 30 million.) Extinction is an intrinsic part of the earth's natural cycles. Few vertebrate species have lasted more than ten million years, which is a relatively short span on an evolutionary time scale. The large mammals (bears, tigers, rhinoceroses, etc.) that we seek to preserve from extinction today, for example, had their ecological niche cleared for them by the late Cretaceous mass extinction that ended the reign of the dinosaurs. The obliteration of species, from an evolutionary perspective, is the harbinger of new forms of life.

The danger of maintaining such a grand evolutionary overview is that it may dull concern for the here and now. Aldo Leopold wrote that the earth is "vastly greater than ourselves in time and space—a being that was old when the morning stars sang together, and, when the last of us has been gathered unto his fathers, will still be young."[17] Leopold is an esteemed environmental thinker, and he provides us here with a worthy reflection. Yet from such a temporally expansive perspective, one might question whether environmental protection ultimately is in vain. The earth will in all likelihood outlast the human race regardless of our misdeeds or our efforts to protect her. In any case, the earth itself will eventually be consumed in flames when, in a few billion years, the sun becomes an expanding fireball known as a red giant. Why, then, should we bother with environmental protection? The danger of expanding one's time horizon beyond a few hundred or thousand years is that it underlines the insignificance of all human action.

Indeed, grand theological overviews that dwarf efforts of environmental stewardship have been used as ideological weapons by anti-environmentalists. James Watt, Ronald Reagan's Secretary of the Interior, belittled environmental concerns because he believed that everything would be destroyed by God in the coming apocalypse. In the same vein, a Congressional representative arguing for increased rates of land development more recently suggested that "preservationists get uptight because they don't believe in the hereafter. Because that means that what's here is all there is."[18] Expanding one's time horizon from a theological, evolutionary, or cosmic point of view may shrink rather than extend one's ethical purview and lead to environmental disregard born of fatalism.

Regardless of what humans do to destroy the earth, its eventual ecological recovery and restabilization is pretty much guaranteed. But that, one might say, is beside the point. Such ecological recovery and restabilization, Stephen J. Gould has noted, would "occur at planetary, not human, time scales—that is, millions of years after the disturbing event." As Gould concluded, "Our only legitimate long view extends to our children and our children's children's children—hundreds or a few thousands of years down the road. . . . A potential recovery millions of years later has no meaning at our appropriate scale."[19] Individual organisms live to die, and die that there might be more life. The same might be said, in evolutionary time scales, of entire species. Environmentalists acknowledge this insight. Yet they do not allow it to undermine their sense of obligation to future generations. They manage their time horizons skillfully. Depending on the context, time horizons should not become too extensive. Environmental destruction that will not be mended for eons remains, for all intents and purposes, unacceptable.

E. O. Wilson has noted that "[t]he one process now going on that will take millions of years to correct is the loss of genetic and species diversity by the destruction of natural habitats. This is the folly our descendants are least likely to forgive."[20] This folly, Wilson predicted, "will be remembered by generations a hundred years from now, a thousand years from now."[21] Norman Myers likewise observed that "[i]n addition to eliminating large numbers of species, we are also causing evolution to lose its capacity to come up with large numbers of replacement species. . . . [W]e are effectively saying that we are absolutely certain that people for the next 5 million years can do without maybe half of all today's species. That's far and away the biggest decision ever taken by one generation on the unconsulted behalf of future generations since we got up on our hind legs."[22] While the earth's recovery from ecological devastation is inevitable in some form, given a long enough period of respite from human interference, current environmental degradation does undermine the planet's capacity to serve as a hospitable and beautiful home to human beings and millions of other species. Environmentalists accept the responsibility to protect the bounty of the earth's natural systems and the evolutionary history invested in each of the earth's species. This applies to the planet's protective ozone layer, its level of greenhouse gases, and its capacity to absorb pollutants and waste, no less than to its level of biodiversity. Wholesale neglect in any of these areas would not destroy the earth. But the planet might quickly be made inhospitable to forms of life that have made it their home for millions of years.

The point is well illustrated by the depletion of the ozone layer. Chlorofluorocarbons (CFCs) were invented in 1928. They quickly came to serve numerous purposes, though their large-scale production and industrial use did not occur for some decades. In 1974, it was discovered that CFCs significantly contributed to the destruction of stratospheric ozone, a band of gas 9 to 22 miles above the earth that protects the biosphere from harmful ultraviolet radiation (UV-B). In the United States, CFCs were banned in nonessential aerosol products in 1978. Concerted action to stem CFC production, however, was not taken until 1987, two years after scientists had discovered a growing "ozone hole" over the Antarctic. Advanced industrial nations eventually committed to producing no more CFCs beginning in 1996.

Production of CFCs fell 76 percent between 1988 and 1995, and restrictions on the production of halon and other ozone-destroying chemicals have been strengthened. Still, the overall global decrease of stratospheric ozone has reached 7 percent. Ozone depletion has reached 60 percent over Antarctica seasonally, creating a hole over the continent more than twice the size of Canada. The problem is predicted to get worse before it gets better, which means that photosynthetic processes will be disturbed, much aquatic plankton will be killed, and many of the earth's creatures, including human beings, will experience higher rates of skin cancer, eye cataracts, and other ailments before levels of stratospheric ozone are normalized in the middle of the next century. The international agreement concluded in Montreal in 1987 to phase out ozone-depleting substances (originally signed by twenty-four nations and later strengthened and ratified by more than 150 countries) demonstrated how a global environmental consciousness can achieve institutionalization and lead to significant collective action. It is an inspiring example of international cooperation and foresight. Nonetheless, the fact remains that it took only half a century for the human race to damage significantly a life-protecting atmospheric layer that the earth required 1.9 billion years to produce.[23]

The Sierra Club motto is One Earth, One Chance. To be sure, there is only one Earth. From a evolutionary point of view, however, the planet has had many chances. The biosphere has recovered from numerous geologic and cosmic traumas worse than those humans have inflicted or could ever inflict. Whether the distant future bears an ecologically rich diversity of life or a much more barren landscape is a human concern. Nature, one might say, is indifferent to the outcome. Radical biocentrists admit as much. Earth First!er Christopher Manes has acknowledged that "[t]he life-forms that will predominate in the future may be grizzly bears and bison or scorpions and roaches, as far as evolution is concerned. For our children's sake, however, we do not have the luxury of indifference."[24] While environmentalists direct an eye to eons of planetary history, they also retain sight of a not-too-distant human future, the future of their children and their children's children. An extended backward glance at billions of years of evolution is balanced with a preview of the next few human generations. That is to say, the human species has but one chance to safeguard the earth's evolutionary odyssey for its decendents.

Environmentalists must deftly manage their time horizons. As a rule of thumb, they suggest that our ecological responsibility should extend across time in direct proportion to our capacity to affect the lives of progeny. The Iroquois people made their decisions with the interests of seven generations in mind. That was a time scale quite sufficient to compensate for any environmental degradation that could be produced by their level of technology and population density. It was also a time scale sufficiently restricted (it went no further than the great grandchildren of great grandchildren) to keep responsibilities within a feasible purview. It did not presume that the present generation had the power or prerogative to control the indefinite future. Green activists hope to add a "Seventh-Generation Amendment" to the U.S. Constitution. Their efforts parallel other attempts to sponsor seventh-generation legislation that would impose personal liability for any "injury, destruction or loss of vital natural resources" that endures from 30 to 210 years (a generation is commonly pegged at thirty years). Similar state initiatives are also in the works.[25]

These are ambitious proposals. Yet they may not adequately offset our potential for mischief. Today we have the capacity to affect the interests of more distant progeny. Most radioactive waste must be isolated for over 10,000 years to prevent toxic contamination of the ground water, the earth's surface, and the air. This is about twice the length of recorded history. Plutonium-239, with a half-life of 24,000 years, takes a quarter of a million years to decay to safe levels, which is to say, about fifty times longer than any civilization has yet survived, longer even than *Homo sapiens* have been on Earth. To produce plutonium today, without adequate methods of safe disposal, is quite literally to endanger thousands of future human generations. When we experiment with substances such as iodine-129, which has a half-life of sixteen million years, our responsibility extends beyond imaginable boundaries.

Many of the most familiar environmental concerns—radiation poisoning, depletion of the ozone layer, the greenhouse effect, the depletion of resources, and the eradication of species and habitats—have ramifications that will exhibit their greatest force long after present generations are gone. Faced with this reality, environmentalists suggest that the task at hand is not to extend our vision indefinitely into the future. The task is to restrict our current practices so that overly extensive calculations and projections become unnecessary.

Caring for Future Generations

Collective action generally occurs when three prerequisites are met. First, a sense of agency has developed, making concerted action to alter the situation at hand feasible. Second, a collective identity has been formed, allowing a relatively clear perception of the constituency to be mobilized into action. Third, an injustice that requires redress has been perceived. Often the perception of injustice proves key to developing and integrating the other prerequisites for collective action. The perception of injustice allows a sense of a victimized "we" to form in collective opposition to a perpetrating "they." The perception of injustice, in other words, gives rise to the identity of being a victim. In turn, the moral indignation stemming from victimization prompts the belief that the removal of the cause of the indignity is possible and necessary.[26] It follows that environmental collective action is unlikely to arise unless particular perpetrators of injustice can be clearly identified.[27] Indeed, empirical studies demonstrate that the perception of injustice is required to sustain high levels of environmental activism.[28]

Environmentalists assert that we are all responsible for environmental problems such as pollution, overpopulation, and overconsumption. Often a specific perpetrator of injustice is not identified. Yet collective action is stimulated. How is this possible? The answer seems to be that environmentalists effectively transform a victimized "we" into a perpetrating "they" by way of an extended time horizon. The present generation identifies future generations as victims. At the same time, the present generation naturally experiences a solidarity with progeny. Hence, environmentalists take on the dual identity of perpetrator and victim. As one activist stated: "We are at war. Global war. We have identified the enemy and the enemy is ourselves. The sides are clearly defined. On one side we have the collective selfish aspirations of one species—our own, *Homo sapiens*. On the other side we have at stake the future of planet Earth and all of her species, a future which includes our own children and grandchildren. . . . Ecologically responsible democracy must consider the rights of the true majority—those billions of people as yet unborn."[29] The struggle between victims and perpetrators is effectively internalized by environmentalists. Collective action is generated in the pursuit of intergenerational justice.

The unwillingness to become a perpetrator of intergenerational injustice stimulates environmental activism. Paul Harrison suggests in *The Third Revolution* that "[t]he time is near when every child will ask its parent 'What did you do in the environment war, mum and dad? Were you one of those who helped to destroy my future? Or were you one of those who helped to save it?'"[30] Like many who write about environmental issues, Al Gore muses on the home his children will inherit and states simply about *Earth in the Balance*, "I really wrote this book for them."[31] A Greenpeace activist claimed that his own commitment to environmentalism was grounded in a form of farsighted reason that was oriented to ensuring "the ability for my kids and my kids' kids to survive." The concern for intergenerational responsibility and the pursuit of intergenerational justice are pervasive. A sense of moral responsibility for future generations is a mainstay of environmental activism.

Environmentalists are concerned about their own progeny. But their vision is not so narrowly focused. The moral purview is wider. A Sierra Club activist observed that she felt a responsibility not primarily as a parent but "as a human . . . to leave the earth in a better shape than I found it." A staff member of WWF maintained that the defining feature of the environmental movement was its cultivation of a "responsibility to the future" and that her own activism was grounded in broadly based "ethics of human relations within and between generations." Caring for their children is a motivating force for many environmentalists, assuming they have children. But the sense of responsibility to future generations is not the monopoly of parents.

Rachel Carson remained childless. In fact, the chemical industry made a point of questioning her prerogative as a "spinster" to raise concerns about future generations.[32] Her antagonists missed the mark. Environmental concern, and more particularly the sense of responsibility to future generations, is widely held among childless individuals and couples. Indeed, a sense of responsibility for future generations is often the impetus behind decisions to limit family size or not have any children. Future focus is also strongly evident in youth who do not yet have or foresee having children.

Surveys indicate that the desire to protect the environment for future generations is "a nearly universal American value." A significant degree of future focus is so prevalent among the general public that researchers find it "curious that this value has not played a more central role in public discussion and scholarly investigation regarding global environmental change."[33] David Durenburger (R-Minnesota), a member of the Senate Committee on Environment and Public Works, was probably correct when he observed that "[m]ost Americans don't embrace the environmental ethic out of a concern for public health statistics. . . . Their highest value is intergenerational, passing on a world at least as good as the one they received."[34]

In the context of the near-universal concern for the future, environmental organizations investigate and publicize the means by which people can fulfill their intergenerational responsibilities. Much effort is focused on highlighting the responsibilities of individuals and particularly the responsibilities of individual consumers. The "three R's" of sustainable living—reduce, reuse, and recycle—are widely propagated as means to limit consumption and its counterpart, the production of waste. Environmental publications also urge readers to do their part as individuals by forgoing the use of pesticides or synthetic fertilizer on lawns and gardens, by buying organic foods and products, by patronizing green businesses, by using public transportation or bicycles rather than relying on automobiles, by investing in socially and environmentally responsible stocks and mutual funds, and by limiting family size.

The concern with individual responsibility is growing. At the same time, environmental organizations remain as much if not more concerned with collective action. The reason is simple. In the absence of collective action and institutional change, isolated individual efforts— "green lifestyles"—will prove insufficient. Economic, political, and social structures often undercut individual initiative and responsibility. The patronage of green businesses is a case in point. Individuals can be encouraged to "buy green" for the sake of future generations. Yet the relatively small demand for green goods generated by environmentally conscious consumers is unlikely to allow a sufficient supply of competitively priced products. Brian Tokar has aptly remarked that "as 'natural' products have become a niche market for those affluent enough to pay for them, the goods available to everyone else are even shoddier and more toxic than before."[35] A focus on the responsibilities of individual consumers does not address the larger economic system of production. Yet, as Tokar observed, that system constitutes the root of the problem.

Collective action grounded in reformed public values and institutions is indispensable to effective environmental care. A reliance on individual responsibility is insufficient. The editor of *Audubon* explains the problem in the context of overfishing: "It's unrealistic to expect any one fisherman to just say no to a bluefin [tuna] bonanza for the long-term benefit of the species and of fishermen yet unborn. A fairer kind of restraint is called for, a *collective* restraint born of popular acceptance. . . . To accommodate all those future citizens, whose love for the natural world will be no less keen than ours, requires . . . a [collective] willingness to forgo using up the world now."[36] To avoid the limitations of uncoordinated and purely voluntary individual actions, collective responsibility for the future must be cultivated and institutionalized.

For this reason, environmentalists' efforts to address issues of intergenerational justice are frequently channeled into the political arena. They seek from governmental bodies the institutionalization of future-focused values. Beyond lifestyle changes, then, environmental organizations ask their members to vote on the basis of candidates' environmental records or platforms, to write to or lobby political representatives, to attend hearings, to educate the general public, to oganize and mobilize.

Environmental organizations take the lead in initiating such collective action. In response to Congressional efforts to roll back environmental protection in 1995, the Sierra Club, Public Interest Research Group, Natural Resources Defense Council, National Audubon Society, Greenpeace, Friends of the Earth, a dozen other national environmental organizations, and 450 local groups circulated an Environmental Bill of Rights. The goal of acquiring a million signatures was easily met within six months. Congress was served with the petition to "preserve America's national heritage, wild and beautiful, for our children and future generations." Zero Population Growth (ZPG) kicked off its own petition campaign to coincide with the inauguration of the 104th Congress in January of 1995. Appealing to the popularity of the Republican Party's "Contract with America," which was aimed at rolling back government regulation, ZPG publicized the need for a "Contract with the Future." "There's a lot of talk about contracts these days," a ZPG newsletter reads, "but the most important contract lawmakers need to make is with the *future*. And that means making a promise to act in ways that assure a high quality of life not only for ourselves but for generations to come."[37]

The Cousteau Society, in turn, has circulated a petition for a proclamation of the "Rights for Future Generations." It has been signed by more than nine million citizens of 113 nations. Approved by the Executive Committee of UNESCO in 1995, the petition was presented to the General Assembly of United Nations. The hope was that the U.N. Charter might officially acknowledge that "[f]uture generations have a right to an uncontaminated and undamaged Earth."[38] Such petition drives at national and international levels are only the most visible of efforts to stimulate collective action with a future focus. In fact, most of the lobbying and litigating that environmentalists engage in is also aimed at fulfilling obligations to future generations. By focusing on these obligations, environmental organizations stimulate collective environmental responsibility and lay the groundwork for collective restraint. The watchword of this effort is sustainability.

Sustainability

Within the environmental community, the effort to concretize a future focus has been subsumed under the rubric of sustainability. Sustainability simply refers to the capacity of a practice, relation, or process to be carried on indefinitely without undermining the environmental conditions of its viability. Sustainability is most easily defined in the negative. America's aquifers, on average, are being depleted 25 percent faster than they are being recharged. That is not sustainable water use. Currently, cropland erosion of soil is occurring between thirteen and thirty times faster than the replacement rate. Over three million acres of prime farmland in the United States are lost annually to development or erosion.[39] That is not sustainable agriculture. Globally, over two-thirds of marine fish stocks are being harvested faster than they can reproduce themselves. That is not sustainable fishing.

The demands of sustainability are relatively straightforward, at least in situations where the depletion of resources can be easily measured. Sustainability dictates that the use of renewable resources, such as topsoil or clean, fresh water, should not exceed their rate of natural recovery. Those resources that are irretrievably diminished by our use of them—namely, all extracted nonrenewable resources such as minerals and fossil fuels—should only be exploited at such a rate that the discovery or development of viable substitutes for them might be expected before they are practically exhausted.[40] Those natural resources that have no imaginable substitutes, such as particular species of flora and fauna, must not be destroyed or used up.

Living sustainably also means refraining from producing things whose deleterious effects or side effects accumulate to harmful levels. Waste, pollution, human populations, and commercial development should not exceed an ecosystem's capacity to absorb them without becoming degraded in its life-sustaining capacities. The mining and burning of the earth's vast reserves of low-grade coal, for example, scar the land and pollute the air. The use of this abundant nonrenewable resource is probably not sustainable owing to its destruction of renewable resources such as air quality. To live sustainably, therefore, we may have to refrain from extracting and using certain nonrenewable resources even though they presently exist in abundance.

Sustainability is a powerful concept in search of concretization.[41] While environmentalists have no trouble defining sustainability in the abstract, they often falter in the attempt to explain how it ought to be individually or collectively realized. Most begin by grounding themselves in their own efforts to lead sustainable lives. One Greenpeace member and Sierra Club activist defined sustainability as an "equilibrium" between what one takes out of the environment and what one puts back into it, adding, "It's very important, of course, but it's not terribly easy to do." His own attempt to live sustainably was exemplified by efforts to xeriscape his yard, using indigenous plants that tolerate periods of drought to reduce or eliminate the need for irrigation. Were everyone in his hometown to water their lawns, he observed, the aquifer from which water was drawn would be severely depleted. At some time in the not-too-distant future, there would be insufficient water to go around. Practicing sustainable yard care, therefore, required some level of xeriscaping.

Though its currency is growing, the significance of sustainability is not widely appreciated within the general public. Nonenvironmentalists, in particular, have little if any idea what the word means. Few recognize the term. For environmentalists, on the other hand, the importance of sustainability is universally acknowledged. Many consider it to be the most important environmental concept. Lindsey Grant, writing for Negative Population Growth, has insisted that "[s]ustainability . . . is perhaps the single best word to express environmentalists' goals."[42] An activist with the Environmental Defense Fund (EDF) and the Wilderness Society maintained that sustainability is "the only philosophy we can survive under."

Few people spoke of sustainability before the 1980s. Only in the fourth wave of environmentalism has it become a driving force. Reflecting on the history of American environmentalism, Gaylord Nelson, the founder of Earth Day, stated that "[t]he idea of sustainability as an issue of general discussion is new. The issue of sustainability, looking at the whole, total picture is quite new. The fundamental issue from now on is sustainability, everything else is subsidiary to that. Sustainability encompasses everything."[43] Developing an ethic of sustainability and forging the institutions that would facilitate the widespread adoption of this ethic constitute the core of environmental work today.

Environmentalists organize most of their thinking around the idea of sustainability even when the situation does not obviously call for its use. Asked how much freedom she would be willing to give up to ensure a healthier environment, one Sierra Club member replied that she would forgo "enough so that the need to give up freedom is spread equally among generations. In other words, the next generation shouldn't have restrictions on their freedom because I wasn't willing to give up any of mine." Here the concept of sustainability is put to work to adjudicate individual liberties in the same manner that it is employed to adjudicate appropriate levels of resource consumption. In both cases, the question of sustainability determines the parameters of legitimate action.

To live sustainably is to maintain the integrity of the environment such that the opportunities for future generations to live well are not diminished. Sustainability is about preserving the quality and diversity of life over time. A member of EDF observed that environmentalists concerned with sustainability extol the duty to leave future generations the "possibility for a quality of life that isn't just mere existence." Environmentalists protect the opportunity of future generations to live lives of at least the same quality as is enjoyed by present generations.

To advocate sustainability, it follows, is to engage in a form of paternalism. Living sustainably does not entail knowing what specific needs and values future generations might have. Environmentalists assume, however, that our own basic needs—health care, nutritious food, decent housing, clean air and water—will also be the basic needs of future generations. In turn, environmentalists believe that future generations deserve to enjoy the same opportunities that current generations enjoy: the opportunities to experience unpolluted landscapes, open space, scenic beauty, and wilderness.

"I have a love of beauty and want to see that perpetuated," a member of Citizens Clearinghouse for Hazardous Waste observed. He identified his love of the "mystery" of life as the value he chiefly wanted to pass on to descendants. Another environmentalist suggested that "keeping at least a few pieces of shaggy frontier for Americans yet unborn to roam in" is the only means we have of preserving for our progeny the opportunity to "understand freedom in its truest form."[44] Under the logo of Do It for the Children, the Wilderness Society urges the preservation of wildlife and wildlands, insisting that "no child should grow up without knowing the beauty and mystery of the natural world."[45] Environmentalists safeguard the opportunity for future generations to value these goods as much as they do. As one Green theorist wrote, "it is only if we value what is sustained, that we can answer the question *why* it ought to be sustained."[46]

Mark Sagoff acknowledges the inherent paternalism of the ethic of sustainability. He has argued that the future focus characteristic of environmentalism is not so much "a responsibility *to* the future as it is a responsibility *for* the future." Sagoff wrote:

> The major decisions we make determine the identity of the people
> who follow us. . . . Our decisions concerning the environment will
> also determine, to a large extent, what future people are like and
> what their preferences and tastes will be. If we leave them an en-
> vironment that is fit for pigs, they will be like pigs; their tastes will
> adapt to their conditions. . . . Our obligation to provide future in-
> dividuals with an environment consistent with ideals we know to
> be good is an obligation not necessarily to those individuals but to
> the ideals themselves. . . . In short, it is a concern about the char-
> acter of the future itself. We want individuals to be happier, but we
> also want them to have surroundings to be happier about. We want
> them to have what is *worthy of happiness*.[47]

We cannot know for certain that future generations will cherish diverse
species of flora and fauna. They may prefer simulated nature in virtual
reality machines. They may also prefer commercial development to nature preserves and trips to the mall to wilderness hikes. In the high-tech,
synthetic world of the future, there may be an abundance of every sort
of commodity but a near-complete absence of nature. Were that to come
to pass, environmentalists would admit that they had failed in their task,
for an important feature of environmentalism is the cultivation of a love
of nature. Critics might bristle at the inherent paternalism of this perspective. Yet environmentalists insist that we are morally censurable if
we strip from future generations the very *opportunity* to cherish the
natural world.

Dave Foreman has pessimistically observed that "ours is the last
generation that will have the *choice* of wilderness."[48] The present generation cannot ensure that its descendants will appreciate biodiversity
and the wonder of nature. To live sustainably, however, is to ensure, at
a minimum, that descendants will have the choice of wilderness. Practically speaking, the only way to protect that choice is to cultivate environmental values among peers and progeny.

From a planetary perspective, all environmental destruction is potentially reversible given enough time—except the extinction of species.
The ethic of sustainability thus dictates that the present generation
should never act so as to eliminate a future generation's opportunity to
experience particular plants or animals and their ecosystemic relations.
Species are unique. When they are gone, they are gone forever. Practices
that reduce biodiversity, regardless of whether they are the unintended
by-products of land development, resource extraction, or pesticide use,
are, in this sense, inherently unsustainable. There is no doubt that we

could get along with less biodiversity. Yet future generations would then be deprived of the opportunity to share the world with certain plants and animals, to use them or their products sustainably, and to experience them in their natural habitats. Sustainability entails the safeguarding of such opportunities.

Living sustainably, to restate, means that our present practices should not impinge on the opportunities of future generations to engage in similar practices. The modifier "similar" is crucial. Time changes everything. Strictly speaking, future generations will never have access to *our* experiences and practices. Descendants shall not have the opportunity to make our historical or scientific discoveries, to produce our cultural or technological innovations, or to understand and experience the world as we did before these discoveries or innovations were made. Environmentalists who advocate sustainable living are not suggesting that we petrify human existence. Living sustainably simply means that we preserve for future generations the chance to engage in the kinds of practices that we have enjoyed and, more generally, that we preserve the conditions that will allow a similar range of experiences. Environmentalists define this as the "conservation of options."[49]

Interdependence and the Role of Caution

Foresightful calculations are obviously in order for those who seek to live sustainably. One must not only gauge present rates of resource depletion but also predict future trends. One would need a crystal ball to perform this speculative work flawlessly. In the absence of prophetic devices, living cautiously and increasing our knowledge of the social and ecological effects of our actions are imperative.

Garrett Hardin maintains that the first law of human ecology is, "We can never do merely one thing."[50] Observing that this ecological law constitutes the "central axiom of environmentalism," Lester Milbrath has spelled out its practical ramifications: "We must learn to ask, for every action, And then what?"[51] The centrality of this question is a direct function of our temporal, spatial, and ecological interdependencies. Applying pesticides to crops, for instance, is not simply doing one thing. Though it may, in the short term, control, the targeted insects or weeds in a certain locale, it also ramifies across spatial, temporal, and ecological horizons. It may contaminate watersheds, decimate migratory bird populations, and negatively affect human health far into the future. Situated within extensive webs of interdependence, virtually no human actions are without repercussions to humans distant in time and space and to other species.

In large degree, environmentalism is a matter of expanding one's awareness of interdependence by asking, And then what? before one acts. Owing to the intricacies of interdependence, even mundane practices, when diligently questioned, provide an environmental education. Not wanting to spoil the aesthetic quality of the landscape by littering, one may have acquired the habit of putting trash in garbage receptacles. But what happens to the garbage after it is put in the receptacles? To answer this question, we must investigate the local landfill or incinerator. This, in turn, might lead one to question the threat posed by manufactured goods that leach toxins into groundwater or pollute the air when burned. One might also question the commercial trends that stimulate excessive waste from packaging and the overproduction of disposable goods. The focus on unsustainable wastestreams might lead one to be concerned with overconsumption and resource depletion. A relatively sophisticated personal ethic of reducing, reusing, and recycling and an orientation toward (macroeconomic) institutional change might be the result—all from repeatedly asking the simple question, And then what?

In many respects, this storyline parallels the development of mainstream environmentalism. The mass movement was inaugurated in the 1970s with campaigns to end littering and stem pollution. A quarter century ago the chief environmental message was "Don't be a litterbug." This approach, to employ Allan Schnaiberg's apt term, cast environmentalists as "cosmetologists." The problem with litter is cosmetic, and the solution is to make better use of trash cans.[52] There was little insight into the deeper causes or effects of environmental degradation. Few people asked, And then what? once the trash was in the can. Keep America Beautiful, a coalition of companies involved in glass, aluminum, paper, and plastic production and use, ran a popular TV spot in 1971 showing Native American actor Iron Eyes Cody with a tear running down his cheek at the sight of roadside litter. The advertisement received tens of millions of dollars in free air time. Meanwhile, Keep America Beautiful was actively engaged in opposing a national bottle bill that would have mandated a recycling deposit for glass bottles.

A recent Greenpeace television advertisement marked the transformation in consciousness to a sustainability ethic. When we next throw away a piece of trash, the Greenpeace advertisement suggested, we should ask ourselves the question, "Where is 'away'?" Filled trash cans may be the end of one problem, but they are the beginning of another.

Landfills and incinerators are not environmentally benign. If trash cans are being filled at too rapid a rate, then obviously enough things are not being recycled or reused. More important, rapidly filled trash cans indicate a society intoxicated with overconsumption. That is ecological thinking. It is radical rather than cosmetic because it engages the question of whether our individual actions can be sustained in the aggregate and over the long term.

Invoking the Iroquois concern for the seventh generation, a member of the National Wildlife Federation (NWF), maintained that "[i]t's all about being aware of one's current situation, being able to place it in a universe that is immense." Continually asking the key ecological question, And then what? is a means of becoming aware of one's current situation while simultaneously placing it in a context broad enough to facilitate sustainable practices. Asking these sorts of questions both instills caution and promotes scientific investigation. As the NWF member concluded, "We need to conserve resources and increase knowledge."

Knowledge is a double-edged sword. It brings awareness of the complexity of natural systems. Alternatively, it may leave us self-assured of our power and prerogative to alter them. Knowledge, in other words, may foster prudence or arrogance. Environmentalists celebrate the former and warn of the latter. They insist that science should be tempered with the insight that we may not yet be ready for many of the lessons that nature has to teach us. We are told that "[n]ature is not only more complex than we think, but it is more complex that we can ever think."[53] Another environmental writer has suggested that "[t]he wilderness holds answers to more questions than we have yet learned to ask."[54] The web of life is so complex that no investigation can fully reveal the intricacies of its patterns or the (long-term) consequences of our interactions with it. Under these conditions, caution should be the order of the day.

If we spent all of our time posing and attempting to answer an extended chain of "And then what?" questions, we would have no time left to act. That would be a problem for environmentalists. With this in mind, Richard Norgaard argues that living sustainably does not entail exact prediction and control of the indefinite future. Such a level of knowledge and power is beyond us. A coevolutionary perspective, grounded in an understanding of the dynamic interdependencies of life, links sustainability not to inaction but to cautious living.[55] To practice sustainable living is to pursue knowledge while practicing prudence.

Norman Myers has observed that "[p]robably the biggest environmental problem of all on the horizon will turn out to be that of the interactions between lesser problems."[56] This "synergistic connection" can never be fully anticipated. Unfortunately, there is already a long history testifying to the dangers of synergism. DDT, environmentalists discovered in the 1960s, not only killed mosquitoes, its intended target, but also threatened entire populations of raptors. That is because pesticides such as DDT (as well as mercury and certain other heavy metals) are prone to what is known as bioaccumulation. Each higher species on the food chain suffers an increasing level of contamination. Whereas DDT was found in negligible quantities in the waters of treated areas, it was present in greater concentrations in the plankton in these waters, and in increasingly greater concentrations in the small fish that ate the plankton, the larger fish that ate the small fish, and the ospreys and other raptors that ate the large fish. In fact, ospreys were found to have levels of DDT many million times greater than that of the water over which they fished. Current studies also suggest that many pesticides, though relatively benign to nontarget species when applied in isolation, may be a thousand times more disruptive of hormone and reproductive systems when organisms are exposed to two or more of them over time—as often occurs in the environment. Along with chemical compounds such as PCBs, dioxins, and furans, these pesticides, known as "endocrine disrupters," may be highly dangerous to human fetuses and the young.[57] The ecological downside of synergism is also patent in the arena of global warming, where the interactive effects of climate change threaten regional drought and flooding, agricultural setbacks, habitat loss, species extinction on land and at sea, and pest infestation.[58]

Wendell Berry wrote that, ecologically speaking, the effects of our actions are "invariably multiple, self-multiplying, long lasting, and unforeseeable in something like geometric proportion to the size or power of the cause."[59] Ecological linkages may be direct or mediated. The results of severing them cannot be fully anticipated. The disappearance of moose from Vermont in the nineteenth century, for example, was largely a product of the overharvesting of beavers. The moose, it turns out, depended on the vegetation that grew in beaver ponds for their summer foraging. Likewise, the decline in songbird populations in the United States was unexpectedly exacerbated by the extermination of large carnivores, such as cougars and wolves. Raccoons, foxes, and possums eat songbirds and their eggs. With the overhunting, poisoning, and trapping of large predators, which hunt mid-size predators such as raccoons, foxes and possums, the populations of the latter exploded. With so many egg and bird eaters around, songbird populations declined.

Since human health and livelihoods depend on the viability of ecosystems, tinkering with nature may have disastrous consequences. The
fact that the destructive effects of our actions only manifest themselves
decades or centuries later, environmentalists maintain, should not lessen
our concern. Russell Train, WWF's chairman of the board, warns that
"[w]e still do not fully understand the implications of human-induced
disruptions to natural ecosystems, but there is growing evidence that we
may be tampering severely with our children's—and their children's—
capacity to survive."[60] Speaking at the ozone treaty negotiations as head
of the U.S. delegation, a senior fellow of the Conservation Foundation
and the WWF stated, "If we are to err, then let us, conscious of our
responsibility to future generations, err on the side of caution."[61] That
is a universal sentiment within the environmental community.

The pursuit of sustainability is the art and science of cautious living.
Yet a certain level of ecological disruption is inevitable, regardless of
our circumspection. We cannot live—certainly not with our present level
of population, economic production, and technological development—
without affecting future generations in ways that remain largely unpre-
dictable. Nonetheless, we need not present progeny with irreversible
faits accomplis, which is the case, for example, every time this generation succeeds in exterminating another species. As one Sierra Club activist insisted, living responsibly means that we do not burden progeny
with environmental problems that are "unsolvable to them." Likewise,
a member of TNC maintained that the onus is on present generations
"not to do anything that they [future generations] will not be able to
undo."

It is difficult if not impossible to predict what problems will prove
unsolvable to future generations. We might therefore understand our
responsibilities as twofold. First, we must increasingly investigate the
social and ecological effects of our actions. Scientific inquiry is indispensable. Second, we must cautiously rein in those practices that threaten
to rob future generations of opportunities that we ourselves have enjoyed. Our first responsibility is to increase knowledge; the second entails
fostering prudence. A rule of thumb for balancing these responsibilities
was offered by a twelve-year-old girl at the plenary session of the Earth
Summit in Rio de Janeiro. She put the point succinctly in addressing her
adult audience: "You don't know how to bring back an animal now
extinct. And you can't bring back the forests that once grew where there
is now desert. If you don't know how to fix it, please stop breaking
it."[62]

To wait until an ecological crisis is upon us before responding is to act imprudently. In many cases, it is to act too late. Things broken cannot always be fixed. With this in mind, the Brundtland Report suggests that "[e]conomic development is unsustainable if it increases vulnerability to crises."[63] Prudence is not the art of resolving crises but the art of avoiding them. Many ecological problems arise from the solutions that we have invented to remedy earlier problems. Sustainable living, it follows, avoids unnecessary risk taking and preempts the need for crisis intervention. This characteristic of sustainability finds its conceptual formulation in the "precautionary principle." The precautionary principle stipulates that "[w]hen an activity raises threats of harm to the environment or human health, precautionary measures should be taken even if some cause and effect relationships are not fully established scientifically."[64] Effectively, the precautionary principle shifts the burden of proof. No longer would a potentially victimized public have to prove that an activity was unsafe. Instead, the proponent of a potentially dangerous activity would have to verify its harmlessness before acting.

The endorsement of the precautionary principle is not restricted to environmental purists. Indeed, Robert Costanza suggests that "the precautionary principle is so frequently invoked in international environmental resolutions that it has come to be seen by some as a basic normative principle of international environmental law."[65] Principle 15 of the Rio Declaration maintains that "[i]n order to protect the environment, the precautionary approach shall be widely applied by States according to their capabilities. Where there are threats of serious or irreversible damage, lack of full scientific certainty should not be used as a reason for postponing cost-effective measures to prevent environmental degradation." The Intergovernmental Panel on Climate Change concluded in 1995 that human activities might raise average global temperatures 1.8 to 6.3 degrees Fahrenheit and raise sea levels by 6 to 38 inches in the next hundred years.[66] With these dangers in mind, a recent World Bank paper on global warming states: "When confronted with risks which could be menacing and irreversible, uncertainty argues strongly in favour of prudent action and against complacency."[67] Most of the signatories to the international accord on climate change reached at Kyoto, Japan, in December 1997 also adopted a precautionary perspective, though the final agreement was far from optimal. Today's emitters of greenhouse gases will be long gone when future generations face disrupted weather patterns, hotter climates, heightened species extinction, submerged islands, and devastated coastlines. Climate change increases our vulnerability to ecological crises. Knowing that is enough to justify action.

With this in mind, we might amend our definition of sustainable living to include the following clauses: Living sustainably means living cautiously. Living cautiously entails reducing vulnerability to ecological crises whenever possible and responding with diligence to threats whenever they arise.

Sustainability and Economics

In addressing the root causes of our ecological straits, Al Gore has lamented that "the future whispers while the present shouts."[68] With similar concerns, Robert Heilbroner has spoken of the "inverted telescope through which humanity looks to the future."[69] The metaphors are apt. The impression the future makes on us is faint.

The voice of the present is loud. It demands attention. It can be virtually despotic. Nowhere does what Gro Harlem Brundtland calls the "the tyranny of the immediate" play a larger role than in economic affairs. A future focus provides environmentalists with a means of framing and arbitrating conflicts between economics and ecology. Financial gains made at the expense of future generations who must pay ecological reparations for our current prosperity are considered illegitimate.

The tyranny of the immediate played a central role in America's national history, which is, among other things, the history of a frontier conquered for quick profit. In the mid-1700s, a Scandinavian naturalist traveling in America recorded that "the grain fields, the meadows, the forests, the cattle, etc. are treated with equal carelessness; and the characteristics of the English nation, so well skilled in these branches of husbandry, is scarcely recognized here. We can hardly be more hostile toward our woods in Sweden and Finland than they are here: their eyes are fixed upon the present gain and they are blind to the future."[70] A century later, in the mid-1800s, the "Gold Rush" erupted in the West. A frantic race for buried treasure left ecological scars on the nation's land and waters. In the mid-1900s, America produced wealth in greater quantities and with greater speed than any other nation in history, and with an ecological destructiveness that sparked a mass environmental movement. Today, the United States is notorious for its culture of conspicuous consumption and for its world record rates of resource depletion and waste production. Though constituting less than 5 percent of the world's population, Americans consume between 20 to 40 percent of world's resources and produce an equal if not higher percentage of its waste and pollution.[71] Asked if the political system in the United States was conducive to environmental care, one activist with Florida Defenders of the Environment responded that "it's not so much the system . . . as the national culture: we don't have a long-term view." The tyranny of the immediate is no stranger to American life.

Economists accept the tyranny of the immediate as a basic feature of human nature. They call it a "positive time preference." Goods available immediately are valued more than the same goods available at a later date. Likewise, problems that affect us today receive immediate attention while those that will affect us tomorrow, or will affect our descendants at a later date, are more easily ignored. The future, to employ the jargon of economic theory, becomes "discounted." The farther an event is displaced in time, the more its value diminishes. The farther down the road that a problem will manifest itself, the weaker its whisper.

Discounting the future is economically rational. What sensible person would not rather have a dollar today than a dollar a year from now. A bird in the hand is worth more than one in the bush. A bird in the hand is more secure. It cannot fly away or be captured by some passerby. A bird in the hand also becomes immediately available for use. It can lay eggs, effectively earning interest for its owner. For these reasons, it makes good economic sense to discount future gains at some ascertainable rate that reflects the loss of compound interest and the inherent insecurities of all future endeavors.

The problem is that good economics often translates into bad ecology. What makes dollars and sense for the economic actor bent on maximizing current returns on investments may easily lead to disaster for society and the biosphere. As John Dryzek wrote, "[A] system may be judged economically rational while simultaneously engaging in the wholesale destruction of nature, or even, ultimately, in the total extinction of the human race. The latter result holds because of the logic of discounting the future."[72] At a standard 10 percent discount rate, a project that would yield a $10 million depletion of ecological resources in 100 years would be economically viable as long as it produced a $725.00 profit today. In standard economic practice, the vision is even more constrained. Financial gains that cannot be reaped for more than twenty years are generally discounted to zero.[73] When costs of resource depletion, pollution, and habitat destruction are shifted to future generations, and the concerns of these generations are discounted by present-day decision-makers, economic rationality portends social and ecological catastrophe.[74] If the discount rate is high enough, and the foreseen consequences of environmental exploitation are far enough away, economic reason may justify current pleasures at the cost of future extinction.

Individuals who produce goods or services may accrue greater profits from present sales by "externalizing" or "socializing" ecological costs. Externalizing costs really means fobbing off expenses on others. Financial benefits from resource exploitation are immediately realized by private individuals or corporations while the ecological costs of this exploitation are distributed across society and future generations. A corporation that manufactures paper, for example, might choose not to bear the price of cleaning up the dioxin-laden wastestreams of its factories. Rather than see its profit margin shrink, it expels untreated effluent into a nearby waterway. This corporation has effectively passed along the cost of paper production to those residents living downstream who suffer from the waterway's degraded state.

In like fashion, costs may be externalized to those distant in time. The nuclear energy industry is widely criticized on this count. The life span of a nuclear reactor is about forty years, after which it must be decommissioned and entombed in concrete or dismantled and disposed of as toxic waste. Decommissioning and disposal costs begin at a few million dollars and may reach hundreds of millions of dollars per facility. Disposing the low-level radioactive waste that is produced by operational nuclear energy plants is also costly and requires special landfills. High-level radioactive waste (spent fuel rods), of which each reactor produces about thirty tons a year, is typically stored above ground in steel drums, awaiting more permanent forms of disposal. Currently there are over 30,000 tons of spent fuel rods awaiting permanent disposal in the United States. By the year 2010, it is estimated that the amount of high-level radioactive waste in need of a home will double. It will rise further as older reactors are shut down. Disposal costs for this waste will be very high, assuming a safe form of permanent disposal is ever discovered. (After searching for more than forty years, industry and government are no closer to finding a solution.) Permanent disposal costs will certainly dwarf the already significant temporary storage costs.

In 1982, the federal government instituted a "user fee" on nuclear energy (1/10 cent per kilowatt hour), which was intended to offset the costs of developing a permanent storage site for high-level radioactive wastes. That site, however, has not been built to date, and it is uncertain whether the fees collected will cover the bill when it comes due. In turn, the cost of dismantling nuclear power plants (not to mention the hundreds of millions of taxpayers' dollars spent each year in fission research and development, some $34 billion since 1948[75]) is not factored into the price consumers pay for nuclear-generated electricity. Future generations will thus bear the monetary and ecological costs of cleaning up nuclear power sites without having had the benefit of a single kilowatt of the energy produced. In short, the nuclear energy industry achieves economic viability by way of externalizing its ecological production costs to those distant in time. (There is also the ever-present threat of an accident that would submit present generations to radioactive contamination. The nuclear energy industry has externalized these costs in space. They have been granted limited financial liability for any mishaps that occur through the Price-Anderson Act. In the event of a catastrophic accident, taxpayers would foot most of the bill for medical costs and cleanup.)

Another means of externalizing costs is exhibited through the depletion of natural resources. Resources that could be considered the common holdings of the earth's inhabitants are used up without compensation to future generations who might otherwise benefit from them. This applies not only to the exhaustion of nonrenewable resources, but also to the depletion of certain renewable resources, such as groundwater, whose regeneration rates are generally slow. In both cases, present generations are cashing in the inheritance of future generations. When the bird is stewing in someone's cooking pot today, those who might have collected its eggs tomorrow will go without.

E. F. Schumacher first observed in the early 1970s that the "modern industrial system . . . lives on irreplaceable capital which it treats as income."[76] The metaphor has become mainstream. As a Sierra Club activist observed, sustainability simply means "you don't spend your principal." The unsustainable depletion of natural resources amounts to treating natural capital as if it were income and treating principal as if it were revenue. Once the capital or principal is depleted, income or revenues necessarily dry up. If we view the capacity of the biosphere to absorb pollution as a natural resource itself—Schumacher called it "the tolerance margins of nature"—then the saturation of the biosphere with pollution also constitutes a using up of natural capital. Future generations are effectively subsidizing current resource use whenever the long-term costs of resource depletion or environmental pollution are ignored. Progeny will pay the externalized costs of today's productivity.

John Locke, the seventeenth century British philosopher, is the conceptual father of the economic practice of externalizing the costs of resource depletion. Locke's labor theory of value maintained that natural resources assume worth only by way of the labor that humans put into them. In turn, Locke proposed that a person gains property rights to that which he labors to produce. Thus, land becomes property and achieves value when plowed for the first time, just as acorns become property and become valuable once they are gathered. For Locke, the natural world only gains worth once it has been tilled or mined or harvested, and then it becomes the sole property, effectively the income, of the tiller, miner, or harvester.

In an inexhaustible and largely unpopulated world, the type of world John Locke assumed America to be at the time of its colonization, counting natural capital as income would be legitimate. If no depletion of resources occurs, then we may consider these resources to bear economic value only to the extent that labor is invested in them. Perhaps the salt in the sea is an example of a virtually inexhaustible resource. The worth of sea salt is approximately equal to the economic costs of its extraction. Few things in this world are as inexhaustible as sea salt, however. Environmental economist Herman Daly has suggested that only in an "empty world," where resources vastly outstrip demand, would we be justified in counting the consumption of natural capital as income. "In today's full world it is anti-economic."[77] On a finite planet, the extraction of resources represents not only labor expended but natural capital depleted. One might think of it as capital taken from future generations. If we assume the earth's resources to be the common stock of humankind, then it is capital taken from one's contemporary neighbors as well.

The economic free market might be coaxed, prodded, and lured into assuming more environmental responsibility. Currently, however, is not well attuned to the ecological costs of doing business. The invisible hand guiding the market decidedly lacks a green thumb. That was the conclusion reached by the National Commission on the Environment, convened in 1990 by the WWF. Its report, entitled *Choosing a Sustainable Future*, warns that "[t]he nation must stop stealing the environmental capital of future generations and must live instead on its fair share of nature's interest."[78] An activist with a local organic farming project who held sustainability to constitute the "underlying principle" of environmentalism defined the term as "controlling your greed for the long-term benefit of the next generations." Not spending nature's capital and controlling one's greed to make do with a fair share of nature's interest are the central tenets of a sustainable economics.

The freedom of the market often saddles future generations with the burden of making ecological reparations. To add insult to injury, the subsidization of resource depletion by future generations is often carried out in tandem with its subsidization by present-day taxpayers. Current public land use practices provide a good example of this double externalization of costs.

On much western public domain, livestock grazing has eroded the land, degraded riparian areas, and reduced biodiversity. Yet current grazing prices on public lands are far less than those charged on private lands. In 1992, for instance, grazing fees on public lands administered by the Bureau of Land Management were just over one fifth the fees charged on private lands, a public subsidy for ranchers that effectively cost taxpayers $55 million. All in all, private livestock grazing on public lands is estimated to cost taxpayers between $200 million and $400 million a year.[79] When indirect costs, such as wildlife eradication programs and water projects are accounted for, the figure rises significantly.[80] The term "welfare ranching" has been employed by environmentalists to describe the situation. To make matters worse, only 12 percent of the grazing permit holders are small ranchers. The top 10 percent of permit holders control half of the nation's public grazing land.[81] These large business interests are well organized, and in 1995 defeated a two-year effort by Interior Secretary Bruce Babbitt to raise grazing fees on public lands.

In a similar vein, the Forest Service has an ongoing practice of selling federally owned timber, including old-growth trees, to private companies at prices well below the costs that the Forest Service incurs building logging roads, mapping and managing sales, and reseeding logged-out areas. The Sierra Club estimates that between 1908 and 1991, the logging program in national forests operated at a net loss to taxpayers of $7.3 billion. From the late 1970s until the early 1990s, average yearly forest timber program expenditures exceeded receipts by $293 million. Losses have risen steadily since. According to the General Accounting Office and the White House Council of Economic Advisers, the Forest Service lost over $1.2 billion in its timber program between 1992 and 1995. In Alaska's Tongass National Forest alone, the world's largest remaining temperate rain forest, the federal government spent $389 million in six years servicing clear-cutting operations that provided only $32 million in revenues.[82] These massive taxpayer subsidies of the forest industry do not reflect the economic losses caused by logging that are incurred by other tradespeople, businesses, and communities that depend on the forest ecosystem, such as salmon fishers whose natural resources are threatened by logging practices that degrade rivers and streams in which salmon spawn.

The "principle" of public lands, according to NRDC, is that they are intended to be "held in sacred trust for all future generations."[83] Yet national forests and other public lands are frequently managed in an unsustainable manner. Profit takes priority over preservation. These private profits are effectively siphoned from the public trough. Jon Roush, president of the Wilderness Society, has estimated that "[t]axpayers have subsidized the destruction of about ninety percent of our ancient forests, our riparian habitat, and our native grass."[84] This represents the height of externalizing costs in space and time.

Unfortunately, the unsustainable and publicly subsidized exploitation of natural resources is not limited to national forests and grasslands. A perennial target of environmentalists is the Mining Law of 1872. This legislation, originally signed by President Grant, permits hardrock mining on public lands without fair market royalties being paid or the costs of land rehabilitation figured into leasing agreements. Currently, mining companies pay no royalties to the federal treasury for the $3 billion worth of minerals taken from public lands each year. Total royalties forgone amount to over $240 billion. Moreover, the taxpayers' bill for the cleanup of many of the half a million abandoned hardrock mines in the United States is estimated to run between $30 billion and $72 billion.[85] Almost all environmental groups also call for an end to the generous government subsidization of water projects, which typically deplete water tables and lead to the salinization of irrigated lands. In the United States, irrigation is subsidized at a cost of $2.5 billion per year. Environmentalists also oppose the subsidization of fossil fuel and nuclear energy production, currently running at $27 billion per year. This subsidization effectively stimulates the production of toxic wastes and pollution and, in the case of fossil fuel energy, greenhouse gases. (Worldwide, water projects consume $50 billion in subsidies each year while annual global energy subsidies amount to $450 billion.)[86] Another focus of concern is urban and suburban sprawl, which paves over green spaces, destroys wetlands, and deprives wildlife of habitat. Yet real estate developers are routinely subsidized by taxpayers, who pay for the roads, schools, fire and police protection, and other services required by new housing and commercial projects. This subsidy may amount to $10,000 per new suburban home.[87] Once again, resource depletion is effectively subsidized twice: once by current taxpayers and once again by future generations who will bear the costs of a degraded natural environment.

According to the Office of Management and Budget and Congress's Joint Committee on Taxation, American taxpayers in 1994 contributed $51 billion in direct subsidies to corporations and forfeited another $53 billion in corporate tax breaks.[88] Not all of these subsidies and tax breaks promote unsustainable practices, but many do. Beginning in 1993, Friends of the Earth and the National Taxpayers Union have worked together in a coalition called Citizens United to Terminate Subsidies (CUT$). The coalition has spearheaded the "Green Scissors" campaign, which unites free-marketers, fiscal conservatives, environmental progressives, and taxpayer activists in an effort to end "polluter pork." The campaign is backed by over two dozen other environmental organizations and citizen advocacy groups. In each of the last three years, the Green Scissors Report challenged Congress to snip between $33 billion and $39 billion from the federal budget and terminate between thirty-four and fifty-seven environmentally harmful federal programs.[89] In addition to the recommended cuts in spending and subsidies, Friends of the Earth has also exposed environmentally destructive tax loopholes, mostly given to extractive industries, that cost the country's taxpayers $4.5 billion each year.[90]

These subsidies and tax breaks, to use Daly's expression, are anti-economic. Yet they constitute an integral part of the standard approach to natural resource exploitation and economic development. They are intrinsic to what Garrett Hardin calls the CC-PP game: Commonized Costs–Privatized Profits.[91] That is not to say that industries and enterprises that remain unsubsidized by public monies are, on that count alone, environmentally sustainable. Cutting "polluter pork" in itself does not ensure that future generations will receive fair treatment. Many unsustainable practices are carried on without state support. Environmentalists point out, however, that abolishing antiecological and antieconomic subsidies would sharply decrease the injury done to future generations while leaving current generations feeling less insulted come tax time.

Greening Business and Getting the Price Right

Given the tyranny of the immediate that reigns over most economic thought and practice, the term *sustainable economics* might seem oxymoronic. Nonetheless, developing the art and craft, and perhaps the science, of a sustainable economics remains the primary task for many environmental organizations.[92] Some, such as Co-op America, do nothing else. Supporting and promoting "green businesses" and cutting "polluter pork" are two of the more common efforts environmentalists engage in to develop a sustainable economics. A more ambitious and elusive goal is *full-cost accounting* or *full-cost pricing*.

The idea behind full-cost accounting is deceptively simple: the price we pay for goods and services should factor in their long-term, social and ecological costs. Full-cost accounting reflects the costs of natural resource depletion, pollution mitigation, health maintenance, and waste disposal, costs generally hidden from consumers as externalities. Stephan Schmidheiny, who led the Business Council at the 1992 Earth Summit, observes that the key to ensuring sustainable business practices is "getting the price right." Schmidheiny wrote that "[u]nless prices for raw materials and products properly reflect the social costs, and unless prices can be assigned to air, water, and land resources that presently serve as cost-free receptacles for the waste products of society, resources will tend to be used inefficiently and environmental pollution will likely increase."[93] Some of the more straightforward proposals for full-cost accounting are geared toward the levying of "green taxes." Many environmental groups endorse a tax on gasoline, for instance, that would pay for the costs of building and maintaining highways and fighting the pollution caused by the exhausts of internal combustion engines. Pollution taxes that would achieve similar results for industrial facilities are also widely endorsed, as are taxes on resource and energy use and waste production. These green taxes are generally proposed in conjunction with graduated decreases in income, property, labor, savings, or investment taxes. The intent is to ensure that there is no net increase of the overall tax burden on individuals and to compensate low-income individuals and families for any of the increased costs passed on to them by industry.

Business critics frequently complain about the cost of environmental regulations, usually in the politically strategic but misleading antinomy of jobs (or profits) lost in a regulated market versus jobs (or profits) gained in an unregulated market. One of the benefits of full-cost accounting is that it achieves environmental protection within a market paradigm. If full-cost accounting were adopted, businesses could be less regulated—in the sense of having fewer governmental restrictions imposed on their means of production—because they would simply pay for the social and ecological costs of production up front and then directly pass on these costs to the consumer. In an era of increasing resistance to bureaucratically administered government regulation, full-cost accounting may be a viable means of promoting a green market economy. Without restricting consumer choice, it employs market-based incentives to ensure sustainability.[94]

In general, support within the business community for full-cost accounting has been tentative and reserved. Few corporations have attempted to integrate its principles into their business practices because of its perceived threat to profit margins. At best, these principles receive lip service. One hears William D. Ruckelshaus, CEO of Browning-Ferris Industries, acknowledge that "economic activity must account for the environmental costs of production" and that "the market has not even begun to be mobilized to preserve the environment; as a consequence, an increasing amount of the 'wealth' we create is in a sense stolen from our children."[95] In the same vein, the Declaration of the Business Council for Sustainable Development states that "the prices of goods and services must increasingly recognize and reflect the environmental costs of their production, use, recycling, and disposal. This is fundamental, and is best achieved by a synthesis of economic instruments designed to correct distortions and encourage innovation and continuous improvement, regulatory standards to direct performance, and voluntary initiatives by the private sector."[96] Such statements are inspiring. As always, however, actions would speak louder than words. The above declarations of support for full-cost accounting and other economic devices designed to encourage sustainable business practices were written or endorsed by the CEOs of some of the largest and most environmentally destructive corporations in the world, such as Mitsubishi, Chevron, Royal Dutch/Shell, DuPont, Dow Chemical, and Browning-Ferris Industries. By heartily endorsing the theory of full-cost accounting, it seems, many businesses hope to escape the task of putting it into practice.

Ernst U. von Weizsäcker observed that "[b]ureaucratic socialism collapsed because it did not allow prices to tell the economic truth. Market economy may ruin the environment and ultimately itself if prices are not allowed to tell the ecological truth."[97] In the long term, telling the ecological truth will prove necessary. But it is not as easy as it sounds. Full-cost accounting entails the discovery and disclosure of the complete life cycles of products, including an estimation of the social and ecological costs of the extraction and depletion of raw materials, the manufacturing process, the services involved in distribution and marketing, and the products' use, recycling, or disposal. Acquiring the data and conducting the studies needed to get the price right present hefty challenges.

The full-cost price of commercially produced fruits or vegetables, for instance, would include (1) the ecological costs of their agricultural production, such as soil erosion, the eutrophication of water sources caused by the use of synthetic fertilizer, and the bioaccumulation of contaminants caused by pesticide applications; (2) the ecological costs of their delivery to consumers, such as the pollution (and the highway maintenance costs) caused by their trucking to market and the energy depletion caused by their refrigerated transport and storage; and (3) the ecological costs of managing the wastestreams of their discarded packaging and noncomposted remains. So calculated, the full-price cost of monocropped fruits and vegetables produced by agribusiness and sold on supermarket shelves would rise well above that of much organic produce purchased at farmers markets or retail outlets. Currently, organically produced food only accounts for 1 to 3 percent of the U.S. food sales. Its higher price—a function of the costs that its nonorganic competitors externalize—limit its demand.

The present total unpriced costs of the U.S. food system are estimated to be between $150 and $200 billion per year. Whereas traditional accounting methods reckon an average $80 profit per acre, full-cost accounting indicates a $26 per acre loss.[98] Full-cost accounting of imported foodstuffs may demonstrate an even greater disparity. It has been estimated that the full-cost price of a hamburger produced on pasture cleared from tropical rain forests might be over hundred times its current market price.[99] Were full-cost accounting introduced, therefore, one might expect sustainable production and marketing practices to replace less sustainable ones—purely for economic reasons.

The full-cost price of automobiles and gasoline, to take another example, would rise significantly, making public transport much more economical. Present gasoline taxes, vehicle taxes, and road tolls cover less than two thirds of the total capital and operating costs of highways. The tens of billions of dollars of operating costs not covered each year by such user fees generally come out of local taxes. In turn, the indirect costs associated with automobile travel, such as highway law enforcement, tending to accident victims, and smog abatement runs up to hundreds of billions of dollars each year. Were these costs figured into the price of motor vehicle fuel, the cost of gasoline would rise by $3 to $7 a gallon.[100] It would rise further if other collateral effects were figured in. The destructive effect of car emissions on the nation's wheat, corn, soybean, and peanut crops, for instance, amounts to losses of $1.9 to $4.5 billion annually.[101] Motor vehicle pollution, in turn, costs Americans between $40 and $50 billion in annual health care expenditures, and causes as many as 120,000 unnecessary or premature deaths. With these expenditures assessed, the full-cost price of gasoline might rise to $11 per gallon.[102] Pricing reform, environmentalists have concluded, would have the effect of "taking cars off welfare."[103]

Along with public transport, alternate forms of energy production would become more economically feasible were full-cost accounting applied to fossil fuel consumption. Some 64,000 deaths from heart and lung disease are caused each year by particulate air pollution in the United States. If the cost of caring for these patients and the cost of lost productivity and premature death were added to the price of burning coal or oil, these fossil fuels would become more expensive than many forms of solar power.[104] Full-cost accounting, in short, represents a future-focused economics that is geared to sustainability. In many cases, the time horizon need only be expanded a few decades to demonstrate the antieconomics of current practices.

Today semitrailers carrying fruits and vegetables from corporate farms in Florida and California pass each other on highways en route to retail outlets in each other's state. This system of production and allocation seems irrational when compared to one based on locally grown and consumed produce. The typical American meal travels 1300 miles from its various points of production to the dinner plate. The economic system that encourages such long-distance nutrition may yield short-term profits. The long-term view is much less reassuring. The point is not that interstate (or international) business activity should be stopped or unnecessarily hampered. The point is simply that the goods and services we buy and sell need to be priced to account for the long-term environmental costs of their production, delivery, use, and disposal. Otherwise, our present savings are made at the expense of future generations.

Full-cost accounting is intrinsic to the field of "industrial ecology," also known as the "science of sustainability," which first served as a topic of discussion at a colloquium held by the National Academy of Sciences in 1991. Industrial ecology employs life-cycle analysis and integrates "design for environment" strategies into the choices that businesses make about material and technology use, and product design. Industrial ecology seeks to "move beyond the regulatory and organizational barriers that single-media, single-species, single-substance and single-life-cycle-stage approaches create to a more holistic and longer-term consideration of environmental degradation."[105] As a preparatory exercise, Jonathan Lasch of the World Resources Institute co-chaired the President's Council on Sustainable Development. Participating environmentalists pushed the council to request the 100 largest U.S. corporations to prepare "sustainability reports" that would provide life-cycle analyses of their major raw materials and products "from the point of extraction from the earth through to ultimate disposal."[106] While the report endorsed life-cycle analyses, the request to manufacturers was never made.

Before components of a sustainable economics could gain a conceptual let alone practical foothold, the common measures by which our standard of living is assessed would have to change. Currently, quality of life is almost universally measured in terms of gross national product (GNP) or gross domestic product (GDP). Yet GNP does not take any of the collateral effects of economic activity into account. It has no future focus. Gross national product is simply a measure of market activity, of money changing hands. Environmental organizations were quick to publicize the fact that the *Exxon Valdez* disaster actually raised the nation's GNP. That is because the cleanup efforts were formally recorded as contributions to the nation's economic productivity. Certainly the quality of life in America was not enhanced by the release of eleven million gallons of oil onto Alaska's seas and shores. Yet standard economic indicators would have us believe so. Less controversial but no less significant examples abound: walking, biking, or taking mass transit to work contributes less to the GNP than the private use of an automobile. Wearing a sweater on winter nights contributes less than raising the thermostat. "For all practical purposes," Al Gore notes in *Earth in the Balance*, "GNP treats the rapid and reckless destruction of the environment as a good thing."[107] As Schumacher intimated and Daly suggests, "greening the GNP" would necessitate our counting the depletion and degradation of natural resources as a debit, not a credit.

When a forest is felled, the loss of trees, wildlife habitat, and healthy riparian areas is not recorded in the nation's accounting books. The sales of board-feet of lumber or cellulose pulp form the only columns in the ledger. The same occurs whenever topsoil is eroded in agricultural enterprises, oil is pumped from the ground for automobile consumption, or aquifers are depleted for irrigation. Gross national product, in short, is not an accurate measure of quality of life, and certainly not of a sustainable quality of life. An executive with Audubon gestured at the change needed in our assessments of national welfare when she said, "One of the most important figures in the newspaper are housing starts, they're an indication of how well we're doing; I'd like to see figures like how many new trees were planted." Indeed, a more comprehensive measure of quality of life would include numerous indicators of social and environmental welfare, such as human morbidity and mortality rates, ecological health indicators for agricultural land, forests, rivers, lakes, and coastal regions, levels of air and water purity, educational standards, presence of wildlands, and biological diversity.

Only with such measures included would the implementation of full-cost accounting become a feasible means of assessing economic indicators. Toward this goal, an Index of Sustainable Economic Welfare has been developed.[108] In sharp contrast to the rosy indications of the GNP, the index has shown a gradual decline in the nation's welfare of about 12 percent since 1976. At the urging of the NRDC and other environmental groups party to the President's Council on Sustainable Development, a recommendation was made to change the way the GNP is calculated to include the value of natural resources and the costs of their depletion.[109] Whether this recommendation is put into practice remains to be seen.

In the United States, green economic indices have not been widely accepted by the business community or the political establishment. In other countries such as France, Germany, Japan, the Netherlands, New Zealand, and Norway, alternate measures, such as the Net National Welfare or the Net National Product, are more widely employed. These indices are calculated by subtracting from the GNP the money spent protecting against and repairing environmental damage, the monetary equivalent of environmental degradation remaining after such protective and reparative expenditures have been made, and an allowance for the depletion of natural resources. With an eye to promulgating these practices, NWF and other environmental groups co-sponsored a conference with the World Bank called Accounting for the Future. As a result, the World Bank, International Monetary Fund, and a number of United Nations agencies have explored means of recalculating GDP to account for environmental losses and gains.[110]

Most environmental organizations endorse the principles behind full-cost accounting. Few have fully explored its potential or concrete application. For some, the difficulty is simply pragmatic. Getting the price right entails extensive data gathering and a good bit of extrapolation. For others, the problem is more principled and strategic. Full-cost accounting ultimately boils down to an effort to translate every human and nonhuman good into its monetary equivalent. Yet many goods are not (easily) measurable in economic terms. Perhaps many goods should not be so measured. In turn, things that can easily be measured in economic terms, such as the costs to industry of environmental regulation, tend to be disproportionately weighed against things that are less easily measured, such as public health and welfare and biological diversity. Efforts to promote full-cost accounting may thus play into the hands of antienvironmentalists who are eager to subject all environmental concerns to the laws of the market.

Environmentalists worry that the use of economic devices such as "cost-benefit analysis" may actually undermine the intent of full-cost pricing efforts and waylay efforts to institute environmental regulations and legislation. Cost-benefit analysis is an attempt to weigh the economic costs of proposed regulations (to industry and consumers) against the economically quantified benefits that would be achieved were those regulations enforced. Faced with the problem of weighing concrete economic costs against more ambiguous predictions of health and human welfare benefits, the cost side of the equation often wins out. Projected economic costs, however speculative or inflated, generally carry the most weight, leading to the "dwarfing of soft variables." Environmentalists contend that the effort to quantify environmental protection may prove its undoing.

Take the example of air pollution. Industry argued that conforming to the 1990 Clean Air Act, specifically the reduction of sulfur dioxide emissions from smokestacks, would cost up to $1500 per ton of emissions. Industry claimed that the high costs and negligible benefits made this regulation unwise. Government estimated the cost at $600 per ton, and figured this an acceptable price. The actual price paid in 1994 was $150 a ton and by the spring of 1995 it had dropped to $128 a ton, widely perceived to be a pretty good bargain for cleaner air.[111] Even in the relatively simple case of reducing sulfur dioxide emissions, highly inflated estimates of the economic costs were quickly made. The benefit to citizens of reduced pollution in terms of increased good health, bluer skies, and cleaner rivers and lakes is patent. Yet these benefits remain difficult to assess in monetary terms. Consequently, their value is easily deflated.

Cost-benefit analysis is grounded in the assumption that everything has a price and is for sale. Not everything of value in life is bought and sold in the marketplace, however. To determine the dollar value of nonmarket goods, economists have developed the practice of "shadow pricing." In determining shadow prices, economists examine "bundled goods" traded on the market that vary only by one or a few nonmarket goods. Thus, peace and quiet is not a marketed commodity, but houses near airports or freeways are. Economists therefore assume that the market value of tranquility is equivalent to the price difference between similar houses that are located near and far from airports or freeways. Here, however, economists may not be discovering the value most people place on peace and quiet. Rather, they may be measuring what those people who are *least* sensitive to noise are unwilling to pay for less noise.[112] The answer, presumably, is not too much.

The effort to assign shadow prices to certain goods may also be demeaning of those goods. Imagine having to put a price on the love, health or life of a child or friend. We speak of these goods, as "priceless." Many environmental goods, such as wilderness, also resist direct translation into market terms. Presumably we value wilderness for the same reason that we value love or friendship—because in a significant sense it is, or should be, beyond economic calculations.

Cost-benefit analysis of public health concerns raises similar issues. In the face of pollution-induced disease we can measure the increased costs of medical care and the costs of human resources lost to sickness and early death. Yet spending money to treat disease or to compensate bereaved families and businesses does not produce the same level of human welfare as is obtained by preventing disease in the first place. Moreover, pricing out the reparative costs of environmental hazards or destruction does not address the potential infringement of the rights of those harmed. As a writer for NRDC has observed, the cost-benefit calculus "sidesteps" basic questions of justice: "Imagine a law requiring federal agencies to measure the worth, in dollars, of free speech or civil rights before acting to enforce them: any attempt to weigh an intangible public good against hard currency will shortchange the former, whether it is the right to vote or the value of protecting a human life against cancer."[113] The point is well taken. Yet public goods are not necessarily "intangible." They are simply difficult to measure in economic terms and may be demeaned when so measured. Calculations become even more problematic, if not impossible, once the welfare of future generations is taken into account.

The problem of assessing the costs and benefits of protecting endangered species is a case in point. The preservation of the nation's native species is not as great a drain on tax dollars as is often suggested. The average annual allocation for the Endangered Species Act (ESA) has been between $40 million and $60 million. This approximates the cost of constructing a mile of urban interstate highway.[114] The ESA, moreover, has not proven much of a threat to property rights. The number of residential, commercial, agricultural, or industrial projects actually stopped because of the ESA ranges between 0.03 and 0.06 percent of all projects surveyed. This means that of the hundreds of thousands of projects evaluated since the ESA came into existence, an average of under five per year have been stopped. It has been said that developers stand a greater chance of an airplane crashing into something they have built than having their projects terminated because of the ESA.[115]

Still, if very few projects have been terminated, thousands have been altered because of the ESA. If that were not the case, the ESA would not be worth the paper it is written on. As one might expect, some of these alterations have proven costly. Southern California scrub land, for instance, is home to gnatcatchers, a threatened species of bird. At the current level of development, all gnatcatcher habitat would be destroyed within twenty years. It has already diminished from 2.5 million acres to less than 400,000 acres since 1940, as the human population of Southern California quintupled. The gnatcatcher's habitat is prime real estate. It can be sold for between $200,000 and $3 million an acre. The economic cost of not developing the land is therefore easily determined in the real estate marketplace. The benefits to society of preserving the gnatcatcher's habitat are not so easily figured.

How does one "get the price right" for the preservation of this habitat? It is very unlikely that an ecological collapse would result from the demise of the gnatcatcher. Yet even from a purely human point of view, forgetting for the moment the potential "rights" of gnatcatchers, saving what is left of natural southern California has its obvious public benefits. The natural beauty of the land, after all, is one of the reasons people had for moving to southern California in first place. How would one begin to assess the economic coefficient of this aesthetic benefit? How, in turn, would one assess the ecological benefit of the land for its capacity to preserve biodiversity?[116] Finally, how would these aesthetic and ecological benefits be extrapolated to account for the welfare of future generations? As Ophuls and Boyan have suggested, "even with perfect information, the economists could not answer most of these questions, for they involve political, social, and ethical issues, not the issue of efficient resource allocation that neoclassical, marginalist economics was designed to handle. They are 'trans-economic' questions."[117] Faced with transeconomic questions, a cost-benefit assessment would likely favor the real estate developers who have figures at their fingertips.

Forests viewed in terms of their potential board-feet of timber bear easily ascertainable economic values. In contrast, the value of preserving forests for their capacity to clean the air, filter water, conserve soil, aid pollination, benefit fisheries, regulate climate, shelter animal and plant life, harbor genetic and medicinal resources, maintain biodiversity for scientific and educational purposes, and provide recreational opportunities for hikers, campers, and sightseers is impossible to calculate accurately. The task is made even more difficult if we wish to assess the value of these services for innumerable future generations. When assessments are made of the dollar value of nontimber forest products, keeping the trees standing often comes out ahead.[118] Yet much of the value of forests is not economically quantifiable. Transeconomic values, environmentalists insist, should not be dismissed.

In sum, there are at least half a dozen reasons why a reliance on cost-benefit analysis as it is currently practiced is problematic. First, it is a notoriously imprecise endeavor. Second, cost-benefit analysis tends to undermine the idea that certain goods are basic rights that should not be subject to economic trade-offs. Third, cost-benefit analyses that do assess the value of nonmarket goods through shadow pricing tend to undervalue these goods and court the danger of demeaning them through their reduction to a monetary equivalent. Fourth, cost-benefit analysis is methodologically predisposed to underweight costs borne by future generations. Fifth, cost-benefit analysis easily becomes a means of reinforcing rather than restraining technological and economic growth. As such, it undercuts the search for alternative, more ecologically sound means of conducting our lives. Sixth, cost-benefit analysis as it is currently practiced is not democratic. As one critic observed, "The main cumulative impact of cost-benefit analysis may be in legitimating the idea that public policy formation is a matter for technical, expert choice and not a question on which non-specialists such as elected officials, still less any broader public, have any rightful say."[119]

In light of these concerns, K. S. Shrader-Frechette argues for the establishment of a technology tribunal, where a citizen jury would make informed decisions about the costs and benefits of such matters as the production and use of synthetic chemicals or the initiation of environmentally dangerous industrial or technological processes. The idea might be extended to encompass greater democratic control over environmentally threatening commercial development.

Democratic influence is a necessary but not a sufficient condition of good public policy making. It should accompany, not replace, science and technical expertise. Shrader-Frechette has argued that democratic decision making need not produce poor policy recommendations because "good policy is not merely a matter of discovering highly technical *truths*, but also a question of attempting to guarantee *justice*. If one recognizes that the purpose of adversary proceedings is not primarily to establish some empirical point, but instead to attempt to provide conditions under which policy can be established in a fair, orderly, timely, and representative manner, then there seems to be no strong case for claiming that such democratic procedures are likely to yield poor policy."[120] Environmentalists worry that technological and economic growth is currently treated as an end in itself, rather than a means to an end. A technology and economic development tribunal would provide one means for allaying these fears.

Environmentalists cannot wholly avoid the treacherous waters of cost-benefit analysis. They must participate in the effort to get the price right. If they spurn involvement in such activities, the likelihood is that cost-benefit analysis will continue to be employed in a very suboptimal manner, which is to say, in a way that undervalues or ignores ethical and environmental concerns.[121] Inevitably, however, environmentalists who critically engage cost-benefit analysis will find themselves pitting the ambiguous benefits of environmental protection against an arsenal of figures detailing economic costs. The primary means at their disposal to combat the tyranny of the immediate that rules over such calculations remains a future focus. This future focus is best informed by scientific knowledge, a generous dose of caution, and sensitivity to those enduring values that cannot be measured in dollars and cents.

Jobs versus Spotted Owls

Nowhere has the purported trade-off between economic gains and environmental goods been more controversial than in the "jobs versus the environment" debate. Indeed, the threat to job security posed by environmental protection has assumed mythical proportions. As many as one in three Americans believes that his or her job is put at risk by environmental regulation.[122] The possibility of losing jobs to environmental protection has proven itself a bugaboo that makes even ecologically oriented politicians cringe. Elected representatives are very reluctant to endorse any environmental proposal that threatens job losses, even when these

losses are relatively minor or impermanent. The vociferous debate over
the spotted owl in the Pacific Northwest is a case in point.

Each year, over three million Americans lose their jobs. The vast
majority of these job losses have nothing to do with environmental regulations. Based on Labor Department figures, the Economic Policy Institute calculated that between 1987 and 1990, only 0.1 percent of job
losses resulted from environmental regulations. Change of company
ownership caused thirty-five times more layoffs.[123] Between 1992 and
1996, AT&T cut 123,000 jobs, IBM dropped 122,000 workers, GM
laid off 99,000 employees, and Boeing trimmed its workforce by
61,000.[124] In contrast, it has been estimated that only 9,000 to 12,000
logging-related jobs will be lost in the Pacific Northwest over the next
fifteen years owing to the protection of the spotted owl and its old-
growth habitat. Industry figures are considerably higher, but they do not
even begin to approach the number of jobs already lost to automation
or corporate "down-sizing."

Meanwhile, other industries in the Pacific Northwest, such as tourism, are likely to produce more jobs as a result of enhanced resource
protection. In some states, income from recreational tourism surpasses
that from logging and mining combined.[125] The Forest Service has calculated that recreation yields four times the economic benefits of timber
cutting and creates sixteen times as many jobs. In an independent study,
John Berger has concluded that recreationalists, hunters, and anglers
contribute thirty-seven times more income to the economy and produce
thirty-two times as many jobs than does logging in the national forests.[126]

Curtailing logging operations to protect the habitat of endangered
species or to preserve old-growth forest will lead to short-term job losses.
Undoubtedly, such losses will be painful to many individuals, families,
and communities. These economic and social costs should not be denied.
Yet one must approach the debate of jobs versus the environment with
an expansive temporal framework. As John Adams of the NRDC argued, "In the long run, every job in this country depends on our natural
resources. And that is the central thrust of NRDC's work: we take the
long view. Our role and our obligation are to look after the long-term
needs of society."[127] Like Adams, most environmentalists respond to the
jobs versus the environment standoff by arguing that the conflict dissi-
pates as soon as a sufficiently long-term perspective is employed. "Ultimately, if we don't protect the environment, then we're going to destroy
the people who live in the environment," one Sierra Club activist ob-

served. "I don't think there's a real conflict here [between jobs and the environment], rather it's a question of long-range versus short-range interests." A staff member for WWF likewise observed that "you have to take a look at that [problem] in two different time frames, the near-and long-term. In the long-term, the objective that we have is to provide a permanent alternative to [having to make] the choice [between jobs and environmental protection]." With the American economy increasingly oriented to services rather than resource extraction, environmentalists note, the choice is becoming even clearer. It is not a matter of the environment *or* jobs. The choice is between the environment *and* jobs or very little of both.

One activist with the Greens involved in a community organic gardening project described her activities thus: "I want to be associated with projects that show that human development and environmental protection go hand in hand, that they depend on one another." The gardening project, she went on to note, demonstrated that small-scale, organic agriculture was both "good for the earth and economically rewarding." Ever since the organization Environmentalists for Full Employment was formed in 1975, the claim has been made that in the long term, and often in the short term, environmental protection creates jobs and stimulates the economy. Whereas early efforts typically focused on the benefits of recycling over incineration or landfilling (recycling creates more than four times as many jobs as incineration and nine times as many jobs as landfilling), environmentalists today have considerably expanded their analyses. The economic viability of environmental care has recently been demonstrated by elaborate economic studies and concrete projects taken on by businesses.

Energy efficiency and waste reduction can be profitable. The 3M Company, a leader in environmentally responsible business, estimates that its "Pollution Prevention Pays" program has saved $537 million and reduced air pollution by 120,000 tons, wastewater by one billion gallons, and solid waste by 410,00 tons over a fifteen-year period. Another $650 million was saved in energy conservation. The company has set itself the goal of zero emissions and expects to reap further savings.[128]

Anecdotal evidence and theoretical econometrics have produced conflicting reports about the effect of environmental protection on the economy as a whole. Fortunately, empirical studies have been more uniform and more uniformly favorable to environmental protection. Over the last two decades, those states with the most ambitious environmental programs in the nation have demonstrated the highest level of economic growth and job creation. On a global scale, findings are similar. Those countries with the most stringent environmental laws also demonstrate the highest rates of economic growth and job creation. Having lax environmental standards does not facilitate the creation of jobs or stimulate economies. Rather, it temporarily insulates wasteful, inefficient, and polluting industries while penalizing those firms that do innovate.[129] Still, the misleading rhetoric is in place. In 1995, Republican Speaker of the House Newt Gingrich labeled the Environmental Protection Agency a "job-killing machine" in an effort to justify its fiscal gutting.[130]

Business interests claim that environmental regulations cost them $125 billion a year.[131] That is a hefty sum. Yet it represents only 1 to 2.5 percent of the total costs of production.[132] A study by the National Federation of Independent Business indicates that small businesses rank compliance with environmental regulation eighteenth on their list of the financial problems that they face, well below the costs of health care, federal taxation, liability insurance, and workers' compensation.[133] That is not surprising, given that the average price effect of current environmental regulations on consumers is only 1.5 percent.[134] In turn, only 2 to 3 percent of taxpayers' total federal contributions goes to environmental and natural resource protection as a whole. This compares to almost half of federal taxpayers' money that is sunk into defense-related spending.[135]

Notwithstanding glaring exceptions such as Superfund (an egregious two thirds of its expenditures pays for consultation and litigation rather than actual cleanup efforts), most monies devoted to environmental protection have proven cost-effective. The EPA recently weighed the economic costs and benefits to industry and private citizens of its enforcement of the Clean Air Act between 1970 and 1990. About $20 billion was spent by industry to conform with the regulations in 1990. Yet $400 billion were saved by the nation in lower hospital costs, decreased medical expenses, fewer lost workdays, and greater productivity than would have resulted without the abatement of smog and polluted air. The rate of return for air pollution control, in other words, is $20 in savings for every dollar invested.[136]

Gregg Easterbrook devotes most of his book *A Moment on the Earth* to criticizing the excesses of environmentalism. Yet he acknowledge that "[e]nvirommental protection has brought to the public visible benefits within the lifetimes of the taxpayers who made the investments. Unlike many other government programs, almost every dollar invested in the environment has yielded clear improvements, usually at a reasonable price." Environmental protection, Easterbrook goes on to say, is responsible for "saving capitalism from itself by compelling industry to function cleanly and with resource efficiency while there was still time to make this transition smoothly."[137] Unfortunately, the transition has not gone as smoothly as it might have. One of the prices paid for the Reagan administration's environmental deregulations, which were revisited by the Republican Congress of the mid-1990s, is the loss of American industry's competitive edge in the global market for environmental technology. The beneficiaries of that loss, to the tune of billions of dollars in annual revenues, are primarily the corporations of Japan, Germany, Canada, and the Scandinavian countries.[138] The world market for environmental goods and services, one of the fastest growing markets today, is estimated at $4 trillion. Currently, the U.S. share of this market is declining.[139]

Environmentalists have long argued that spending money to protect the environment is good for the economy. Economists are starting to agree. Environmental protection, economic opportunity, and market competitiveness can go hand in hand.[140] Conservation measures taken by industries are proving profitable. Life-cycle analysis, for instance, has become a cost-effective money-saver for business.[141] Environmental protection is widely considered a growth industry. By 1992, the environmental protection industry had generated $355 billion in total industry sales and $63 billion in federal, state, and local government tax revenues. It employed four million Americans. Depending on how one defines the sector, the environmental industry in the United States constitutes up to 3 percent of total employment. By 2005, 5.4 million Americans are expected to be employed in the area of environmental protection.

Analysts rightly warn that environmental protection can be disruptive to specific industries and that poorly designed and implemented environmental protection programs, laws, and regulations often do more harm than good. Standard forms of regulation often set a floor as well as a ceiling to permissible discharges, undermining incentives for industries to pollute less than the laws permit.[142] There is a greater need for "innovation-friendly regulation" that fosters creative means of pollution prevention rather than technology-specific, end-of-the-pipe compliance. Regulation should stimulate the development of clean technology instead of simply forcing industries to invest in cleanup technology.[143] Market-oriented, incentive-driven, innovation-stimulating approaches to environmental protection should be experimented with and selectively employed as alternatives to the traditional "command and control" regulation.

After an extensive survey of the available literature, Roger Bezdek concluded that "[d]espite one's personal opinion of the environmental movement or of particular environmental programs or activists, environmental protection, rather than being a drag on the economy, can represent a major profit-making, job-creating, tax revenue-generating opportunity for this nation. Further, until the debate shifts from artificial tradeoffs, such as 'jobs versus spotted owls' and 'profits versus energy efficiency,' the United States will be at a competitive disadvantage because other nations have already identified environmental technology as a 21st-century winner and are investing in it heavily."[144] The economic profitability of conservation is demonstrable. The strength and viability of the national and international environmental business sector are evident. The overall economic impact of environmental protection is positive.

Environmentalists cannot and should not avoid debating the economic costs and benefits of environmental protection. The debate is necessary because economic welfare and steady employment are goods that people value and are reluctant to risk. Environmentalists must address these issues squarely. In doing so, they should not uncritically celebrate all tax revenues, profits, or jobs generated by environmental protection. A significant portion of these goods are generated from after-the-fact pollution cleanup and habitat or species restoration projects. Uncritical celebration of these revenues, profits, and jobs may undercut more systematic efforts of pollution prevention and habitat preservation. Moreover, conservation measures and cleanup technology may prove insufficient to preserve the environment if growth in production, consumption, and population continues unabated. Paul Hawken's gloomy assessment is that "[i]f every company on the planet were to adopt the best environmental practices of the 'leading' companies—say, the Body Shop, Patagonia, or 3M—the world would still be moving toward sure degradation and collapse."[145] Unlimited growth of even an environmentally friendlier economy remains unsustainable. In the end, there is no substitute for a full-cost accounting of the economy.

The chief obstacle to implementing such accounting is structural. Those who exercise the most power in society and might effectively promote a truly sustainable economics are typically those who have become adept at externalizing the costs of their enterprises. The rules of the game, in other words, do not bode well for sustainable business. As Stephan Schmidheiny wrote, "The painful truth is that the present is a relatively comfortable place for those who have reached positions of mainstream political or business leadership. This is the crux of the problem of sustainable development, and perhaps the main reason why there has been great acceptance of it in principle, but less concrete actions to put it into practice: many of those with the power to effect the necessary changes have the least motivation to alter the status quo that gave them that power."[146] The problem is not simply that those with power or wealth seek to preserve their privilege. The problem is that our economic and political system as a whole is structured in such a way as to impede a future focus.

Schmidheiny observed that it is difficult to demand of political leaders "who rely on the votes of the living to achieve and remain in high office, that they ask those alive today to bear costs for the sake of those not yet born, and not yet voting. It is equally hard to ask anyone in business, providing goods and services to the living, to change their ways for the sake of those not yet born, and not yet acting in the marketplace."[147] Businesses are chiefly concerned with the bottom line. Marketplace competition often forces businesses to be environmentally myopic to ensure their (short-term) economic survival. Likewise, the electoral pressures faced by national political representatives generally militate against the adoption of long-term concerns and perspectives. The temporal horizon of politicians is often as short as the electioneering sound bites they generate.

Environmentalists contrast their long-term perspectives with the abbreviated temporal horizons of the powers that be. The Nature Conservancy most sharply distinguishes itself from business organizations, John Sawhill has noted, because "when we talk about the long term, we don't mean next week or next quarter."[148] Likewise, Sierra Club executive director Carl Pope contrasts his organization's perspective with "Beltway myopia." The severely constrained temporal vision of Congress means that the task of repairing environmental degradation is left to "future generations: an irrelevant class of people who can't vote, aren't consumers, and don't have political action committees."[149] Since environmental organizations are not tenure-limited bodies, they are well disposed to confront the long-term concerns that business executives and elected representatives often find at odds with their economic or political survival.

Environmental organizations face a bottom line. Like politicians and business people, environmental activists depend on public support and are averse to alienating their constituents. Yet, unlike most politicians and business people, environmental organizations remain primarily concerned with the long-term effects of their actions. An activist who identified most strongly with Earthwatch explained his work as "helping the earth speak." When one speaks for a planet four and a half billion years old that must provide a home for an indefinite number of future generations, one's time horizon necessarily expands.

Counting on Posterity

Oscar Wilde once said that a cynic was someone who knows the price of everything but the value of nothing. The free market economic system is cynical. It is very good at setting short-term prices but is not up to the task of safeguarding long-term social and ecological goods. Understanding the value of these goods is central to the tasks of living sustainable lives and running sustainable businesses. As important as weighing long-term costs and benefits is, however, the quality of life cannot be reduced to economic calculations. How, for instance, would one begin to quantify in dollars and cents the key ingredient of the good life: human happiness?

Happiness, for most people, is a function of both the pleasure they *take* from life and the meaning they *give* to life. The pursuit of pleasure may prompt one to discount the future, to live for the day and care not for the morrow. Maximizing pleasure, for this reason, is not usually a sustainable practice. In contrast, the pursuit of happiness, which grounds a meaningful life, is much more conducive to sustainable living. Making life meaningful, environmentalists suggest, entails contributing to the preservation of a cultural and evolutionary legacy.

When asked about the need to balance the protection of the environment with economic concerns, one activist member of various environmental groups responded that short-term compromise was often necessary. In the long-run, she insisted, "the more successful way to live would be to live in harmony with [the environment]; one would live a happier, fuller life in harmony with the natural order rather than in conflict with it." For environmentalists, cultivating a future focus becomes a means of settling value conflicts in the here and now. Their happiness is tied to the meaning they give to life, and earthly life becomes meaningful chiefly through its preservation and enhancement.

Children are dependents. They are wards of adults. The relationship, at times, seems rather one-sided. Children gain everything from us, including life itself. Yet what do they give back? Certainly they give love. Perhaps later in life they might contribute to one's care and financial security. But what about grandchildren and great-grandchildren? What about those descendants whom we shall never know? As Trumball once famously asked, why should we care a whit about posterity, for what has posterity ever done for us?

Economically speaking, the answer is nothing. "It is an economic fact that posterity never has been, and never will be, able to do anything for us," Ophuls and Boyan wrote. "Posterity is, therefore, damned if decisions are made 'economically.' "[150] Genetically speaking, the answer is different. Posterity extends our genetic life, and thus offers us a kind of biological immortality. Culturally speaking, progeny are no less important—for a similar reason. They carry on and further develop our technological, intellectual, aesthetic, and ethical achievements and thus lend us a sort of cultural immortality. Posterity carries us into the future biologically and culturally, just as ancestry carries us into the past. Through our relation to ancestry and posterity our lives gain continuity and meaning. We become part of a greater whole. Perhaps for this reason most people reject efforts to discount the future when assessing long-term environmental issues.[151]

In his discussion of the relation of time to environmental politics, John O'Neill wrote:

> Our primary responsibility is to attempt, as far as possible, to ensure that future generations do belong to a community that has a narrative continuity with ourselves—that they are capable, for example, of appreciating works of science and art, the goods of the nonhuman environment, and the worth of the embodiments of human skills, and are capable of contributing to these goods. This is an obligation not only to future generations but also to those of the past, so that their achievements continue to be both appreciated and extended; and to the present—ourselves. The future matters to us now.[152]

Past, present, and future generations are linked. They demonstrate relationships of interdependence.

The future focus propagated by fourth-wave environmentalists establishes and expands the sensibility of generational interdependence. We may always ask, What can posterity do for us? For environmentalists, however, the crucial question is, What can we do—or perhaps more importantly, what can we leave undone and unspoiled—for posterity? The point is well captured by a popular slogan: Environmentalists Make Great Ancestors! Living up to this reputation requires the diligent pursuit of intergenerational justice. That is a life's work.

The Quest for
Environmental Justice

Social Interdependence
across Space

"In just about any environmental issue," an activist with Florida Defenders of the Environment observed, "there are some people who
benefit from environmental destruction. That's why it's going on:
whether it's mining companies, oil companies, or people [burning down
rainforests] in the Amazon. If you look at the whole world, however,
it's probably 99 percent of the people who are being harmed [by environmental destruction], especially if you look at the long term." The
activist's statement testifies to a common characteristic of fourth-wave
environmentalists. Their ecological perspective produces not only an expanded temporal horizon but an expanded spatial horizon as well.

While the sensibility of temporal interdependence remains the most
salient feature of contemporary environmentalism, the affirmation and
cultivation of spatial interdependence are pervasive. The task ahead for
environmentalists, Dianne Dillon-Ridgley, president of Zero Population
Growth (ZPG), remarked, is to "recapture a moral vision for our society—a vision that satisfies human needs and is both inclusive and sustainable."[1] The concern for sustainability bespeaks a future focus. The
concern for inclusivity bespeaks an expanded sociogeographic perspective. To think inclusively is to think of the world as a whole, despite its
national, ideological, racial, economic, and political cleavages. The pursuit of intergenerational justice thus finds its complement in the global
pursuit of social justice, also known as environmental justice. While intergenerational justice and social justice remain conceptually distinct, for

fourth-wave environmentalists they are largely inseparable in practice. Sociogeographic inclusivity and temporal sustainability go hand in hand.

"If you think in the long term, you've got to think globally," one Greenpeace activist stated succinctly. With this in mind, Lester Milbrath combined the popular adage "Think globally, act locally" with the corresponding environmental dictum, "Think tomorrow, act today."[2] Future focus begets a global focus, and vice versa. The term *global*, importantly, refers not only to an extended geographical sphere, but also to an extended socioeconomic and political sphere as well. Global sensibility, for fourth-wave environmentalists, indicates an acknowledgment of environmental responsibility that breaches standard socioeconomic and political boundaries.

Social Interdependence

The justification for the environmental rights of future generations is difficult to separate, logically speaking, from the justification for the environmental rights of those who inhabit other nations or other classes or races within the same nation.[3] In upholding the rights of future generations, we acknowledge an obligation to people who remain incapable of actively representing their own interests. Their lack of participation in decision-making, for temporal reasons, does not absolve us of our duties to them. If we accept the legitimacy of obligations to future generations, then it is difficult to justify the denial of similar obligations to those who are incapable of actively representing their interests for geographic, socioeconomic, or political reasons. The effective lack of self-representation of non-nationals or marginalized citizens of certain races or classes does not absolve us of our responsibilities to them. In both cases, moral obligation has reached beyond narrowly constrained temporal or sociogeographic boundaries. As the Brundtland Report stipulated, the "concern for social equity between generations . . . must logically be extended to equity within generations."[4]

Environmental protection engaged in on behalf of future generations is also difficult to separate from environmental protection engaged in on behalf of local and global neighbors for practical reasons. The complementarity of these tasks is grounded in the growing interdependence of the world's peoples. To protect the environmental health of our children and grandchildren, it is increasingly necessary to protect the environmental health of neighbors in the "global village." The welfare of parts has come increasingly to depend on the welfare of adjoining parts and on the environmental health of the whole. "In today's interdependent world," Sharon Pickett of ZPG has stated, "cooperation across borders to address social, economic and environmental issues has become not only beneficial, but imperative."[5] The world is shrinking, ecologically speaking, and our duties have correspondingly expanded. More and more of the environmental concerns we have for future generations can be adequately addressed only through environmental protection that is socially and geographically inclusive.

"In the end," Aron Sachs of the Worldwatch Institute wrote, "environmental justice is such a powerful concept because it brings everyone down to the same level—that of a shared dependence on an intact, healthy environment. . . . Protecting the rights of the most vulnerable members of our society, in other words, is perhaps the best way we have of protecting the right of future generations to inherit a planet that is still worth inhabiting."[6] Ensuring a healthy environment for progeny within a neighborhood or nation entails the extension of ecological care and the promotion of ecological values beyond its borders. In the same vein, more and more of the environmentally destructive practices that we engage in locally or nationally deleteriously affect those who live beyond our neighborhoods and nation, just as they will harm those who live beyond our lifetimes. That is the upshot of interdependence.

A member of a regional lake management society remarked that he was "overwhelmed with all the interconnectedness" demonstrated among the countries of the world in environmental affairs. The linkages are indeed overwhelming. Pollution does not stop at national borders, a fact that was well demonstrated in the fall of 1997 as smoke from brush-fires set by commercial loggers in Indonesia poisoned the skies across much of Malaysia, Thailand, Brunei, and the Philippines. Other environmental threats are no more respective of national borders. When Americans release carbon dioxide by burning fossil fuels, they threaten the lives and livelihoods of Polynesians on the other side of the globe whose islands may be submerged as global warming melts polar icecaps and expands the oceans' surface waters. (That is why the Alliance of Small Island States has strongly urged industrialized nations to cut their greenhouse gas emissions.) Were the Chinese to manufacture large amounts of CFCs to supply their burgeoning economic development, the inevitable release of great quantities of these ozone-destroying chemicals would threaten the health and welfare of Swedes, who would suffer increased cases of skin cancer and cataracts. When increasing population and poverty force Amazon dwellers to slash and burn rainforests, Russians who might benefit from pharmaceuticals derived from forest flora and fauna suffer for this loss of biological diversity. Ecologically no less than technologically, the world has shrunk. More so than ever, the fates of neighbors in the global village are joined. The technology, demographics, and economics of most countries today cast "ecological shadows" that extend well beyond their own borders. There are no more spatial sinks that can harmlessly absorb pernicious ecological externalities. The global frontier has been closed.

Over 95 percent of all human generations have lived by hunting and gathering wild game and plants. They led relatively isolated lives within small kinship groups and tribes. Their actions seldom affected their distant neighbors. Obviously there is no going back to these times. The total area of the planet is under 200 million square miles. A hunting and gathering lifestyle requires about a square mile per hunter-gatherer. Two or three population doublings ago, humanity forever forfeited the option of relatively isolated lives for its members.

A Sierra Club activist insisted that one of the "icons of beliefs" for environmentalists is respect for "the cultural identity of people." The world should be uncrowded enough, he observed, that different peoples can "go about their stuff" without being unduly interfered with. Yet, he mused, a changing world has made such isolationism impossible. Today, our levels of population, consumption, and agricultural and industrial production affect our neighbors across the street and around the globe. "The single worst thing about the inexorable pressure of population is that our gonads have driven us into a corner," an activist stated. He observed that we have arrived at the point where we have no choice but to get involved in the affairs of other people. In the face of this necessity, environmentalists prefer that action involving other nations and cultures be oriented to education rather than more intrusive intervention.

Contemporary environmentalists ideally see themselves as global citizens whose chief obligation is to live in a way that, if practiced universally, would remain sustainable. A member of various local and regional environmental task forces and commissions remarked, "My primary identification is as a creature on the planet. It's important to me to live as if every action were going to be imposed globally." Earth Share, an organization of over forty environmental groups formed in 1988 to gain access to workplace fund-raising, has formulated a shorthand version of this dictum. Its motto reads Its a Connected World. Do Your Share.[7] An executive of a local Sierra Club group observed that learning about the "complex interdependencies" in the natural world stimulates one to assess the social world from an increasingly global perspective. "Being an environmentalist," he said, "guarantees that you think this way eventually."[8] The question that environmentalists always ask, And then what? pertains as much to the repercussions of our actions on the sociogeographically distant as on the generationally distant. The nonlocalizable nature of many environmental problems has effectively expanded the moral universe in space no less than in time. To be an environmentalist today is to cultivate a sense of global interdependence.

The Rise of Globalism

By the beginning of the second wave of environmentalism in the 1960s, Americans were becoming aware of the global nature of environmental issues. By the end of the third wave, this awareness was pervasive. The globalization of environmentalism over the last quarter century is undeniable. In the forty years from 1930 to 1970, only forty-eight international conventions or treaties on the environment were signed. In the following decade, from 1971 to 1980, another forty-seven were signed. By the end of the 1980s, more than 250 international environmental agreements were on the books.[9] In 1971, only twelve countries including the United States had instituted environmental regulatory agencies. Over 140 countries had such agencies operating by the mid-1980s. In 1972, the United Nations Conference on the Human Environment, held in Sweden and known as the Stockholm Conference, was attended by delegates from 113 nations. Relatively few heads of state participated, and it was boycotted by the Soviet Union and the governments of Eastern Europe. Representatives from 134 nongovernmental organizations (NGOs) participated.[10] In June 1992, more than 110 heads of state and representatives from 178 nations attended the Earth Summit in Rio de Janeiro. Over 15,000 participants representing 1400 NGOs from 165 nations also took part.

It has been suggested that the growth of environmental concern between the Stockholm and Rio summits represents a "paradigm shift" in thinking. Rather than perceiving a world of independent states, policy makers and the general public increasingly view the planet "as a single integrated system with complex linkages among large-scale ecological systems of land, oceans, atmosphere, and biosphere."[11] Unfortunately, the global sensibility displayed at Rio has not issued in a commensurate level of concrete initiatives on the part of the participating nation-states. Defining interdependence as the "new kid on the global block," Norman

Myers observed that "[n]ot even the most advanced nation can insulate itself from [global] environmental impacts, no matter how strong it may be economically or how advanced technologically or how powerful militarily. . . . Yet however much interdependence is a built-in fact of our new world, we have yet to mobilize the political collaboration to reflect it: we simply do not recognize it as a strict fact of life."[12] Translating the awareness of global interdependence into concrete practice is the task fourth-wave environmentalists have set themselves.

For many environmental organizations, global activities mark a significant shift in their orientation. The Sierra Club, for instance, has recognized that "the planet is smaller and more interdependent than when Muir roamed the Sierra. . . . We can no longer afford to be concerned only with what happens in our backyards."[13] Sierra Club vice president Michelle Perrault has listed as a central goal for her organization the effort "to train more members to speak up not only on behalf of their local communities, but for the whole earth."[14] Many prominent U.S. environmental organizations formed in the 1980s and 1990s (for example, Rainforest Action Network, Rainforest Alliance, Conservation International) have explicit global mandates. Most if not all of their work is conducted outside of the United States. Even those organizations that have always maintained an international overview, such as Greenpeace, locate their period of truly "going global" in the last decade.[15] Today, in fact, there are few national environmental organizations that do not have global perspectives and international agendas.

Political scientist Richard Rosecrance wrote that "because many, perhaps most, of the desirable features of modern life depend upon what other countries do, one government cannot assure economic security or welfare for its citizens. It can only negotiate with these objectives in mind. . . . The nation has been reduced to a 'mediative state,' an instrument that balances the pressures of international life on the one hand with domestic life on the other." Rosecrance concluded that "the government has to obey two masters but the electorate would wish that it were only one."[16] Regarding environmental affairs, Rosecrance got it half right. International interdependence is a reality, to be sure. Yet in many cases citizens, not state officials, acknowledge and affirm this interdependence foremost. A recent survey, for example, indicates that 87 percent of Americans believe the environment to be a global issue that demands an organized, global response. Eighty percent believe that NGOs, such as environmental groups, should take the lead in confronting these environmental threats.[17]

The environmental movement has regularly taken the initiative of persuading national leaders that world-order environmental problems must take precedence over the interests of particular domestic constituencies. World citizenship is, for environmentalists, an attitudinal reality with concrete ramifications. The point is not that nation-states have ceased to command the lion's share of political loyalties in today's world. States still command these loyalties and likely will continue to do so for the foreseeable future. Allegiances, however, are quickly multiplying. Environmentalists have taken the lead in cultivating this global expansion of allegiances and are negotiating its shifting, complex terrain.

National governments maintain that they are responsible first and foremost to their own citizens. Their capacity and willingness to concern themselves for the welfare of the peoples of other lands are limited and problematic. As the nuclear catastrophe at Chernobyl demonstrated, however, national borders can no longer mark the borders of responsible government. The extensiveness of environmental interdependence today presents a challenge to the 300-year-old notion of national sovereignty, which holds the state to be the supreme arbiter of what takes place within its borders. Noting that "environmental matters don't end at any kind of political boundary," some environmentalists take the position that "national sovereignty is a harmful fiction." Others argue that for all the environmental ills associated with it, national sovereignty remains the prerequisite for the effective enforcement of almost all environmental regulations today. In either case, obligations grounded in interdependence, we are assured, do not end at national borders. The question is simply whether nation-states or other organizations and agencies should be primarily responsible for seeing these obligations fulfilled.

Most environmentalists do not judge national sovereignty to be an obstacle to their work. A staff member at Greenpeace maintained that the "principle" of national sovereignty practically never conflicted with the mission of his organization. A Sierra Club member agreed that national sovereignty in itself was not an impediment to environmental protection. It may actually be an incentive, he suggested, assuming that "we can develop within each country a pride in its natural history like we have pride in other national histories."[18] An activist with the Greens observed that whether national sovereignty is good or bad for the environment depends on the country in question and its commitment to environmental protection. She went on to state that "[t]he globalizing things like NAFTA [North American Free Trade Agreement] and GATT [General Agreement on Tariffs and Trade] are more of a threat than national things. Some countries don't have environmental protection policies. When you globalize that and get to the least common denominator of environmental protection—then you have the worst situation." This viewpoint has become widespread within the environmental movement, as national environmental laws come under fire from international trade agreements. The World Trade Organization (WTO), for instance, has attempted to outlaw the green labeling of products (such as "dolphin-safe" tuna) as an illegal barrier to trade. The World Trade Organization also denies nations the right to embargo natural resources that are caught or harvested beyond their borders in ways that cause harm to endangered species. In March 1998, for example, the WTO ruled that provisions in the Endangered Species Act requiring turtle excluder devices to be used by wild shrimp netters who want to sell their shrimp in the U.S. was a barrier to free trade. The turtle-excluder devices are more than 95 percent effective in preventing the entanglement and drowning of sea turtles, many species of which are endangered. The ruling will force the U.S. to change the law or face severe economic sanctions. Likewise, the Multilateral Agreement on Investment (MAI) would make it easier for foreign investors to buy and develop tangible assets, such as factories, timberland and mining rights, and to sue local, state and national governments whenever environmental restrictions or regulations negatively affects expected profits from these assets. Greens have characterized the Multilateral Agreement on Investment as "NAFTA on steroids."[19] In the face of economic globalization, national sovereignty and environmental protection often stand hand in hand.

Environmentalists look forward to a more integrated world. Yet they are less interested in undermining the nation-state system than in employing it strategically whenever possible and circumventing it whenever necessary.[20] There is, at present, no effective global governmental organization dedicated to environmental protection. Effective bilateral and multilateral agreements and agencies do exist, but their effectiveness ultimately remains grounded in state-sponsored actions. The 1987 Montreal protocol on the reduction of ozone-depleting substances and the 1997 Kyoto accords on the reduction of greenhouse gases are cases in point. Environmentalists understand the need to work within the nation-state system. At the same time, they consistently applaud the globalization of environmental stewardship, even when this globalization erodes national sovereignty. Thus, environmentalists celebrated the first use of the International Court of Justice in the Hague, also known as the World Court, to settle an environmental case. In 1997, the International Court of Justice began its investigation of whether Slovakia had the right to divert the Danube for hydroelectric power, an action that was destroying Hungary's wetlands downstream and threatened its supplies of agricultural and drinking water. In preparing its judgment, the World Court entertained arguments from environmental groups such as Greenpeace, the Sierra Club, and the World Wildlife Fund (WWF). These environmental organizations argued that the Szigetkoz ecosystem on the borders of Austria, Hungary, and Slovakia represented a global heritage that was not the property of any particular national government and deserved, for this reason, international protection.[21]

Unlike nationally mandated governments, environmental organizations are particularly well suited to pursue a global approach to environmental threats and environmental care. Just as they are often better positioned to look out for the long-term interests of society than tenure-limited elected officials, so their transnational mandates and, in many cases, their transnational memberships make them well suited to address global issues and pursue global strategies.

The crucial role played by NGOs in global affairs has been widely recognized. National governments increased the monies channeled through NGOs from $100 million in 1975 to over $2 billion by 1988.[22] The 1987 United Nations World Commission on Environment and Development, comprising representatives from twenty-two nations on five continents, concluded in its report that further increasing the role played by NGOs should be a "high priority." These organizations were recognized to serve as an "efficient and effective alternative to public agencies in the delivery of programmes and projects." The report acknowledges that "an increasing number of environmental and development issues could not be tackled without them."[23] The importance of environmental NGOs was also acknowledged at the Earth Summit in 1992, which mandated the creation of a new United Nations body, the Sustainable Development Commission. One of the primary roles of this commission is liaison with NGOs. Today, a number of environmental NGOs lay claim to outperforming many nation-states in the fields of global environmental oversight and protection. The largest environmental organizations spend more on environmental protection than most countries do, and have consistently spent more than the United Nations Environmental Program.[24]

At the 1994 Cairo Conference on population and development, the importance of environmental NGOs was again underlined. Over half of the delegates from the United States to the governmental conference were representatives of NGOs. In total, there were 4000 representatives of 1500 NGOs from 133 countries in attendance. The conference produced a twenty-year Programme of Action designed to stabilize population at 7.8 billion by 2050. Whether this blueprint will be financially and politically backed by the signatory states remains to be seen. If post-Rio compliance is any indication, we likely will be disappointed again. Yet, as one observer noted, "if the Cairo Conference did nothing else, it bore witness to how surely NGOs are moving the world forward, piece by piece."

Environmental organizations facilitate international conferences and mediate environmental disputes between nations. They also circumnavigate national governmental bodies altogether, achieving their ends directly through transnational public pressure. Greenpeace provided a controversial example of this latter tactic in its oppostion to the ocean disposal of offshore oil rigs and storage tanks. In the spring of 1995, the Royal Dutch/Shell Group announced that it would sink its Brent Spar oil-storage rig in the Atlantic ocean. Greenpeace activists initiated a campaign to prevent its sea disposal. In typical Greenpeace fashion, activists took direct action by attempting to physically occupy the structure. Royal Dutch/Shell resorted to using water cannon to fend off Greenpeace helicopters that were depositing activists on the rig. As always, the Greenpeace cameras were rolling.

The oil company claimed that sea disposal was no more environmentally harmful than land disposal. The fact that the financial cost of dumping the Brent Spar at sea was less than one fourth the cost of its land disposal, however, made environmentalists suspicious of the oil company's "scientific" assessment of disposal methods. Independent analysts estimated that the ecological damages caused by land disposal were relatively close to those caused by sea disposal. This prompted some to characterize Greenpeace's efforts as "a circus and sideshow that distracted from the big environmental issues affecting the world."[25] Yet Greenpeace was interested in more than the disposal of a single rig. If the Royal Dutch/Shell Group's attempt to use the ocean as an inexpensive dumping ground were successfully blocked, activists hoped, oil companies would have to internalize the ecological costs of ocean drilling. They would have to take the high price of decommissioning and disposing of ocean rigs on land into account before making the decision to engage in future offshore drilling. That is to say, Greenpeace was attempting to force oil companies to engage in full-cost accounting. Activists were arguing in word and deed that geographically distant space, in this case the open sea, could no longer cheaply and uncontroversially be used as a sink in which the costs of economic enterprises could be externalized.

Equally significant is the fact that Greenpeace attempted to achieve its ends through transnational means. Shell U.K. was in charge of the rig and adamantly refused to entertain the option of land disposal. The British government endorsed this decision. An international boycott called by Greenpeace caused gasoline sales of Shell subsidiaries to plummet, not only in the United Kingdom but also across the continent. Shell's German sales, for instance, declined by 30 percent. In the end, the British company was pressured by European subsidiaries of Shell to reverse its decision, despite the intransigence of the British government. Greenpeace's victory was globalist in its means as well as its ends. The environmental organization employed a transnational forum to effect transnational change in a pitched battle against a transnational corporation.

Many less dramatic but no less significant struggles are waged by other groups on other issues. The Environmental Defense Fund (EDF), for example, has pursued a globalist strategy regarding the problem of climate change. Their effort to curb greenhouse gas emissions calls for international energy taxes and emission limits. Under the EDF plan, national governments would establish emission allowances to accommodate various business and community needs. In the United States, EDF proposes that all firms that import or produce fossil fuels be required to offset the carbon content of their fuels that exceeds 1990 levels of emissions. Individual fossil fuel companies would be free to meet these requirements in any number of ways, such as investing in improved fuel efficiency in the auto industry or in new low-carbon technologies. All emission offsets would be tradable across industries and, eventually, across national borders.

An EDF publication explains that a "U.S. oil company, for example, might offset some of its emissions by subsidizing the manufacture of more efficient refrigerators in India and China, or by a carefully monitored tree planting and maintenance program—next door, in the next state, or in another country."[26] The EDF has been a strong voice for this sort of collaborative, market-based environmental regulation. It is a controversial approach that many environmental organizations disclaim. Worries about regional disparities abound, as was demonstrated at Kyoto. The problem is that companies might escape responsibility for cleaning up toxic "hot spots" that impair the health and welfare of citizens in certain low-income or minority communities or in less developed countries by buying pollution credits or engaging in pollution mitigation elsewhere. Without reentering the debate of the merits and dangers of such market strategies, we may observe an important development: the primary orientation is global. The attempt is made to gear the (geographic) scope of the solution to the nature of the problem.

A Sierra Club activist has predicted that a close-knit community of nations will form over the next decade. This integration "will occur as a result of people becoming aware of the interconnectedness of countries. . . . The planet is small, and people will begin to realize this." The prediction of a close-knit community of nations seems wishful thinking. With the advent of the new millennium, and with many of the more prominent ideological struggles of the past on the wane, environmentalists are perhaps justified in believing that peoples across the globe will adopt a worldview that is, for the first time in history, truly a *world*view. But the change probably will come slower than many would hope.

Political ideologies are often animated by an antagonism between ingroups and outgroups. Chauvinistic nationalism and fascism are obvious examples. The quip about a nation being a people with nothing in common but a common enemy illustrates the point.[27] Environmentalism, to the extent it may be called an ideology, marks a definite departure from this orientation. Green consciousness is inherently globalist.

Environmentalists seldom voice prerogatives, rights, or needs in exclusive terms. The politics they practice tend to be global in intent, if not always in origin and effect. The notion of competitive struggles between ingroups and outgroups is generally considered counterproductive, certainly in ecological terms if not also in social terms. That is because military or economic battles typically bear within them the seeds of ecological defeat for the victors no less than for the vanquished. Consequently, environmentalists are interested in fostering nonexclusive transnational and transpersonal sensibilities.[28] If there is talk of opposing sides staked out in battle, then the reference is generally to a global struggle against the common enemy of environmental degradation.

In this vein, Michael Oppenheimer of EDF has argued that the resources that previously went into fighting the cold war should now be diverted to an increasingly hot war—a war that needs to be fought against global climate change.[29] Al Gore has suggested a similar transformation of former cold war efforts into planetary environmental stewardship.[30] Tom Athanasiou likewise argues that we face a level of ecological deterioration that will soon require an all-out "war to save the earth."[31] The intention is clearly to foster globalist sensibilities and translate the environmental struggle into what William James termed the "moral equivalent of war." These efforts have been widely applauded within the environmental movement, although some remain wary of the militaristic overtones.

Daniel Deudney worries that exploiting the emotive power of militarism to mobilize environmental awareness and action will actually undermine a needed globalist sensibility. Deudney wrote that "[h]arnessing these sentiments for a 'war on pollution' is a dangerous and probably self-defeating enterprise . . . The movement to preserve the habitability of the planet for future generations must directly challenge the tribal power of nationalism and the chronic militarism of public discourse. . . . For environmentalists to dress their programmes in the blood-soaked garments of the war system betrays their core values and creates confusion about the real tasks at hand."[32] Deudney is concerned that the medium may betray the message. Metaphors of war are generally not conducive to sensibilities of global inclusivity.

Environmentalists observe that ecological degradation and resource scarcity have become sources of social instability, civil strife, and even violent conflict within and between states.[33] As fresh water becomes an increasingly scarce commodity, for example, the potential for international conflict over its use rises. (Almost three quarters of the world's major river systems, which provide nearly 40 percent of the world's population with water, are shared by two or more nations.) Many environmentalists believe that dwindling reserves of nonrenewable resources or resources not quickly renewable, such as oil and fresh water, might result in savage "eco-wars." Though they employ the threat of civil or international strife over dwindling natural resources as a means of garnering support for increased ecological stewardship, environmentalists also guard against inadvertently fostering the "tribalist" sentiments that might accompany natural resource shortfalls. They are wary of the "greening of hate" and the "eco-fascism" that may result from environmental crises if a people's awareness of their ecological vulnerability remains unaccompanied by a globalist sensibility.[34]

Life on Spaceship Earth

Globalism has been defined as "a heightened sensitivity to the fragility of the life-support system of the planet and a sense of human solidarity in a world of increasing interdependence."[35] Within the environmental community, globalism gained full momentum in the 1980s, when notions of the "common heritage of humanity" and the "global commons" became widespread. The onset of the sense of planetary interdependence, however, goes back at least to the 1960s, at the beginning of the second wave of environmentalism. In 1965, Adlai Stevenson popularized the notion of Spaceship Earth in a speech delivered to the United Nations Economic and Social Council (UNESCO). "We travel together, passengers on a little space ship," Stevenson said, "dependent on its vulnerable reserve of air and soil; all committed for our safety to its security and peace; preserved from annihilation only by the care, the work, and I will say, the love we give our fragile craft."[36] On Spaceship Earth, the message goes, ecological interdependence sows a common fate and a common task for humankind.

The cultivation of a sense of "shared fate" is crucial to the environmental movement.[37] This sense of shared fate is fostered by the Spaceship imagery. Despite its popularity, however, the Spaceship Earth metaphor has been increasingly subject to criticism. There are chiefly two reasons for this critical challenge.

First, by proposing that everybody is "in the same boat," the spaceship image would have us assume that a universal equality exists. Yet not all passengers on Spaceship Earth enjoy the same privileges or suffer the same deprivations. Those with little power and wealth often sweat and starve in the smoke-filled engine room. Others dine and delegate on the air-conditioned bridge. Many of the larger environmental organizations, owing to their Western heritage and upper-middle-class orientations, have been justly accused of obscuring the nature of Spaceship Earth's inegalitarian community. They speak of the need for everyone to make sacrifices for the earth but ignore the fact that most of the earth's peoples have little to sacrifice.

Economically secure Westerners often ignore the conclusion of Stevenson's UNESCO speech, where he said of Spaceship Earth that "[w]e cannot maintain it half fortunate, half miserable, half confident, half despairing, half slave-to the ancient enemies of man-half free in the liberation of resources undreamed of until this day. No craft, no crew, can travel safely with such vast contradictions. On their resolution depends the survival of us all." Environmental organizations with third world constituencies and grassroots domestic organizations representing disadvantaged groups brought these issues into focus during the 1980s and 1990s.[38] This challenge to the supposed inclusivity of Western environmental concerns produced a significant increase in efforts to address issues of environmental justice in national and international contexts.[39]

The National Wildlife Federation (NWF) for example, recently reformulated its mission statement to include a reference to global environmental justice. The organization now bears a mandate "to educate, inspire and assist individuals and organizations of diverse cultures to conserve wildlife and other natural resources while protecting the Earth's environment to promote a peaceful, *equitable* and sustainable future."[40] Likewise, the Wilderness Society announced in 1994 that its guiding vision was of "a world in which diverse human populations and the cultural and political institutions that guide them adhere to the principles of sustainability and social justice."[41] The sensitivity to diverse cultures and to the mutually supportive goals of equity and sustainability reflects the mandate of environmental justice. Achieving social justice on Spaceship Earth, movement organizations maintain, is the first step to making ecological sustainability something that is itself politically, economically, and culturally sustainable.

The second challenge to the Spaceship Earth imagery concerns its tendency to promote political centralization. During the second and third waves of environmentalism, the notion that a central authority was necessary to prevent global environmental collapse developed. The Spaceship Earth metaphor bolstered this belief. Centralism of power has proven itself alluring to many environmentalists. When organizations such as Friends of the Earth characterize the task for the Rio Summit as setting "a new agenda for life on earth," critics worry that environmentalists are engaging in "totalitarian thinking" and desire "powerful, centralized, and intrusive institutions of governance either at the global level or within individual societies."[42]

In 1968, Hardin published his famous article, "The Tragedy of the Commons," in the prestigious journal *Science*. The article made an instant impact and gained long-standing fame. It has been reprinted over fifty times in edited books. Relying on a quasi-historical example of environmental degradation, Hardin argued for environmental protection through centralized control. He observed that individuals who grazed their livestock on common fields in days gone by may have destroyed their resource base for lack of a central authority. A livestock owner, Hardin argued, would rationally seek to maximize his personal benefits by grazing more and more of his own animals on the commons. In the aggregate, such lack of self-restraint eventually produced a commons eroded beyond further use by overgrazing—to everyone's detriment. Extending his example, Hardin argued that various common resources may be depleted to the point of collapse in the absence of a political authority to restrain individual acts of exploitation.[43] Hardin was concerned not only with the overuse of land, but with the availability of clean water and air. His foremost concern was the depletion of planetary resources caused by overpopulation.

If humans reproduce in greater numbers than the earthly commons can sustain, Hardin insisted, there will be a tremendous collective loss despite the (perceived) gain each individual attributes to having more children. The tragedy of the commons is the ecological decline of a planet in an age of burgeoning populations and overconsumption. Hardin concluded that "[r]uin is the destination toward which all men rush, each pursuing his own best interest in a society that believes in the freedom of the commons. Freedom in a commons brings ruin to all."[44] The only solution, Hardin maintained, was "mutual coercion, mutually agreed upon." Environmental protection, it appeared, must have an authoritarian face. The belief that a central authority was necessary to prevent global environmental collapse drew wide attention and significant support from second-wave environmentalists.[45]

Despite their global sensibilities, fourth-wave environmentalists have become wary of the centralized administration that would be required to oversee global environmental protection. "The Earth is one but the world is not," wrote the authors of *Our Common Future*. The report underlined the need for the world's nations to unite in an effort to protect the environment.[46] The problem with any such unity, however, is that it threatens to centralize power. A special issue of *The Ecologist* in 1990 challenged the Brundtland Commission's report for just this reason. The commission used the environmental threat to world security, critics charged, as a means to consolidate the power of the Western industrialized nations at the expense of local initiatives and grassroots movements. While the commission spoke of "our common future," the rhetorical question that critics asked was, Whose common future is really being secured? Who is being protected by centralized control over environmental affairs, the local dwellers of the land or the bureaucracies and corporations that rule over them?[47] Once again, globalist thought was running afoul of its own imagery.

Spaceship Earth globalism may promote a political quietism and a centralization of power that remains at odds with an active citizenry. Marshall McLuhan's adjustment of the metaphor well illustrates the issue. "There are no passengers on Spaceship Earth," McLuhan insists;

"everybody's crew." The Spaceship Earth metaphor, McLuhan has suggested, may be counterproductive to the extent that it promotes elite-directed politics and faith in traditional authorities rather than citizens who take on the responsibility of being crew members. The metaphor cuts against the hands-on stewardship of local habitats and grassroots activism that are the hallmarks of contemporary environmentalism.

Fourth-wave environmentalists reject the idea that centralism is necessary for environmental protection. Empirical and historical research has demonstrated that commons areas and common-pool resources can be actively and carefully regulated, and have been for hundreds of years. They need not demonstrate the same inherent tragedy in the absence of centralized control as the "open access" resources described by Garrett Hardin. Relatively small groups (under 15,000 members) often enjoy a good deal of success at collectively managing common resources. Indeed, neither private property nor state property, Hardin's alternatives, has a particularly good track record regarding natural resource and wildlife preservation.[48] The "tragedy of the enclosure," despite its centralized control, often proves as detrimental to the environment as the "tragedy of the commons."[49]

While environmentalists are suspicious of centralized authority, they do not reject it altogether. Most are relatively content to exist within an elite-directed world.[50] The key question is whether power is politically accountable. A decline of centralized yet politically accountable national or international authority over environmental protection may prove catastrophic if it translates into an increase in decentralized yet politically unaccountable corporate power. While environmentalists reject the idea of a centralized "globeacracy," they worry more about a decentralized world run by corporate giants. The revenues of transnational corporations such as the Ford Motor Company are larger than the national economies of Norway and Saudi Arabia. Philip Morris's annual sales exceed the gross national product of New Zealand. Indeed, the dozen largest corporations have individual sales revenues that are greater than the gross domestic product of 161 of the world's nation-states. Over half of the world's hundred largest economies belong to multinational cooperations.[51] Aware of these facts, environmentalists insist that any further decrease in national sovereignty and the powers of international governmental bodies would bode ill for political accountability and for the environment.

The same argument holds for decreases in federal power at the national level. In the face of growing corporate power, political decentralization may be environmentally disastrous. Capital is fluid. It tends to flow to wherever short-term profits are highest. Labor is less fluid and ecosystems cannot move at all. In the absence of nationwide environmental laws and regulations, short-term profits will be sought in states and regions where labor can be cheaply exploited and where nature can be used as a sink for waste. Audubon president Peter Berle observed the results: "No governor can stand up to the CEO of a big corporation who threatens to take major facilities and thousands of jobs to another state unless environmental standards are relaxed."[52] At both the national and international levels the corporate strategy of divide and conquer may be employed.

Environmentalists look with trepidation at "the emerging global economic order."[53] Globalization, in this case, describes a world chiefly characterized by the borderless flow of capital. It is a world dominated by transnational corporate powers that "do not consider it their responsibility to ponder the long-term political, social and ecological consequences [of their activities], much less deal with them." The problem with economic globalization, in short, is that it "doesn't have a conscience."[54] Brent Blackwelder of Friends of the Earth has warned of the "corporatocracy" that emerges as Western multinationals gain "carte blanche to the world's resources and have become the economic engine that is driving a growing and unsustainable global marketplace."[55] As the United Nations' 1997 *Human Development Report* indicates, the biggest "winners" of economic globalization are corporations that engage in trade and finance. The biggest "losers" are the world's poor.[56]

Globalism is a two-edged sword. The imperative to look beyond one's neighborhood and nation too often becomes an excuse for abdicating local and national environmental responsibility. With these concerns in mind, over two dozen environmental organizations currently participate in the International Forum on Globalization. The forum organizes ongoing "Global Teach-Ins" to educate the public about the "economic, environmental, social, cultural and democratic threats of corporate economic globalization."[57] Observing that the "current trends toward globalization are neither historically inevitable nor desirable," forum members advocate diversified, locally controlled economies attentive to "ecological limits."[58] In a similar vein, Earth Island Institute promotes Sustainable Alternatives to the Global Economy (SAGE), a project led by people of color that exposes the destructive effects of economic globalization on urban minority communities.[59] Noting that trade agreements such as NAFTA and the World Trade Organization penalize governments for enforcing environmental regulations, Sierra Club executives observed that "[t]he globalization of trade is as much about eroding democracy as it is about increasing trade." Edgar Wayburn, Chair of the Sierra Club's International Committee, stated that "[t]he pressures unleashed by the accelerating economic globalization of the past several decades threaten the values underlying all of the Club's work."[60]

Environmentalists who look favorably toward the emergence of a global community insist that it must be diverse. They are averse to the emerging "globo-culture" and lament the cultural imperialism of the West that has created a "McWorld" geared to Western consumer standards. The goal for many environmental organizations, as a Rainforest Action Network publication suggests, is to "redefine globalization."[61] Fourth-wave environmentalists distinguish between a globalist sensibility grounded in citizen activism and local caretaking from economic and political globalization grounded in a homogeneous consumer culture and the worldwide exploitation of resources by transnational corporations. The latter form of globalism is the legacy of Western-style economic development, that is to say, capital intensive, top-down industrialization. It is incompatible with sustainable development.

"Properly speaking," Wendell Berry wrote, "global thinking is not possible. Those who have 'thought globally' (and among them the most successful have been imperial governments and multinational corporations) have done so by means of simplifications too extreme and oppressive to merit the name of thought."[62] Globalism may bespeak a green, planetary consciousness that valorizes local caretaking and cultural diversity. Economic globalism, however, threatens to overpower the local ecological knowledge and solicitude that are the sine qua non of environmentalism.

Buckminster Fuller's 1969 monograph, *Operating Manual for Spaceship Earth*, identifies the planet as "a "mechanical vehicle, just as is the automobile" and suggested that the proliferation of computers would dissipate all future ideological differences.[63] Fuller's vision was grand. By describing the planet and human society in mechanical terms, however, and by celebrating technology as a panacea, Fuller failed to understand the ties to land and community that foster much environmental protection. Fourth-wave environmentalists underline the importance of these organic relationships. They favor increased citizen participation and local control and observe that love for and knowledge of one's ecological locale stimulate its caretaking.[64] Wendell Berry's twist to the Spaceship Earth metaphor illustrates this orientation. "The only true and effective 'operator's manual for spaceship Earth,' " Berry wrote, "is not a book that any human will ever write; it is hundreds of thousands of local cultures."[65]

The responsibility of being part of the crew of Spaceship Earth, one might conclude, derives from the nature of the threats being faced and the available solutions. While many of the vulnerabilities on Spaceship Earth are global, remediative actions and objectives will generally remain geographically circumscribed. As Berry wrote, "The real work of planet-saving will be small, humble, and humbling, and (insofar as it involves love) pleasing and rewarding. Its jobs will be too many to count, too many to report, too many to be publicly noticed or rewarded, too small to make anyone rich or famous."[66] Environmental protection is chiefly grounded in the actions of countless people looking after their own human and biological communities.

A good rule of thumb for environmentalists is "[n]ever globalize a problem if it can possibly be dealt with locally."[67] That is the message environmentalists propagate by their frequent use of the slogan Think Globally, Act Locally. Local activism must work in tandem with, not become subservient to, global thinking. To the extent that ecological care begins at home, relatively small, active, self-responsible communities of citizens are required. Cultivating such communities proves difficult within large nation-states and would be even more difficult within a global regime. For this reason, environmentalists tend to be situation specific in their policy recommendations. The correct balance between centralization and decentralization, between globalism and localism, is a matter of context. Different problems call for different solutions.[68] In many respects, however, the correct balance between local action and global thinking is suggested, if not completely realized, within the environmental movement itself. The transnational, global affiliations of environmental organizations, coupled with their regional chapters and community-based projects, make for a pragmatic linkage of local engagement and global awareness.

Forging the Local-Global Linkage

While mounting human populations, rates of consumption, and industrial and technological development have made a global sensibility necessary, the only viable solutions to ecological problems, environmentalists maintain, will include homegrown and community-centered remedies. While acknowledging the "desperate need for world-wide change," a local activist and writer on environmental affairs worried about globalist efforts to "legislate the law of other places."[69] The attempt to devise a single, uniform solution to all our problems may actually be the main problem, he suggests, for we will not arrive at appropriate solutions to particular problems by displacing ourselves from them. With this concern in mind, the role taken on by fourth-wave environmental organizations is increasingly that of a facilitator of local initiatives.

A 1991 editorial on the twenty-year history of Greenpeace reads: "As we grow more global in the scope of issues we address and increasingly come to grips with social justice issues, we can no longer afford to be just an environmental organization for the predominantly white, relatively affluent, industrialized world. As Greenpeace increases its work in communities of color and in Latin America, Asia, Africa and the Eastern Bloc, we are simultaneously pledging that this work be led and carried out by Latinos, Asians, Africans and Eastern Europeans."[70] Zero Population Growth's Campaign for a Quality Future has similarly focused its energies on "creating sustainable communities" by organizing town meetings and empowering locals to manage their own growth and development.[71] The initiative, which fosters "local, grassroots efforts and citizen awareness," is grounded in the principle that "sustainable communities will evolve *only* when citizens truly understand the connections between personal, local, national, and global issues."[72] Likewise, The Wildlands Project, which claims that its vision is "continental," promotes "locally controlled" economies as a means of achieving sustainable societies balanced with wilderness protection.[73] John Flicker, president of the Audubon Society, has predicted that "the next generation of the conservation movement will focus on the community level."[74] Thus, while movement organizations are tempted, for publicity purposes, to parade their achievements, they are also moved, for ecological reasons, to share both the responsibility and accolades for environmental protection with local communities. It is at the local level, they realize, where lasting solutions to most environmental problems will be conceived and carried out.

Certain hardy globalists, as one Greenpeace staff member proposed, would "do away with national borders and have some kind of global justice and legal system." This system would be based on "an environmental bill of rights, and labor and social and economic bill of rights that would set down the constraints under which capital could operate." Even in such cases, however, the abolition of national borders and the institution of global rights are proposed not so that transnational power might rule over national governments but out of concern for the "community level." The proposed global bill of rights is meant to "empower people at the local basis," just as national bills of rights empower and protect individual speech and action. Again, global oversight and regulation are intended to stimulate not undercut locally directed environmental caretaking.

In an interdependent world, many seemingly local issues have nonobvious global causes and ramifications. Soil erosion, for example, is typically portrayed as a local, regional, or at most national problem. It stands in contrast to the patently global problems of ozone depletion and climate change. Wendell Berry wrote that "[s]oil . . . is lost a little at a time over millions of acres by the careless acts of millions of people. It cannot be saved by heroic feats of gigantic technology but only by millions of small acts and restraints, conditioned by small fidelities, skills, and desires. Soil loss is ultimately a cultural problem; it will be corrected only by cultural solutions."[75] Soil erosion, Berry reasoned, is best alleviated by fostering ecological mores adapted to local cultures. The global context should not be ignored, however. Soil erosion has, in the aggregate, global consequences. For example, it reduces the amount of food that can be produced and traded internationally and often pollutes lakes and rivers that cross borders. Soil erosion also has global causes. The economic instabilities that arise from the global interdependence of national and local economies may promote the agricultural practices that cause or exacerbate erosion. As local farmers are forced by the demands of international markets into monocropping, allowing few or no fallow years, and using marginal lands, soil becomes systematically eroded. Both in its causes and in its effects, the problem of soil erosion cannot be divorced from global considerations.

The problem of overpopulation demonstrates analogous relationships. Though the world as a whole is overpopulated, not all of its nations, regions or cities are. Overpopulation, one might argue, is not a global problem but a municipal, regional, or national issue. To the extent that the global economy breeds cultures of dependence and poverty, however, population pressures may be exacerbated. Impoverished people, it is well known, often have more children than the economically secure as hedges against infant mortality and economic destitution in old age. Conversely, the effects of overpopulation, such as the cross-border migration of refugees, the unsustainable depletion of natural resources, and the decrease in biological diversity owing to habitat destruction, are global in scope. Yet despite the global causes and effects of over-population, remedies must be attuned to local and national contexts. Contraceptives, for instance, effectively lower fertility rates in industrialized countries. Their success is significantly reduced if they are employed as a singular means of population control in lesser developed, patriarchal societies where a woman's value is largely determined by the number of children she bears. An effort to raise the status of women through education, improved health care, and economic empowerment is an indispensable means of population control here, as was recognized at the 1994 Cairo Conference.[76] Once again, global concerns must be approached in light of local contexts.

Like globalism, localism has its seamy side. Restrictive localist concern characterizes certain forms of environmental activism. The so-called NIMBY syndrome is a case in point. The acronym NIMBY, which stands for Not In My Backyard, was originally supplied by an incinerator industry public relations executive. It was meant to characterize pejoratively the opposition posed by local activists to an incinerator siting in their neighborhood. At that time, grassroots environmentalism was identified as a selfish form of localism whose goal was to shift the burden of waste management to other neighborhoods, towns, or counties. Localism, in this case, was identified with isolationist self-interest.

Without doubt, a certain amount of NIMBY activism deserves this criticism, particularly as it was manifested in earlier waves of environmentalism. The acronym was pejoratively yet accurately employed, for example, to describe the efforts of suburban homeowners to exclude the development of low-income housing in their communities.[77] By the 1980s, however, the grassroots toxics movement was strategically exploiting localist self-interest to stimulate more expansive concerns and action. Indeed, studies reveal that those who oppose hazardous waste disposal facilities in their neighborhoods but are *not* opposed to their being sited elsewhere, the true "NIMBY Syndromers," are not the members of minority and working-class neighborhoods that groups such as CCHW typically mobilize. Rather, they tend to be high-income, well-educated individuals with a conservative political identification.[78] For the grassroots activists of the toxics movement, particularly those who frame their efforts in terms of environmental justice, NIMBY struggles are not primarily about shifting social burdens. They are issues of public health versus private profit, with the self-serving actions of corporate waste companies the object of censure.

The Citizens Clearinghouse for Hazardous Waste describes its efforts as People Working Locally and Connecting Nationally. Standing by its motto, the organization closed all of its field offices in 1994 in order to fund more local groups. Its extensive system of autonomous affiliates is linked nationally through a clearinghouse. This is part of an ongoing effort to translate NIMBY into NIABY—Not In Anybody's Backyard. When individuals first find themselves situated "downstream" of pollution, their struggle to remedy the situation often stimulates concern for other downstreamers. Support for collective action is drawn along a broad front. Though originally acting in their local self-interest, people learn that they are neither the first victims of an environmentally destructive system nor will they be the last. Lois Gibbs observed that as the toxics movement matured, people decided that they would not let "their solution become someone else's problem."[79] A sense of solidarity with other downstreamers was cultivated. Citizens Clearinghouse for Hazardous Waste, which in 1997 changed its name to CCHW: Center for Health, Environment and Justice, describes itself as People United for Environmental Justice. What sustains the pursuit of justice is the threat—actual or potential—of health-impairing environmental degradation faced by each and all. The monthly newsletter of CCHW is called, suitably enough, *Everyone's Backyard*.

While most toxics groups restrict their agendas to the backyards of their own community, state, region, or nation, many of the larger nationals have made the move to NOPE, Not On Planet Earth. Greenpeace stands at the forefront of environmental organizations in this regard. It has a geographically catholic definition of "backyard." The organization has sent its ships after toxic-waste-filled barges and tankers destined for foreign dumping grounds. Arriving before the waste-laden ships in foreign ports, Greenpeace activists have stimulated enough local opposition to the dumping to force the ships to move on or return home with their toxic cargoes intact. Greenpeace also initiated the International Waste Trade Project in 1989. The project sought a global agreement to outlaw cross-border waste transport, which typically flowed from the industrialized north to the less developed south.

Highly industrialized countries generate 90 percent of the nearly 400 million tons of hazardous waste produced each year in the world. It is estimated that thirty million tons of this waste had been crossing national borders annually, with much of it moving from industrialized to developing nations.[80] Following the ratification of the 1994 Basel Convention, for which Greenpeace deserves much of the credit, the trafficking of most forms of hazardous waste across national borders was out-lawed. This agreement on the global waste trade marks a first step to meeting the thirteenth of the "Legal Principles for Environmental Protection and Sustainable Development" adopted by the World Commission on Environment and Development in 1987. This "nondiscrimination" principle maintains that states "shall apply as a minimum at least the same standards for environmental conduct and impacts regarding transboundary natural resources and environmental interferences as are applied domestically."[81] This international accord parallels the expansion of activist concerns from NIMBY to NIABY to NOPE.

The geographic expansion of the concept of one's backyard is reinforced by the dynamics of multinational corporate power. NIMBY activists are often made aware of the global nature of the problems they face when they learn that the decision to locate a landfill or incinerator in their neighborhoods was made by executives of a large corporation headquartered in a distant community, state, or nation. The toxic waste to be shipped into their neighborhoods for disposal, in turn, may originate from counties or states other than their own. The local problem, activists learn, represents a small part of a much larger, unsustainable economic pattern of externalizing ecological costs.

One of the chief tactics employed by local toxics activists is known as "plugging up the toilet." The idea is to make corporate waste disposal neither politically easy nor economically cheap by organizing citizens to oppose local sitings of landfills and incinerators. Plugging up the toilet works because limiting the availability of disposal facilities through public protest raises the cost of disposing waste. "That's the American way," Lois Gibbs observed, "scarcity raises the price . . . That's when real change will come. All they understand is profit and loss. When the cost is high enough, corporations will decide to recycle wastes and reclaim materials, to substitute nontoxics in their products, to change their processes of production."[82] Plugging up the toilet has proven quite effective. As one study concludes, "Until grassroots protests became successful at limiting disposal capacity, neither industry nor government had seriously considered alternatives—like banning hazardous production processes—that might impose redistributive [economic] burdens upon industry."[83]

The strategy of plugging up the toilet seems straightforward enough. Yet it is often misunderstood. Reflecting on the efforts of grassroots activists to stop construction of an incinerator designed to dispose of chemical weapons and, potentially, other hazardous waste, Gregg Easterbrook has naively asked, "But wouldn't that be a good idea? If hazardous wastes really are as dangerous as claimed, their destruction is an urgent need."[84] The problem is twofold. First, the incineration of hazardous wastes produces noxious air emissions. Second, the production of hazardous materials will continue unabated as long as their disposal remains cheap and uncontroversial. Plugging up the toilet does not solve the problem of getting rid of existing toxic wastes. In fact, it aggravates the problem. That is the point. Plugging up the toilet sends a loud, or rather acrid, economic and political message that fewer toxics should be produced in the first place.

Critics worry that NIMBY activists (and perforce NIABY and NOPE activists) who are plugging up the toilet have effectively gone BANANA. These activists want waste disposal businesses to Build Absolutely Nothing Anywhere Near Anything. Clearly as long as industries and municipalities produce unrecyclable, uncompostable waste, business or government will have to find places to put it. The real question is how much waste needs to be produced in the first place and how toxic must it be? If nothing else, NIMBY localism underlines the unsustainability of economic practices that take for granted the availability of geographically distant sinks. On Spaceship Earth, NIMBY activists argue, such sinks no longer exist. The technique practiced by local toxics groups of plugging up the toilet challenges the "out of sight, out of mind" mentality on which ecological cost externalization is based. NIMBYism, in this case, is simply the other side of the globalist coin. Globalism, after all, is not only a concern for the global ramifications of (aggregated) local actions. Globalism is also a concern for the effect of global processes on particular locales.

Political Interdependence

Aristotle observed that democracy is a system of government requiring citizens to rule and be ruled in turn. Democracy is political rule grounded in interdependent relationships among citizens and between citizens and their representatives. These relationships are mediated by constitutional rights and electoral processes. Because it is grounded in and fosters interdependence, democracy has become a natural ally of environmentalism.

Gifford Pinchot was the first environmentalist to observe the connection between democracy and environmental care. In his 1910 publication *The Fight for Conservation*, Pinchot characterized the conservation movement as inherently democratic. He wrote:

Equality of opportunity, a square deal for every man, the protection of the citizen against the great concentrations of capital, the intelligent use of laws and institutions for the public good, and the conservation of our natural resources, not for the trusts, but for the people; these are the real issues and real problems. . . . Natural resources must be developed and preserved for the benefit of the many, not merely for the profit of a few. . . . Conservation is the most democratic movement this country has known for a generation. It holds that people have not only the right, but the duty to control the use of the natural resources, which are the great sources of prosperity. And it regards the absorption of these resources by the special interests, unless their operations are under effective public control, as a moral wrong. Conservation is the application of commonsense to the common problems for the common good, and I believe it stands nearer to the desires, aspirations, and purposes of the average man than any other policy now before the American people.[85]

Pinchot's democratic values have been put into practice by many who argue that the care of nature and the protection of a safe, healthy environment go hand in hand with democratic government. Many argue that environmental protection will be best served by a more participatory democracy, where information and influence are more equitably shared.

In a contemporary statement of Pinchot's thesis, Donald Snow has characterized the environmental movement thus: "Conservation is not only about resource protection, preservation, and science; it is also about good government. Since long before Earth Day, conservation NGOs have been among the leading champions of citizen participation, open government, and access to information. Conservation NGOs are worth special attention because of the crucial role they play in our civic and public affairs. Without them, our political life would be much poorer."[86] Environmental organizations take on the task of providing citizens with the knowledge they require to make informed decisions about their well-being. Groups that engage in grassroots organizing or promote grassroots action directly stimulate more participatory governance.

In its mission statement, Friends of the Earth sets itself the task of "empowering citizens to have an influential voice in decisions affecting the quality of their environment—and their lives."[87] Among fourth-wave environmentalists, the linkage between environmental protection and political life is typically characterized in democratic terms.[88] Environmentalism is portrayed as a democratic struggle for the right of individuals and communities to protect and enhance their health and well-being. The pursuit of environmental justice, in particular, is inherently a political struggle. One observer remarked that "the environmental justice movement is not so much demanding recognition of a new set of substantive rights as ensuring that nations follow through on commitments they have already made."[89] Basic human rights to health and welfare depend on a nontoxic, fertile environment. The provision and maintenance of a life-sustaining and life-enhancing environment, it follows, are part of the political bargain struck between citizens and their government.

Environmental groups understand themselves as democratizing forces. To the extent that environmental struggles yield an organized citizenry, these struggles expand and enrich democratic life. Evidence suggests, in turn, that the reverse is also true. Democratic life may expand and enrich environmental learning and foster environmentally responsible behavior.

A study conducted by Adolf Gundersen demonstrates that democratic deliberation heightens citizens' concern for the environment and improves their ability to put their environmental values into practice. "Given the opportunity to engage in political deliberation on environmental questions," Gundersen observed, "citizens *do* learn. Hence expanding such opportunities holds a very real promise for environmental solutions." Democratic deliberation strengthens individuals' environmental commitments because it heightens what Gundersen calls "environmental rationality." The term describes an appreciation of coevolutionary interdependence. The environmental rationality stimulated by democratic deliberation is chiefly characterized by the adoption of "long-term thinking" (interdependence across time), "collective thinking" that ultimately entails a "global" perspective (interdependence across space), and "holistic thinking" that demonstrates an appreciation of how the various parts of nature and society interrelate and interact (interdependence across species). Environmental rationality is also "evolutionary" in its scope.[90]

Gundersen's study measures the "environmental promise" of democratic deliberation rather than democratic participation. Environmental groups often focus on the latter. A campaign organized by CCHW to end the production of dioxin, for instance, is characterized as a blend of ecological and political advocacy. Lois Gibbs wrote: "Shutting down the sources of dioxin will not be easy, but look at what we stand to win—less cancer, stronger immune systems and fewer birth defects. We will also see the beginnings of a rebuilt democracy based on the coalition efforts of people who have figured out how to limit corporate influence and maximize public participation."[91] Environmentalists' campaigns against corporate power are described as battles "against the alienation and isolation that destroys the promise of democracy." The American people, CCHW reminds its members, "didn't revolt against King George so we could have King Dow [Chemical]."[92] A democratic impulse is a driving force of fourth-wave environmentalism.

The CCHW effort of "rebuilding democracy through the grassroots environmental movement" goes so far as to place the political task at the forefront. Environmental protection ostensibly serves not as an end in itself but as a means of strengthening democracy. "What many of these groups have in common, other than their focus on environmental issues," Donald Snow wrote, "is a strong desire to inject democratization into public decision making; their leaders seldom see environmental issues as an end and a cause in themselves, but rather as an opportunity to enhance the power and authority of individual citizens against the seemingly all-powerful state and the commercial interests that dominate it."[93] Snow's remark aptly describes many environmental justice groups and grassroots organizations. For the large national groups, however, the situation is reversed. Democratic participation clearly serves the end of environmental protection. As Robert Cox of the Sierra Club indicates, strengthening civic life is simply the "best strategy" for preserving wilderness and protecting the environmental health of communities.[94] Whether as a means to environmental protection or an end in itself, increasing democratic participation is a key ingredient of almost all environmental organizations' recipes for success.

Democracy is not a panacea for environmental ills. Indeed, most environmentalists agree that democratic government is a mixed blessing. As an activist with the Sierra Club and Greenpeace observed, "Democracy creates problems but also enables solutions." The promise of democracy is that of an informed citizenry with the power and incentive to look after its long-term interests. The problem is that modern democratic culture is consumeristic and hyper-individualistic. That is not conducive to environmental concern or to collective efforts to protect the environment.

An Audubon activist observed that as an expression of the public will, democratic politics will only lead to environmental protection if the public embraces environmental values. The problem, he believed, was that an individualistically oriented public will shy away from collective action. In turn, democratic-individualist culture, particularly in the

United States, has become closely aligned with the unfettered market and the untrammeled growth of consumption. Environmental degradation is the result. Liberal democracy's celebration of the independent individual, to the extent that it undermines the affirmation of social interdependence and valorizes consumer choice, bears an antiecological potential.[95] For this reason, environmentalists, particularly the more active ones, consistently express reservations about (market-oriented) individualism.[96] Survey research demonstrates that "[t]he general public overwhelmingly believes in markets, business and capitalism, while the [environmental] activists do not. Most Americans do not see a lot wrong with a system that enables one to acquire more goods rather than being content with only a few. For environmental activists committed to an egalitarian culture, living with less is not just a sacrifice for the sake of the environment but a concrete expression of . . . their rejection of the acquisitive life of competitive individualism."[97] As a rule, most environmentalists do believe in market economics; certainly they do not advocate a centralized economy. But environmentalists do not believe that the market should be left wholly unregulated and unaccountable, that business should be allowed to externalize environmental costs, or that governmental action is unnecessary to redress environmental ills created by the market. Hence, environmentalists are more critical of the free market than is the general public, and less optimistic that the market, left to itself, will prove environmentally benign.[98]

A benevolent dictator with sound ecological values might be more conducive to environmental care than a democratic system. Most environmentalists are quick to point out, invoking ecological principles, that this option is unacceptable because, among other things, it cannot be sustained over the long term. As one environmentalist jokingly remarked, "Clearly the best system would be a benign dictatorship. The problem is how to choose my successor!" An EDF staff member expanded the point: "People are affected in their daily lives by environmental problems, and they can express concerns to representatives who are out of a job if they don't comply. Of course a dictatorship or a philosopher king would do things more effectively, but that's not sustainable." A member of Florida Defenders of the Environment, in a gesture common to many, justified democracy by underlining the lack of alternatives. He observed that "we have a world of many nations and the only ones doing any good with environmental protection are the democracies, probably due to freedom of press and speech. [In democracies] complaints can be aired and publicized." In the end, environmentalists' attitudes toward democracy's ecological potential are probably best summarized by Thomas Jefferson's famous justification of public control of government. "I know of no safe depository of the ultimate powers of the society but the people themselves," Jefferson wrote, "and if we think them not enlightened enough to exercise their control with a wholesome discretion, the remedy is not to take it from them, but to inform their discretion."[99] Environmentalists are aware of the ecological merits and dangers of democracy and democratic culture. While they want to make the world safe for democracy, environmentalists remain equally concerned with making democracy ecologically safe for the world. For the most part, they place their hopes in education.

Despite any reservations that environmentalists have about democracy, they consistently advocate greater public participation in government and decision-making. Some recommend that a "Ministry of Citizen Participation" be formed with a mandate to facilitate citizen forums and social movement involvement with federal administrative agencies. Others insist that "fishbowl planning" is needed to ensure public participation and oversight. The World Commission on Environment and Development has recognized that "the pursuit of sustainable development requires . . . a political system that secures effective citizen participation in decision making."[100] Opening the political process to increased citizen participation and influence is not without its costs. Democratizing measures might make the governmental administration of environmental protection a more cumbersome process than it is already. With more voices to hear, there would be more views and interests to sort through before action could be taken. Efficiency may suffer.

Environmentalists justify the unwieldiness of democratic decision-making by relying on an ecological paradigm. Ecosystems thrive because they are based on complex interdependence. Complex, interdependent systems cannot be sustained according to centralized plans and programs. The proper analog of the interdependent political community is not an efficient machine, but a dynamic ecosystem.

Referring to demands by extractive industries for greater "streamlining" of application procedures for public land use, one Greenpeace writer observed that "a streamlined democracy—like a streamlined ecosystem—is a contradiction in terms. To streamline public appeals is to eliminate an essential component of our democratic system of checks and balances, while inviting further exploitation of our public lands by private industry."[101] A Cousteau Society member argued similarly that "healthy government" demands a certain level of democratic "disturbance," just as many forests require periodic fires to replenish themselves.[102] Fourth-wave environmentalists affirm and promote democratic institutions that make the care of the environment a politically sustainable achievement—however imperfect and cumbersome these institutions prove to be.

The environmental movement has been described as a "fractal" organization, signifying that the democratic participation and ecological sensibility that environmentalists seek to stimulate in society at large are also exhibited within their own organizations. The movement constitutes both a means to an end and, in some sense, an end in itself. As social movement theorist Alberto Melucci put it, "The medium, the movement itself as a new medium, is the message. As prophets without enchantment, contemporary movements practice in the present the change they are struggling for."[103] Certainly not all environmental organizations see themselves as the seedbeds of democratic politics. Greenpeace, for instance, gives "checkbook" members virtually no say in the workings or direction of the organization. In part, this may be because the organization wants to engage in actions that are more radical than many financial contributors would personally endorse. The Nature Conservancy (TNC) also allows virtually no membership control. Its president, John Sawhill, has made warm reference to his "corporate brethren" and observed that "at the Nature Conservancy, we have always taken pride in managing ourselves like a business."[104] In the interests of efficiency, most of the large national organizations structure themselves hierarchically. They are managed by professionals and run as business enterprises. Among the grassroots groups, however, the story is different.

Commenting on the "new environmentalism" that erupted from the grassroots in the mid to late 1980s, Jordan and Snow wrote:

> While the mainstreamers tend to rely on hierarchical models of organization, the new environmentalists are more highly democratized. The former tend to see themselves as environmental professionals working to influence the established order of environmental management by working within the system. The latter tend to doubt whether the system, as it is currently constituted, will ever work for them. . . . The grassroots groups are essentially *political*, while many of their establishment counterparts have become essentially *technical*.[105]

Jordan and Snow overstate the case. Many grassroots groups are goal oriented and hierarchical. The critique of the mainstream organizations, however, does not wholly miss the mark.

The NWF, for instance, is far from a democratically run organization. It operates much like any business corporation, with executives who have a firm grip on the reins of power. Yet it "strongly supports the democratization of environmental decision-making" at local and national levels.[106] The same can be said of most of the nationals. They rely heavily on a bureaucratic elite. Yet they endorse increased citizen participation and more democratic politics within society at large. In the fourth wave, even the most conservative and highly professionalized national organizations actively promote democratic means to achieve environmental ends at municipal, state, and national levels of government.

At the grassroots level, protest activity often serves as the substitute for the professionalized lobbying, litigation, and scientific research carried out by the national organizations. This protest activity is conducive to democratic control. Large corporations and irresponsible governments—the actors most responsible for environmental degradation—wield a disproportionate amount of political and economic power in society. The linkage between environmental degradation and centralized political and economic power is painfully obvious to small, grassroots environmental organizations that, unlike the large nationals, do not have much political or economic clout of their own. It follows that democratic politics and environmental protection are natural allies for grassroots groups concerned with environmental justice.

Nonetheless, a substantial degree of organizational hierarchy is the norm for national and local environmental groups, both within the mainstream reformers and among the radicals (with the Greens constituting a partial exception). Moreover, most local groups, like the nationals, do not value democratic political reform above environmental protection. The former is seldom as vigorously pursued as the latter. Increased democratic participation is almost universally endorsed in the abstract. Support for democratic participation tends to diminish and hierarchical structures and decision-making processes reappear, however, whenever the achievement of specific goals demands greater efficiency.[107] Even the most democratic of environmental groups, the Greens, is not wholehearted in its endorsement of greater participation. Greens engage in continuous debate over the extent to which heightened internal democracy undermines or facilitates the achievement of their ecological goals.[108]

Politically speaking, the grassroots toxics movement and the Greens come closest to practicing what they preach. Though nationally organized or affiliated, both remain primarily localist in organization and

democratic in style. Among the large national groups, the Audubon Society is relatively democratic in its structure. The Sierra Club (which
touts itself as "the most powerful grassroots environmental group in the
world") is constitutionally the most democratic.[109] The Sierra Club
membership directly elects its officers. By means of polls and referenda,
the membership at large is also involved in the formation of certain
policies. Under the club's by-laws, for example, direct and binding membership votes can be taken on any issue at the request of 2.5 percent of
the membership (approximately 1300 Sierrans).[110] Nonetheless, the organization cannot be characterized as a truly grassroots group or a participatory democracy. Indeed, what makes the Sierra Club one of the
more powerful political forces in the environmental movement today—
that is to say, what gives it clout in Congress, in the courts, and with
the media—has as much to do with its professionalism as its democratic
processes.

Different organizations respond differently to the challenge of coordinating their ecological goals with democratic means. The tension
between relatively effective, tried-and-true forms of centralized politics
and looser, more democratic forms of participation will undoubtedly
remain an uneasy challenge for the environmental movement as a whole.
Democracy is the form of political life most grounded in an appreciation
of complex social diversity and the social interdependence this fosters.
As such, democracy is easily wedded to ecological thinking. Yet democracy is an unwieldy method of rule. It will remain an ideal for environmentalists that is only ever partially achieved.

Like Pinchot, today's environmentalists find that the economic
power of big business constitutes the greatest threat to democratic efforts
to protect the environment. Comparatively speaking, even the largest of
the national environmental organizations are small fry in this league.
"Greenpeace, compared to its beginnings, is a huge organization," a
spokesperson observed. "Yet compared to the Fortune 500 polluters and
bomb-makers we confront, we're tiny. Our annual budget is equivalent
to four hours of General Motors' [annual revenues]."[111] Such disparity
in resources has its political consequences. One of the five platforms of
the Environmental Bill of Rights circulated by Sierra Club, Public Interest Research Group, Natural Resources Defense Council (NRDC), National Audubon Society, Greenpeace, and a half-dozen other major
groups is to "get the big money out of politics." Campaign finance reform is the proffered solution.

During the preelection period and one-and-a-half sessions of the 103rd Congress, from January 1991 to June 1994, the environmental movement, through its fourteen existing political action committees (PACs), contributed $1.7 million to congressional candidates. During the same period, the waste management industry contributed $1.4 million, mining PACs contributed $1.9 million, timber interests donated $2.3 million, the chemical industry supplied $3.8 million, construction chipped in $7.8 million, transportation (including automotive) concerns contributed $20.9 million, energy and natural resources PACs furnished $21.7 million, and agricultural interests shelled out $22.7 million.[112] These business interests, which do not include all those PACs opposing environmental regulations in the halls of the Congress, contributed a total of $82.5 million, almost fifty times the amount delivered by the environmental movement as a whole. Between 1993 and 1996, corporations and PACs intent on rolling back environmental laws and regulations contributed more to congressional campaigns that any other single cluster of interest groups.[113]

Environmental groups are also grossly outspent in the arena of public relations. The American Petroleum Institute, for example, spent $1.8 million for publicity in 1993 alone, largely in an effort to defeat a proposed tax on fossil fuels. It is only one of fifty-four industry members of the Global Climate Coalition, a leading public relations organization that works to play down the threat of global warming. By comparison, the five major environmental groups that focus on the issue of climate control (EDF, NRDC, Sierra Club, Union of Concerned Scientists, and WWF) were able to devote a combined total of only $2.1 million to the cause.[114]

Environmentalists frequently observe that despite the existence of "free" elections, the power of monied interests undermines the democratic potential of our political system. With the average winning Senate candidate spending $4.6 million and the average winning House candidate spending $516,000 for his or her election campaign in 1994, catering to special interests with deep pockets has become the sine qua non of gaining and maintaining political office.[115] The costs of "democracy" are still rising. Overall spending for the 1995–96 federal election cycle has been estimated at $2 billion.[116] Of course, environmental groups themselves participate in the system. The League of Conservation Voters prides itself on contributing more money to influence outcomes in elections in the 1990s than ever before. Still, the deck is clearly stacked against environmentalists. In the present political system, money talks. That means that the voice of the earth is often stifled.

Diversity and Social Justice

Diversity is affirmed as a strength by fourth-wave environmentalists. Years of conflict and criticism were necessary before this appreciation of difference arose. Without doubt, its affirmation remains incomplete.

Racism and classism were clearly evident in the early conservationist movement.[119] As late as the 1960s, the Sierra Club effectively discriminated against people of color. This discrimination was vehemently opposed by David Brower, who insisted that the club declare itself open to people of "the four recognized colors." At the time, however, Brower did not speak for the majority. Presented with a resolution calling for an end to the exclusion of minorities, the club's board of directors voted it down, with one board member irately observing that "this is not an integration club; this is a conservation club."[118] As the 1960s and 1970s slowly brought racial integration into the mainstream of American life, the conservation groups followed suit—although with the same reluctance displayed by society at large.

The reluctance was understandably resented by those who bore the brunt of discrimination. Responding to the events of the first Earth Day in 1970, the black mayor of Gary, Indiana, observed that "[t]he nation's concern with the environment has done what [segregationist] George Wallace was unable to do: distract the nation from the human problems of black and brown Americans."[119] Whether this distraction was strategically employed or not, the problem of racial and economic inequities was ignored by most second-wave environmentalists. In a 1973 poll, the Sierra Club found its members voting three to one against a proposal to increase the club's involvement with minorities and the urban poor.[120]

To this day, the middle-class and upper-middle-class demographics of the environmental movement persist. The movement remains predominantly white and wealthy.[121] While environmental concern per se is unrelated to income,[122] environmental activism is not widely embraced by low-income classes. A study by Resources for the Future revealed that 8 percent (and some studies suggest 10 percent) of the general population are members of local or national environmental organizations. At the same time, 27 percent of those who earn over $30,000 are members, six times the rate of membership for low-income individuals.[123] "American

environmentalism," one social commentator concluded, "is a secular religion of the white middle class."[124] The religion is currently spreading
to a wider population, but its social and economic roots are patent.

Peter Berle of the National Audubon Society acknowledged in 1987
that "[n]ot one major environmental or conservation organization can
boast of significant Black, Hispanic or Native American membership."[125]
Within the Group of Ten, by 1989, less than 17 percent of its staff
members were minorities and less than 2 percent of its staff were minority professionals.[126] That year, the ten largest environmental groups
embarked on a minority outreach program. By 1991, however, the situation was not much better. Audubon still had only thirty-five nonwhites in a workforce of 320. Less than 7 percent of the paid positions in
NWF were held by nonwhites, while the Wilderness Society could claim
only 5 percent of its staff as minorities.[127] Up to one third of the mainstream groups had no people of color on staff, and over one fifth had
none on their boards. To make matters worse, the largest share of staff
in these organizations was being hired away from other movement organizations, perpetuating a revolving door of personnel that effectively
restricted the entry of minorities.[128]

Most, if not all, major environmental organizations have made efforts to ameliorate their poor representation of minority interests and
their dearth of minority members and staff. Their success has been limited. Even those groups most anxious to increase the cultural diversity
of their membership (such as the Greens, who list social justice and
respect for diversity among their ten key values) manage to count people
of color as only 10 percent of their supporters.[129] Part of the problem
is that social movement activists, whatever their cause, tend to be more
educated than the population at large and tend to come from the middle
or upper middle classes. The leadership and staff of the "grassroots"
toxics movement are not exempt from this demographic reality.[130] Studies also suggest that only 7 percent of working-class people demonstrate
the characteristics considered crucial to involvement in collective action,
such as a sense of agency, a group identity, and the perception of injustice committed against them.[131] As Philip Shabecoff wrote, "Unfortunately, it is true that the leadership of national environmental groups *is*
largely white, male, and well educated, with incomes above the national
average. This description, however, would fit activists in virtually every
social movement."[132] While Shabecoff's observation is accurate, the unfortunate fact remains that minorities and low-income classes remain
inadequately represented within the environmental movement.

The difficulty environmental organizations originally faced in attracting working-class and minority staff and membership was largely a product of their lack of interest in the concerns of the working class and minorities. In a nationwide survey, 95 percent of the leaders of over 500 conservation and environmental groups agreed that "[m]any, perhaps most, minority and poor rural Americans see little in the conservation message that speaks to them."[133] In fact, people of color continue to receive fewer benefits from the efforts of environmentalists than any other segment of the population.[134]

The restricted focus of certain mainstream groups on the preservation of wilderness may be part of the problem. A majority of Sierra Club members, for example, view "protecting wild places" as the club's top priority. Fewer than half as many list "cleaning up pollution" as a primary concern.[135] Aldo Leopold wrote that "[t]here are some who can live without wild things, and some who cannot. . . . For us of the minority, the opportunity to see geese is more important than television, and the chance to find a pasque-flower is a right as inalienable as free speech."[136] Leopold's sentiments would be echoed by many members of the environmental movement today. While television might easily be given up for the sake of wilderness, however, doing without a job or a healthy neighborhood is another issue.

To care for the long-term preservation of biodiversity, one first has to be able to care for the short term preservation of one's economic security and health. Richard Leakey put the point succinctly when he said: "To care about the environment requires at least one square meal a day."[137] Minorities and the poor generally maintain that the basic needs of healthy food, decent housing, and a toxic-free environment rank above the aesthetic and spiritual benefits of wilderness. Until these basic needs are sufficiently satisfied, the latter goods will not be embraced. Working classes, moreover, often perceive the preservation of natural habitats as a threat to economic development and secure jobs. In this context, nature preservation is understood not only as a luxury for the rich, but also as a threat to the poor.

Many minorities grow up in depressed economic settings where the appreciation of wilderness or an aesthetically appealing environment remains largely uncultivated. They have little contact with nature, let alone wilderness. This problem is exacerbated by a lack of funds for nature excursions or vacations. Priorities and sensibilities are affected.[138] As Dorceta Taylor remarked, "It is unrealistic to expect someone subsisting at the margins of the urban or rural economy, or who is unemployed, to support wildlife and wilderness preservation if she or he has no access to or cannot utilize these resources."[139] Aware of this problem, a number of national organizations have made an effort to forge a relationship between underprivileged urban dwellers and wilderness. The National Wildlife Federation, for example, began NatureLink in 1992. Designed to cultivate environmental sensibility in segments of the population where it has traditionally been lacking, the program brings environmental awareness and commitment to urban dwellers who would otherwise have little exposure to the natural world.[140] Such programs are prudent investments. Exposure to nature and nature experiences, studies consistently demonstrate, are prime predictors of environmental commitment and environmentally responsible behavior.[141] "Building a more intimate relationship between people and land," John Sawhill observed, is the only means of creating "the connectedness—the sense of interdependence—with nature" that serves as the foundation of an environmental ethic.[142]

To gain greater working-class and minority representation, the national groups must do more than offer nature excursions, solicit membership from a more diverse public, and hire a more diverse staff. They must concern themselves with working-class and minority issues and become partners in local communities and workplaces. Until the late 1980s, however, such a concern for social justice did not exist. Overconsumption by the rich, the other side of the social justice coin, was also all but ignored. The Group of Ten's 1985 *Environmental Agenda for the Future*, for instance, was largely a top-down approach to environmental protection that did not broach the issue of social justice. Overconsumption, in turn, was left off its list of eleven key areas of environmental concern. Only a brief mention was made of the need for Americans to use their resources "more efficiently" in a section devoted to human population growth. Yet this document was described as a "consensus view" of ten of the largest national groups.[143]

The formerly patent and still lingering neglect by environmental groups of social justice issues remains undoubtedly connected to the realities of organizational funding. Up to 70 percent of the combined budgets of Washington-based mainstream national organizations arrives through the mail from members and small donors.[144] These members and donors come primarily from the middle class and upper middle class. He who pays the piper generally calls the tune. The environmental concerns of middle-class and upper-middle-class America have therefore been the concerns that environmental organizations primarily support.[145]

National environmental groups, for example, frequently investigate, document, and widely publicize the real, but seldom deadly, threat posed by pesticides to suburban residents and consumers. Until recently, however, these groups remained largely oblivious to the tens of thousands of farm workers, predominantly migrants and minorities, gravely poisoned each year by pesticides. The disproportionate affect of pesticides on non-agricultural people of color has also been neglected.[146]

Allan Schnaiberg wrote that "[i]t is never sufficient to point to *the* environment as having been protected. The question must be asked, for whom and from whom has it been protected. Environmental quality and social welfare issues are not socially or politically separable."[147] Only within the last decade, and then with hesitation, have mainstream organizations acknowledged the inseparability of ecological and social health.[148] For the most part, the mainstream was goaded into action by environmental justice activists. These activists presented the nationals with "the reality of a broader vision and a redefined mission encompassing greater diversity for the environmental community in terms of both issues and organizational structure."[149] One minority critic stated the case succinctly, observing that large environmental organizations see "diversification as bringing more numbers on board to do the same job; we see it as a way to concentrate on a new job."[150] In the 1980s, the nationals approached the issue of inadequate minority and working-class representation as a problem of demographics. By the late 1980s, however, the effort to increase minority members and staff was complemented with the "new job" of integrating minority concerns into the agendas and operations of the national organizations. Today, the struggle for social justice is a vibrant current within the mainstream.

In the early 1980s, NWF president Jay Hair admitted that he did not see an association between environmental issues and social justice questions. By the end of the decade, that association was clear.[151] The NWF eventually included sections on environmental justice in its annual reports, and it now boasts its own publications documenting pollution's disproportionate impact on poor and minority communities.[152] Likewise, in 1995 NRDC reformulated its mandate "to reach broader sectors of society by pursuing integrated solutions to environmental, economic, and social problems."[153] The NRDC has subsequently instituted an Environmental Justice Initiative to address the environmental concerns of marginalized groups.[154]

Brent Blackwelder of Friends of the Earth has insisted that "[y]ou can't set environmental problems apart from social problems, problems

of rich and poor. . . . The environmental injustice of pollution is a fact
of life. Those least politically potent have received the worst pollution.
This is indisputable. It's racial, and poverty based. You see it worldwide.
The question is how to deal with it. First, we must recognize it and
speak out against it. Second, we have to change the economic system
that allows it to occur."[155] Celebrating twenty-five years of environmental activism in 1994, Friends of the Earth maintains that its activities are
now shaped by three considerations that will "redefine environmental
activism" for the next quarter century: first, challenging "the imbalance
of economic power and its devastating effect on our health and environment"; second, "redefining prosperity" so as to "debunk the notion
that the 'good life' in America requires massive over-consumption and
its resulting environmental degradation"; and third, "empowering people and communities" so they may "hold governments and corporations
accountable for their decisions."[156] The overall goal is to protect nature
while redefining what it means to live the good life in a socially diverse
world. This entails the pursuit of integrated solutions to ecological, social, economic, and political problems. As Bunyan Bryant wrote: "We
can no longer afford to champion the rights of trees and nonhuman life
without also championing the rights of all people, regardless of race,
sex, income, or social standing. We can no longer afford to treat certain
categories of people as if they were not part of a biodiverse community."[157]

The greater willingness, and in some cases one might even say the
eagerness, of environmental organizations to integrate the concerns of
minorities and low-income classes into their agendas and to integrate
individuals from these groups into their staffs and memberships has
arisen for one major reason—the understanding that their interests are
interdependent. The linkage between poverty and environmental degradation, like the linkage between race, class, and susceptibility to environmental hazards, has been theoretically sustained and empirically
substantiated. In the mid to late 1980s, low-income neighborhoods and
minorities began organizing against many of the same corporate perpetrators of environmental degradation that the nationals had taken on
for years in the courts. New grassroots groups and national environmental organizations formed alliances. Partly responding to these linkages and partly advancing them, the Congressional Black Caucus has
consistently demonstrated the best environmental voting record of any
bloc in Congress (as calculated by the League of Conservation Voters).
A similar claim is made for the Hispanic Congressional Caucus.[158]

Support for the environment has traditionally been lower among minorities. There are good indications that this may be changing. Pollster George Pettinico recently cross-tabulated responses from national surveys to select out "true greens," identified as people who rate the environment as one of the most important issues facing the nation, who say the environment played a role in their voting preference in the last presidential election, who believe government is doing too little to safeguard the environment, and who favor environmental protection over economic expansion. While 46 percent of all whites could be called true greens, 58 percent of African Americans and Latinos warranted that designation.[159] In recent years, moreover, the number of blacks belonging to environmental organizations as a percentage of the black population has not significantly differed from the number of whites belonging to environmental organizations as a percentage of the white population.[160]

Dr. Ben Chavis, former executive director of the National Association for the Advancement of Colored People (NAACP), has corroborated these findings. He observed that "the environmental issue penetrates even across lines of race, across lines of socioeconomic circumstance. . . . I've been in a number of communities where whites and blacks have never marched together, but they're marching together now around the environment."[161] Robert Bullard likewise observed that

> [m]any environmental activists of color are now getting support from mainstream organizations in the form of technical advice, expert testimony, direct financial assistance, fundraising, research, and legal assistance. In return, increasing numbers of people of color are assisting mainstream organizations to redefine their limited environmental agendas and expand their outreach by serving on boards, staffs, and advisory councils. Grass roots activists have thus been the most influential activists in placing equity and social justice issues onto the larger environmental agenda and democratizing and diversifying the movement as a whole. Such changes are necessary if the environmental movement is to successfully help spearhead a truly global movement for a just, sustainable, and healthy society and effectively resolve pressing environmental disputes.[162]

Though frustrated with the pace of change, Bullard remains sanguine about its direction.

Political observers have testified that the success of the environmental movement largely rests with its efforts to reach a wider, more diverse constituency. Adam Walinsky argued that "[i]f we allow the environment movement to become a pretty plaything of the affluent, and ignore the real environmental problems of the ghetto and the farm laborer and the blue-collar worker in his factory, then we will be choosing political suicide, and the name of our sword will be irrelevance."[163] Other observers concluded that "[i]f environmentalists form new and enduring coalitions with labor, community, or social equity movement groups or organizations, they actually have a greater chance of dominating the agenda, through political veto power over economic elites and their government supporters."[164] In the same vein, Lois Gibbs has maintained that the recent outreach of the large environmental groups to minorities and working-class people has become environmentalism's political lifeline. "Environmentalism's history shows that we succeed when we consciously and systematically focus on building a political base of support within the American public," Gibbs has stated. "This holds true for *all* environmental issues, whether they be national parks, endangered species, toxic waste, or garbage. Efforts to preserve and improve the environment are sure to be set back, if not fail outright, when advocates for the environment forget or ignore the fact that environmental causes are just as political as any other public policy issue."[165] By recognizing and nurturing social and political interdependence as well as ecological interdependence, fourth-wave environmentalists have created a more holistic program of action, gained a more diverse membership, and assured a broader base of support.

Community-Based Environmentalism

The integration of the concerns of the disadvantaged into the agendas of national environmental organizations has been greatly facilitated by the lessons learned in the international arena, particularly in developing countries. Evolving global sensibilities, in this case, have had significant domestic repercussions. In the second and third waves, environmentalists frequently garnered support for their battles against habitat destruction, resource depletion, and overpopulation through their apocalyptic portrayals of the devastating appetites of teeming, impoverished masses overseas. Paul Ehrlich, for example, gained the attention of his American readership by beginning his book *The Population Bomb* with an account of how he himself developed a *"feel* of overpopulation" when, en route to his hotel one night in Delhi, his taxi had to navigate streets "alive" with begging, urinating, and defecating Indians.[166] These sorts of characterizations and the ensuing calls for coercive population control led to criticisms that Western environmentalists were racist. At best, they were blaming the victim.

Only in the fourth wave did environmental organizations systematically target the poverty and powerlessness that contributes to environmental degradation. Beginning in the late 1980s, environmental organizations initiated projects designed to empower those who suffer most from environmental degradation and economic marginalization. As Aaron Sachs wrote, over the last five to ten years "environmentalists learned to address the social and cultural context of their campaigns, they became better able to demonstrate the immediate human value of intact ecosystems. . . . [L]arge groups like Greenpeace and the Environmental Defense Fund lent their political clout to small, local human rights campaigns. . . . The North's aloof tree-lovers became compassionate defenders of local peoples."[167] At this time, many environmental organizations abandoned their top-down approaches, which typically had been focused on government programs. Instead, they began working from the bottom up, grounding environmental protection in the development of economic opportunities for local communities and indigenous peoples.

Endorsements of community and community-based environmentalism, to be sure, tend to be short on critical analysis. They neglect the divergent interests that exist within communities and the political processes involved in conservation efforts.[168] Too often these gestures to democratic values never yield concrete projects. Yet their endorsement marks a positive development. In any case, there are few effective alternatives to community-based environmentalism, especially in developing countries where the failure of centrally organized, top-down efforts is patent.[169]

Community-based environmentalism goes hand in glove with a focus on civil and economic rights. The 1988 murder of the Brazilian rubber-tapper Chico Mendes, who had heroically worked to protect rainforests from destructive cattle ranching, sparked many U.S. environmental organizations into action on this front. In 1989, a number of NGOs, led by the Sierra Club Legal Defense Fund, lobbied to have the United Nations conduct an international study of the relationship between human rights and environmental issues. In 1992, the NRDC joined up with Human Rights Watch to document the threats faced by environmental activists. In 1995, the Sierra Club and Amnesty International issued their first joint letter, addressing the link between human rights abuses and environmental degradation in Nigeria.[170]

The need for this latter effort became patent when Ken Saro-Wiwa, president of the Movement for the Survival of the Ogoni People and a nominee for the Nobel Peace Prize, was executed with eight other activists by Nigeria's military regime in November 1995. Saro-Wiwa had been working to protect his people from the environmental ravages of multinational oil corporations that were "developing" Nigeria, the most prominent among them being Royal Dutch/Shell. Shell has pumped an estimated $30 billion worth of oil out of Ogoni territory over the last forty years. Yet the local people have seen few economic benefits and have suffered much environmental destruction. Meanwhile, the oil company admits paying field allowances to Nigeria's army, and has been accused of intimidation, illegal detention, torture, and colluding with Nigeria's armed forces to suppress dissent (an estimated 2000 Nigerians have been killed and 80,000 uprooted).[171] Referring to the Sierra Club's and Amnesty International's joint letter, one observer remarked that "[n]o grassroots [environmental] campaign can be successful if the basic rights of the individual are not respected."[172]

While top-down approaches may occasionally succeed in developing a country's industry or agriculture, this development seldom translates into greater economic equity and frequently results in increased environmental degradation. Multimillion dollar hydroelectric dam projects are cases in point. The Narmada River dams in India and the Three Gorges dam on China's Yangtze River have disrupted lives and livelihoods and imposed the relocation of millions of villagers and subsistence farmers. They have been consistently opposed by local inhabitants, who may be aware, as is the case in China, that up to 40 percent of the ten million people who lost their homes to earlier hydro projects are still not adequately resettled. Over the past decade, as many as ninety million people worldwide have lost their homes to dams, roads, and other development projects.[173] Rather than developing sustainable practices that would allow the majority of people to make a decent living while preserving their land and waters, these projects aim to increase exports and trade or otherwise raise economic productivity and the gross domestic product. Geared to Western models of economic progress, they largely disregard social and environmental inequities and costs. Such projects, particularly those sponsored by the World Bank and International Monetary Fund, have become frequent targets of Western environmental groups.

A project to develop the Bangladeshi fishing industry illustrates the distinction between top-down development and sustainable development. Project managers decided to develop the Bangladeshi fishing industry by stocking lake waters with large fish, such as carp, that have greater retail value than the indigenous, smaller fish. The aim was to move away from subsistence fishing to an economically more productive and expansive enterprise. The problem is that once stocked, carp generally drive out most smaller fish. These smaller fish have traditionally been the mainstay of the work and diet of the local poor. A writer for NRDC's *Amicus Journal* explained that the stocking project falls into the category of "the Marie Antoinette school of development: 'Let them eat carp.' To the poor of Bangladesh, carp are what cake was to the impoverished masses of eighteenth-century France, a luxury consumed only on special and rare occasions. Under the stocking approach, the poor lose the 'bread' of small fish. The winners are the few and the losers the many, and, as in many other times and places, 'development' lies in the eye—or pocketbook—of the beholder."[174] The moral and ecological bankruptcy of the stocking project's top-down approach was underlined by a project official when he said that "The only problem with the project is the people."

Fourth-wave environmentalists argue that the only long-term solution to environmental problems is the people. This lesson was first learned when Western organizations confronted the problem of environmental protection in developing countries. The political culture of developing countries makes top-down approaches to environmental protection problematic. Citizens of advanced industrial states believe that government should be primarily responsible for environmental care and remains the most effective agent of environmental protection. People of developing nations, in contrast, are more likely to believe that citizens and citizen groups are most effective at protecting the environment and should bear the primary responsibility for it.[175] The difference in attitude likely reflects the perceived ineffectiveness or lack of interest in the protection of the environment demonstrated by governments in many developing nations. It follows that the human resources for environmental protection in the developing world are largely to be found at the community level.

Many local agrarian cultures in developing countries well demonstrate the ecological perspective—an awareness and concern for social and environmental interdependence—that environmentalists of industrialized nations are still struggling to cultivate. One Western student of modern peasant societies wrote:

In our own society, environmentalists are beginning to lead the way toward a recognition of individual/community interdependence that peasants have already achieved in their less complex world. . . . An ecological perspective and an awareness of interdependence, however, come with difficulty and only gradually in urbanized industrial society. . . . Peasants . . . have already achieved the view that our society is only slowly approaching. . . . Far from being an anachronism or a dying class in a modern world, as an adaptable class of survivors, peasants may have a good deal to teach urban society.[176]

In this and many other respects, Western environmentalists are learning lessons in social and natural ecology from indigenous peoples, minorities, and the poor.

In the past, many Western environmentalists were justly labeled "eco-imperialists." Ramachandra Guha observed that in India "[t]he initial impetus for setting up parks for the tiger and other large mammals such as the rhinoceros and elephant came from two social groups: first, a class of ex-hunters turned conservationists belonging mostly to the declining Indian feudal elite, and second, representatives of international agencies, such as the World Wildlife Fund (WWF) and the International Union for the Conservation of Nature and Natural Resources (IUCN), seeking to transplant the American system of national parks onto Indian soil. In no case have the needs of the local population been taken into account, and as in many parts of Africa, the designated wildlands are managed primarily for the benefit of rich tourists."[177] The WWF initially formed as a funding agency for wildlife protection in biologically rich but economically poor countries. Western conservationists, WWF admits, generally "ignored" and "alienated" local peoples. They refused to work "within the context of local cultural values."[178] Recently, WWF's agenda has shifted to a strategy of community empowerment. The WWF adopted a new mission statement in 1989, abandoning its narrow focus on the preservation of charismatic species and committing itself to protecting biodiversity, promoting sustainability, and building "a future in which humans can live in harmony with nature."[179] Concern has effectively been refocused on the welfare of local communities affected by wildlife protection programs. Pressure from local communities was instrumental to this change.

Making habitat and species preservation a paying proposition for locals, WWF learned, is the only workable solution because it is the only socially and ecologically sustainable practice. Addressing human needs,

WWF now acknowledges, "is the key to all successful long-term conservation efforts. . . . [w]e cannot simply defend isolated species and habitats. To conserve the world's natural resources for future generations, we must also help to alleviate the human crises and development pressures that put wildlands and wildlife in danger." The WWF currently works "with local leaders, grassroots groups, governments, and international funding institutions to improve living standards and to integrate conservation into public and private-sector development programs."[180] Armed with this new agenda, the organization launched a multimillion dollar "Education-for-Nature" program to train local conservationists.[181]

Other conservation organizations have undergone similar transformations. In its perennial efforts to raise funds for its purchases of ecologically valuable land, TNC has never been shy about parading its successes (over eight million acres of wildlife habitat preserved) and capitalizing on its simple yet powerful appeal.[182] "The Conservancy gets results you can walk around on," an advertisement script reads over a glossy photograph of an expanse of exotic wilderness.[183] Yet there has been a significant shift in the Conservancy's strategic orientation. In 1990, an ecosystem management policy was first explicitly formulated. The Nature Conservancy then shifted from having straightforward acreage goals and a focus on the conservation of landscapes to a more integrated socioecological approach. The organization now embraces a broad concern for ecosystemic health and the stimulation of local commitment to conservation through sustainable development. Conservation is grounded on the tripod of ecology, economy, and community.[184] This shift not only indicates a growing concern for the rights of the local inhabitants that interact with endangered ecosystems. It also reflects a pragmatic acknowledgment that local economic sustainability is the prerequisite for wildlife and habitat protection. Greg Low, TNC vice president, stated that "conservation works place by place . . . in every ecosystem we're working in, we need long-term community support or we will fail." With this concern in mind, Low encourages people to "think locally, act locally."[185] While the primary goal of TNC remains the preservation of biodiversity, its means have changed to include the aid and protection of local human populations that are acknowledged to be the primary stewards of the land.

This shift in attitude has spread throughout the environmental movement. Among other things, it has promoted many joint efforts with indigenous groups. Native peoples have long criticized environmentalists

for showing little concern for the human inhabitants of threatened ecosystems and favoring state control of sensitive land through "debt for
nature swaps." Instead of centralized state control, which may work to
the detriment of local inhabitants and their livelihoods, "debt for indigenous stewardship" swaps have been proposed.[186] The justification, as
Louis Bruyere, president of the Native Council of Canada, stated, is that
indigenous peoples are the base of "the environmental security system.
We are the gate-keepers of success or failure to husband our resources."[187] Grounding environmental protection in organic human-
nature relations, he argued, will prove more successful than the bureaucratic management of land.

Founded in 1972, Cultural Survival was an early promoter of sustainable development based on the preservation of indigenous cultures.
Only in the fourth wave of environmentalism, however, did this idea
gain widespread support within the movement. The NWF is now involved in designing and funding a program to preserve millions of acres
of the Amazon forest as "conservation zones" that will be managed
sustainably by its own residents.[188] Rainforest Action Network grounds
its work on the understanding that "native peoples who have lived in
the forests for thousands of years are the forests' best caretakers."[189]
The WWF began to work directly with indigenous communities in Brazil
in 1993. Its efforts that same year to coordinate conservation efforts
with the Hoopa Valley people marked the first common venture of a
national conservation organization and a native North American tribe.
The WWF maintains that to protect habitats and species it is imperative
"to preserve the traditions and livelihoods of the indigenous . . . people."[190] The EDF has helped the Panara Indians of Brazil regain their
rainforest homeland, from which they were forcibly evicted in 1974.
Bruce Rich, director of EDF's international program, observed that
"strengthening local tribal and cultural communities . . . is the natural
complement of—and often the prerequisite for—saving many of the
world's most biologically rich ecosystems."[191] Of the twenty-five countries with the most native mammal and bird species, sixteen are also
among the top twenty-five in number of languages and cultural groups.
Preserving indigenous knowledge and culture goes hand in hand with
environmental protection.

Environmentalists are aware that not all contemporary indigenous
cultures are ecologically attuned. While native hunters and agricultur-
alists are less disruptive to ecosystems than commercial resource extract-
ors, they do occasionally exploit natural resources beyond the natural

rates of regeneration.[192] Moreover, many indigenous cultures are quickly adopting less ecologically benign livelihoods. Defenders of Wildlife observes the distinction between "Corporate Natives" (also known as "Progressives"[193]), who unsustainably log their Alaskan properties for short-term commercial profit, and "Subsistence Natives," who live much closer to the land and try to preserve a more traditional way of life.[194] It is problematic for non-native environmentalists to categorize indigenous culture. At the same time, it is clear that many indigenous peoples have over the years lost many of their organic ties to the land. Helping indigenous peoples assert, build upon or rediscover these ties is the delicate task taken on by growing numbers of Western environmental organizations.

Local communities in the United States, like their international counterparts, do not appreciate environmentalists who threaten their means of making a living from the land. Imposed regulations generated by lawyers and lobbyists are generally resented. In the past, environmentalists were frequently criticized by land-based peoples for their "failure to present themselves as anything other than just the latest wave of outsider elitists imposing their will without regard for the needs of others and without a clear understanding of the relationship and the interdependency between the people and the lands in question. Their own conduct invites the designation of outside predators, whose recreation-dominated agenda keeps them ignorant of the effects of their attitudes and actions on local social conditions."[195] This criticism has been taken to heart. Today, environmentalists "must be listeners, organizers, consensus builders and, above all, good neighbors. They are biodiversity's ambassadors—catalysts for community-based conservation."[196] The goal is for local land-based people, both in the United States and across the globe, to become the driving force behind sustainable development.

Defenders of Wildlife, NWF, and other environmental organizations have worked extensively with local communities in their efforts to reintroduce grizzly bears to the Selway-Bitterroot wilderness area of Idaho and western Montana. Locals were mostly concerned with preserving their timber-based jobs. In an effort to address these concerns, a coalition of interests was formed, which recommended that a fifteen-member citizen management committee oversee the reintroduction of the bears. The plan was accepted by the Interior Department, which set a precedent by allowing a citizens' committee to act in lieu of the Fish and Wildlife Service.[197] In a similar fashion, Greenpeace has worked extensively with local East Coast and Alaskan fishermen in efforts to preserve their way of life. Greenpeace representatives and locals are attempting to limit the use of corporate-owned "factory trawlers" that virtually vacuum the sea of target fish as well as unwanted "by-catch." In Alaskan waters alone, factory trawlers may waste over half a billion pounds of marine life taken as by-catch a year.[198] Likewise, the Sierra Club works with local Nevada ranchers and independent prospectors to counter the environmentally destructive activities of large mining interests. The Club also works with small farmers in Kansas and anglers in North Carolina whose ways of life are threatened and whose water and air are polluted by corporate hog and poultry farms.[199] Environmental organizations have made similar efforts in urban settings. Here they set themselves the task of adopting and addressing the concerns of city dwellers, particularly the urban poor and minorities, whose neighborhoods suffer most from environmental pollution but whose economic lives are often tied to the offending industries.

Friends of the Earth admits that in the past "national organizations have focused on the 'big fights' taking place in Washington, D.C., and have too often turned a deaf ear to people engaged in local issues that may never make headlines, but make a big difference in the quality of peoples' lives." To address these concerns, Friends of the Earth has initiated its own Community Support Project, which helps groups and individuals garner the resources needed to fight and win local battles.[200] Taking the same tack, Defenders of Wildlife launched GREEN, the Grass-Roots Environmental Effectiveness Network. Its aim is to help local activists from hundreds of grassroots organizations protect wildlife and habitat by providing them with information and technical support.[201] The NWF, in similar fashion, avidly promotes its own community-based approach to environmental problem solving.[202] John Sawhill, president of TNC, has asserted that "community-based conservation will emerge as the primary vehicle" through which his organization will operate in the future.[203]

Aaron Sachs observed that "[i]f more projects emphasized the full, well-informed involvement of local peoples, governments could no longer treat them—or their environments—as expendable. And if there were no expendable people or ecosystems, development would have to be sustainable."[204] Sachs's statement reflects an orientation that is widespread among environmentalists today. The central message is that "communities will be an integral part of any sustainable development schemes that succeed over the long term."[205] This message is slowly rubbing off on governmental and intergovernmental bodies as well. The United Nations, for example, initiated a program for Earth Day 1996 to found local "Green Brigades" in communities throughout the world.

While environmental organizations are clearly moving in the direction of community involvement and co-management, there remains much room for improvement. Too often locals merely become involved in the implementation of programs that are conceived and designed at distant organizational headquarters. Too often the socioeconomic concerns of the poor are still slighted in favor of the recreational concerns of the rich. Nonetheless, the clear trend is to include locals from the ground up: in the identification and articulation of the problem at hand; in the brainstorming of tactics, strategies, and solutions; in the marshaling of resources; and throughout all phases of implementation and management. As T. H. Watkins of the Wilderness Society wrote, "It is becoming increasingly clear that without a working symbiosis between the blue suits of bureaucracy and the blue jeans of local activism, the century toward which we stagger will be infinitely more bleak than reason tells us is tolerable."[206] The effectiveness of national environmental organizations depends on their willingness and ability to work with local communities.

With the U.S. Congress shifting more regulatory responsibilities to state and local levels of government, national environmental groups will find their domestic goals difficult if not impossible to meet unless they accelerate their efforts to act locally.[207] On a global scale, the task is similar. Developing countries currently account for 98 percent of world population growth.[208] Stemming this growth will entail more than fostering environmental consciousness. It will entail stimulating sustainable economic development from the bottom up. Tom Athanasiou has suggested that "[h]istory will judge greens by whether they stand with the world's poor."[209] History's judgment, more precisely, will be based on whether environmentalists work with the world's poor to make ecological security and economic security mutually reinforcing.

Environmental organizations attending the Earth Summit in 1992—there were well over a thousand of them—insisted that efforts to educate the public about environmental issues should be oriented to the promotion of "equitable sustainability." They determined that "[s]uch education affirms values and actions which contribute to human and social transformation and ecological preservation. It fosters ecologically sound and equitable societies that live together in interdependence and diversity. This requires individual and collective responsibility at local, national and planetary levels."[210] This statement provides a summary of fourth-wave issues and orientations: sustainable development, interdependence, ecological integrity, social equity, local-global linkages, and adaptive change in nature and society. It is a statement of the coevolutionary challenge.

The Quest for
Environmental Integrity

*Ecological Interdependence
across Species*

Environmentalists' affirmation of temporal and spatial interdependence expands the horizon of human caring. Efforts to protect the environment are carried out not only for the benefit of the near and dear, but for fellow humans in future generations and other lands. This expansive concern for human welfare is the primary motivation for most environmentalists. But it is far from the only motivation. Environmentalists protect the environment not only for the benefit it accords human beings in the here and now or across time and space but also for the sake of other species and the habitats that sustain them. Environmental care is actuated by concern for the welfare of the plants and animals that share the earth with us. The struggle to protect this interconnected web of life is best characterized as a quest for environmental integrity.

How significant is concern for the nonhuman denizens of the earth in environmental affairs? Garrett Hardin has offered this analysis of the Greenpeace effort to save whales:

> In large groups, social policy institutions necessarily must be guided by what I have called the Cardinal Rule of Policy: *Never ask a person to act against his own self-interest.* It is within the limitations of this rule that we must seek to create our future. . . . We are told of idealists on board this [Greenpeace] vessel who appealed by megaphone to the captain of a Russian whaler to cease his activities in the interests of the whales and posterity. The captain's reply was, of course, of the sort that we of the older generation call "unprint-

able." And why should it not be? Whatever sneaking admiration we may have for the idealists of the Greenpeace Foundation—and I confess I have more than a little—their program is quixotic because it violates the Cardinal Rule by asking people to act against their own self-interest.[1]

Hardin has suggested a good rule of thumb. By and large, people act in their own self-interest, and environmentalists would be wise to keep this truth in mind. Yet Hardin ignores a crucial fact. Greenpeace activists have been very successful in their antiwhaling campaign, probably well beyond their original hopes. Their success, admittedly, came less from shouting at sea-bound whalers than from inspiring the general public and persuading their political representatives back on dry land. Nonetheless, the activists' appeals were not to the self-interest of individuals. Their appeals were precisely those that Hardin dismisses as quixotic: to the welfare of future generations and, more directly, to the welfare of the whales themselves.

Why and how has Greenpeace been successful in this regard? The answer, in large part, is that the public to which it appeals harbors a love of nature that manifests itself in a fondness for whales. This love of nature, often displayed as affection for nature's more charismatic species, does not obviate predominant concerns for human welfare or narrower forms of self-interest. It does modify these anthropocentric and egocentric orientations. In large part, it does so by redrawing the boundaries of our moral universe. "For those who love the nonhuman as well as the human," Roger Gottlieb wrote, "our casual devastation of other species is felt as a crippling moral failure."[2]

The extension of moral concern to other species largely derives from a love of nature. If love of nature was restricted to environmental activists, however, there would be little hope of the movement's success. Fortunately, it is widespread. Clearly, the love of nature is stronger for some than others. For most, its translation into a moral imperative is sporadic. Nonetheless, it has worldly effects. Environmental groups appeal to and cultivate the general public's love of nature. They aim to translate this emotional attachment into a viable ecological ethic.

Love of nature often harbors an incipient biocentrism. Indeed, the biocentric perspective is more common than one might assume. Fewer than one in four Americans feels that our only responsibility to nature is to make it serve human needs or that plants and animals have no rights of their own independent of human interests. Fewer than one in seven Americans believes that we need not care about the extinction of species that do not serve any economic, aesthetic, or other human interests. Two out of three believe that animals, not only humans, have a right to live free of suffering. More than one in four believes that animals should have the same moral rights that humans do, with fewer than two in four disagreeing with this statement. More than one in five expresses the radical view that if human activities are to cause an extinction, then that extinction by rights ought to be that of the human race. On the whole, the public rejects the biocentric argument that other species are *equally* important as the human species.[3] Yet moderately biocentric convictions are held by significant numbers of Americans, and radically biocentric views are held by substantial minorities.[4] Biocentrism is, of course, more prevalent in the environmental movement than in the public at large. Those environmentalists who hold biocentric values, moreover, tend to be more active within the movement and more rigorous in their practices.[5] Hence biocentrism constitutes a force to be reckoned with.

The Love of Nature

The impetus behind biocentric attitudes is simple yet powerful: the love of nature. Asked to explain her activism, an executive with Audubon answered succinctly: "I want to see the earth continue to be a green place with a diversity of creatures. . . . I have such love for all of this." A member of The Nature Conservancy responded to a question about her activism by stating, "I'm just a person who loves to be outdoors. . . . I need a daily dose of the wild within my system." Such attitudes are pervasive among environmentalists. Environmentalists do what they do, David Orr wrote, because of "an early, deep, and vivid resonance between the natural world and [them]selves."[6] A member of Florida Defenders of the Environment explained that his activism was a direct product of growing up on the edge of the Everglades and witnessing its steady degradation over the years. Seeing the object of his love destroyed brought him to battle. This path to environmental activism is well trodden.

Almost all environmentalists profess an early and enduring love of nature. Indeed, it is one of the primary motivations people have for becoming involved in environmental affairs.[7] Appreciation of the material usefulness of nature alone generally will not stimulate environmental activism. Nonutilitarian values, such as the love of nature, are crucial if activism is to be sustained.[8] Kathryn Fuller, president of the World Wildlife Fund, insisted that "it is this passion for nature—this unstinting, unqualified regard for life in all its variety—that informs everything we do at World Wildlife Fund. . . . We return time and again to the animating impulse that first drew us to conservation and let it guide our steps toward the future."⁹ While political, fiscal, and bureaucratic concerns inevitably structure much the work done by large environmental organizations, the love of nature that prompted the leadership and staff to join these organizations continues to motivate.

A long-standing tradition in the Western world represents the human mind as the mirror of nature. The mind reflects the laws and truths of the natural world. In many respects, however, nature is the mirror of the human mind. It reflects both our desires and our fears. From time immemorial, nature has been anthropomorphized. Stories of satyrs and nymphs, nature gods and goddesses are abundant. Nature has always been and remains today permeated with religious and ethical values. Nature, it has been said, is "the original Rorschach . . . an evocative mix of amorphous stimuli onto which we are invited to project our innermost longings and anxieties."¹⁰ Long before nature was physically altered to any great extent by human beings, it had been conceptually invaded. It was reformulated in our image and in our imaginations and appropriated through myth and ritual. In contemporary times, nature has been further integrated into the human orbit, materially as well as mentally. Indeed, nature has been affected by humans to such an extent that the demarcation between what is "natural" and what is synthetic or "manmade" is difficult to maintain today.

In their exuberance to celebrate and protect nature, environmentalists sometimes fail to question what nature is, or has become. They do not acknowledge that much of what is called nature today is far from a pristine, unadulterated wilderness. Biotic life and habitat across the globe have been significantly transformed through human contact and interaction.¹¹ "In a sense there is no 'nature' left in the world," Walter Truett Anderson wrote. "There is no place on Earth—certainly not on an Earth whose sunlight filters through an ozone layer that has been accidentally altered by human technology—that is truly, as the saying goes, untouched by human hands. Indeed, all the things we do to preserve 'nature,' everything from wilderness management to endangered species legislation, are in one way or another human interventions."¹² In many respects, environmental protection is simply one more form of human intervention. But if, in fact, there is no nature left in the world, what is it that environmentalists love and fear losing?

The meaning of the loss of nature for environmentalists is tackled by Bill McKibben in his book, *The End of Nature*. McKibben wrote:

> How can there be a mystique of the rain now that every [acidic] drop—even the drops that fall as snow on the Arctic—bears the permanent stamp of man? Having lost its separateness, it loses its special power. Instead of being a category like God—something beyond our control—it is now a category like the defense budget or the minimum wage, a problem we must work out. . . . What will it mean to come across a rabbit in the woods once genetically engineered "rabbits" are widespread? Why would we have any more reverence or affection for such a rabbit than we would for a Coke bottle?. . . . Someday, man may figure out a method of conquering the stars, but at least for now when we look into the night sky, it is as Burroughs said: "We do not see ourselves reflected there—we are swept away from ourselves, and impressed with our own insignificance. . . ." The ancients, surrounded by wild and even hostile nature, took comfort in seeing the familiar above them—spoons and swords and nets. But we will need to train ourselves not to see those patterns. The comfort we need is inhuman.[13]

Certain environmentalists, and much of the general public, conceptually divorce the natural from the human. For the most part, however, environmentalists do not ground their work in the naive belief that nature is wholly disconnected from the human world. Indeed, an awareness of the ubiquity of humankind's intrusions on natural processes and biotic communities prompts and sustains environmental activism. While the distinction between the "man-made" and the "natural" is impossible to formulate unproblematically, environmentalists work to ensure that it will not be impossible to experience.

In this context, we ask again, What does it mean to love nature? St. Bernard defined love as wanting something simply to be. Love is a celebration of another's existence. Biocentrically oriented environmentalists want nature, or wilderness understood as largely untrammeled nature, simply to be. While nature seldom if ever remains wholly untouched by the human presence today, the hope is that it can still operate in ways largely unaffected by us. Love of nature is a love of biotic life and habitat regardless of its usefulness and beyond its capacity to be integrated into the human world. Nature is loved, one might say, because of its *otherness*. Writer Barry Lopez has explained that "the exhilaration I experience seeing fresh cougar tracks in mud by a creek is an emotion known to any person in love who hears the one-who-is-loved speak."[14] In most respects, the cougar is useless to Lopez. The cougar, in this case, is not itself seen. It might never be seen. But it is loved nonetheless. Lopez wants the cougar simply to be, apart from any encounter he might have with it. There are few cougars left, and those that remain have had their habitats much altered by human encroachments. Still, cougars represent a desired other. They offer an inhuman comfort.

The pure delight in nature's existence plays a crucial role in sustaining environmental behavior. At times, however, this delight may be frustrated by the exigencies of activism itself. An environmentalist involved in numerous local and regional groups explained that his activism was a product of wanting to share and perpetuate a sense of connection with all living things. Yet his desire to be in nature was often stymied by the demands of his activist work. He observed that one's relationship to nature "can't be a connection that you only carry around in consciousness. It has to be renewed constantly, on an hourly, minutely status. Primitive man never left the presence of other living beings, and because he lived in low population densities, primarily what he did come in contact with were other species. Now, most environmental action, like county committee meetings and congressional stuff, is all urban-oriented. It is totally out of contact with other species of living beings and as such is a dissatisfying situation for me to be in." Many environmentalists lament the tragic irony that their activism deprives them of the time and opportunity to encounter the very object of their devotion. The large, national organizations are particularly prone to sapping the animating force of their staff and volunteers in this manner.

Dave Foreman has suggested that all staff members of environmental organizations should be required to take wilderness vacations as a condition of retaining their jobs. He worries that their love for nature, which in most cases stimulated their choice of careers, might atrophy from lack of exercise. Their commitment and work would suffer for it. Certain organizations maintain such arrangements on an infrequent basis, ensuring that yearly retreats for staff are held in natural settings. Other groups, such as the Audubon Society, World Wildlife Fund, and the Sierra Club, promote nature outings or vacations for their staff and members. The Sierra Club sponsors some 25,000 nature outings a year involving over 130,000 participants.[15]

There are good reasons to promote such excursions into green spaces and wildlands. Positive nature experiences, studies reveal, are the sine qua non of most forms of environmental activism. Even mainstream environmental behavior, such as recycling, taking public transportation, or keeping oneself environmentally informed, is largely stimulated and sustained by nature experiences.[16] In turn, love of nature is a necessary, if not sufficient, condition for almost all forms of environmental activism. Because this love of nature is difficult if not impossible to cultivate and sustain in the absence of natural surroundings, the importance of having regular access to parks, green spaces, and wilderness is paramount.

Love of nature has been identified as an innate characteristic of our species. Harvard sociobiologist E. O. Wilson called this innate love of nature *biophilia* and suggested that it might supply the foundation for an environmental ethic.[17] It is not clear that all humans are subject to biophilia. Biocentrists obviously are. And surely other environmentalists are no strangers to the love of nature, however much their humanistic orientations take precedence. After all, what most environmentalists seek to bequeath to their progeny and share with their neighbors is a world in which nature's beauty and bounty persist. As an activist with a local group averred, "It's because I've been allowed to see so much beauty [in nature] that I'd like to pass it along." Without a widespread love of nature, it is unlikely that collective endeavors to protect and preserve a healthy, beautiful, and biologically diverse natural world would arise or be sustained, regardless of any humanistic concerns people have for progeny and neighbors. Strict biocentrism may be a rarity or even an impossibility for human beings, but a basic love of nature is the cornerstone of environmental caretaking.

For many environmentalists, nature serves as the measuring rod against which all other values are assessed. A staff member with The Wildlands Project outlined this point of view. His most basic assumption was that "nature is good." Nature, he insisted, should serve as the "fundamental measure of goodness." Environmentalists often idealize nature in this fashion. Yet nature lovers do not delight in everything that is natural. Many natural processes produce hardship and suffering. While environmentalism has prompted a greater caretaking of the natural world in recent decades, nature herself has not always returned the kindness. The most deadly earthquake of this century, the 1976 Tangshan quake in China, killed over 240,000 people. In the 1980s, volcanic eruptions ended the lives of over 30,000 people. In 1991, a cyclone killed 140,000 people in Bangladesh. That same year the eruptions of Mount Pinatubo in the Philippines left tens of thousands homeless and spewed twenty million tons of sulfur dioxide into the atmosphere. In 1992, Hurricane Andrew destroyed some 85,000 homes and caused an estimated

$30 billion in damages. Above all this, hundreds of thousands of people die of other "natural causes" each and every day.

These events and processes are not *loved* by biocentric environmentalists. Yet they are perceived as part and parcel of the earthly cycle of growth, demise, decay, and regrowth. To love nature and to believe that nature is "good" is to revel in the wonders of the earth's forces, even though one might suffer from and consequently resist their specific manifestations. Loving nature precludes hating nature, but it does not preclude fearing nature's power and avoiding its dangers. As Aldo Leopold suggested when a bolt of lightning struck near his feet, "[i]t must be a poor life that achieves freedom from fear."[18]

Contemporary environmentalism is sustained by the love of nature. That is not to say that the love of nature is in any sense a recent phenomenon. It predates environmental activism of any sort. Nature was loved by humans far before nature needed to be saved from humans. It is unlikely, moreover, that the depth or breadth of the love for nature that is experienced today is new in any important sense. Sun gods and earth goddesses populate the most ancient of pantheons. Nature worship is as old as the hills. What is new about the love of nature is that it is now informed by a sophisticated understanding of ecological interdependence. In large part, our insight into the intricate patterns of the web of life was stimulated by witnessing our careless severing of many of its strands. Our current appreciation of the beauty and complex dynamics of nature, tragically, was gained by way of her piecemeal destruction.

Biocentric and Anthropocentric Perspectives

The scholarly debate between anthropocentrism and biocentrism has framed most efforts to construct an environmental ethic.[19] An affirmation of ecological interdependence is common to both camps of environmental thought. Yet the love of nature demonstrated by biocentrists and anthropocentrists differs in degree if not in kind. Biocentric environmentalists (also known as deep ecologists, ecocentrists, or radical environmentalists) are thoroughgoing biophiliasts. Anthropocentric environmentalists (also known as social ecologists, reform environmentalists, and humanist environmentalists) have an abiding reverence for nature even if they do not ethically or spiritually identify themselves with nature. An examination of the characteristics of biocentrism and anthropocentrism will highlight the various roles that the love of nature plays in environmental caretaking. It will also demonstrate how distinct philosophical and ethical perspectives may converge in the affirmation of interdependence and the quest for ecological integrity.

An action may be considered altruistic rather than egoistic even though it originates from an individual's will or desire. The effort to save an unrelated child from a burning house is certainly altruistic. Yet it remains the product of human will and desire. Likewise, actions may be considered biocentric rather than anthropocentric even though they originate from the thoughts and actions of human beings. Thus, to say that one acts anthropocentrically is not simply to say that one acts as a human being. To act anthropocentrically is to act in the pursuit of distinctly human interests.

All actions purposefully carried out by humans are, by definition, oriented to satisfying some human value. Minimally they reflect the human actor's desire to act purposefully and effectively. All human values are, in this sense, "anthropogenic." That is simply to say that they have their origins in human beings. Anthropogenic values need not be anthropocentric. They need not place human needs and human welfare above the needs and welfare of all other forms of life.[20] Arne Naess, a Norwegian biocentrist who coined the term *deep ecology* in the early 1970s, paraphrases Protagorus' well-known statement to make this point: "Man may be the measure of all things in the sense that only a human being has a measuring *rod*, but what he measures he may find to be greater than himself and his survival."[21] Many environmental activists expend time, energy, and resources to save distant wildland from destructive forms of development. Yet they may in fact never visit or otherwise directly enjoy or benefit from these wilderness preserves. They value the wild for its own sake, regardless of whether it is personally useful to them.

Much if not all biocentrism is, in practice, anthropocentric to some degree. That is simply because the notion of human welfare can be expanded to include almost anything that humans desire or pursue. One might preserve wilderness for no other reason than because one wishes to contemplate nature in as pristine a state as possible. That is still to preserve wilderness for a human good. Many biocentrists recognize this subtle self-interest in their work. An Audubon staff member explained her activism rather apologetically, saying, "I just love trees and birds and animals and I want to continue to experience it all. I don't want the earth to become covered with concrete and buildings. It's pretty selfish, I guess, it's just the way I prefer the world to be." To preserve nature because one finds nature gratifying from an emotional point of view is, ultimately, to serve a human end.

To be unconditionally biocentric, one would have to preserve nature knowing that no human being would ever benefit from it in any way—economically, recreationally, aesthetically, or spiritually. Presumably, the only purely biocentric act (like the only purely altruistic act) would be self-sacrificial, perhaps to the point of being suicidal or genocidal. Certain biocentrists indicate that they would indeed give up their lives to preserve wilderness. Some insist that they would rather see the extinction of the entire human species than the further degradation of nature. Asked what she would be willing to do to protect the environment, one member of multiple national and local organizations stated, "I would die for it. If 99/100ths of the human population on earth could vanish [and by vanishing save the environment], I would gladly raise my hand [to be one of them]." A Portland-based group, the Voluntary Human Extinction Movement, has the motto May We Live Long and Die Out.[22] An executive with a population control group is not so charitable regarding the value of a long life: "We've reached the point where (with a few notable exceptions) if you are a human and alive today, you are a burden to the planet, and the sooner your life is over the better."[23] A radically biocentric orientation may well demand that one sacrifice one's interests, one's life, and perhaps the lives of one's fellow men and women to keep the biosphere from being further disturbed by human hands.

Most self-declared biocentrists do not embrace such stark, misanthropic perspectives. They adopt a more nuanced approach that does not scorn humanity's presence on the earth but seeks to limit it. Their efforts are aimed at facilitating the conditions that will allow as many species as possible, including the human species, to flourish and evolve.

While biocentrists and anthropocentrists may be distinguished in many ways, the major dividing line is clear. Biocentrists promote a conceptual and ethical shift away from humanism. Anthropocentrists propose a shift to a higher, more enlightened form of humanism. The former seeks to extend our concern beyond human utility; the latter broadens human utility to include the noneconomic valuation of nature. In turn, anthropocentric environmentalists tend to link environmental care to social justice. This typically entails contesting rugged individualist and unrestrained capitalist ideologies. While biocentric environmentalists chiefly promote identification with and reconnection to nature as the remedy for our ecological woes, anthropocentrist environmentalists insist that identification with and reconnection to social and political community is a necessary part of any solution.

Biocentrists maintain that all life is of equal value. Bill Devall and George Sessions, who have perhaps done the most to popularize biocentrism in the English-speaking world, wrote: "The intuition of biocentric equality is that all things in the biosphere have an equal right to live and blossom. . . . This basic intuition is that all organisms and entities in the ecosphere, as parts of the interrelated whole, are equal in intrinsic worth."[24] Anthropocentrism, from this viewpoint, unjustly subjugates nonhuman life to human standards, values, uses, and, inevitably, abuses. Anthropocentrism takes the first step toward the domination, exploitation, and destruction of nature, biocentrists insist, when it bifurcates the world into (higher, rational) human subjects and (lower, nonrational) natural objects. The problem here is "speciesism," defined as the belief that one is entitled to treat members of other species in a way that would be improper to treat members of one's own species. Were we to extend the rights presently accorded to human beings to all other species, biocentrists frequently argue, the degradation of our ecosystems could be halted and reversed.[25] "In the first Copernican Revolution," Earth First!er Christopher Manes wrote, "humanity was forced to abandon the erroneous but gratifying view that we and the planet we walked upon were the center of the universe—geocentrism. In the second Copernican revolution, we may be forced to abandon the even more self-aggrandizing belief that we are the center of the moral universe and have a special, privileged status in the biosphere—anthropocentrism."[26] Biocentrism is a radical, and potentially revolutionary, perspective. Advocates believe that it constitutes a turning point for humanity.

Humanist environmentalists maintain that human life deserves a unique and privileged status within the biosphere. Human beings bear the rights and responsibilities of stewards of the earth. They also have first claim to its resources. Gifford Pinchot's efforts to conserve natural resources for the greatest good of the greatest number of his fellow citizens exemplify a populist form of humanism. Most environmental organizations, with the exception of certain groups such as Earth First!, are openly humanist and utilitarian. The "Editorial Creed" of the National Wildlife Federation's magazine presents an expansive version of Pinchot's position: "To create and encourage an awareness among the people of the world of the need for conservation and proper management of those resources of the Earth upon which our lives and welfare depend: the soil, the air, the water, the forests, the minerals, the plant life and the wildlife."[27] Even those organizations whose work is wholly oriented to wildlife preservation generally maintain such utilitarian, humanist standards. Defenders of Wildlife, for example, insists that the Endangered Species Act should be enforced on the basis of "a thorough and objective scientific evaluation of the rate at which species are being lost and the degree to which these losses threaten human health, food sources and other economic resources."[28] Again, human lives and welfare are taken to be paramount concerns (if only for the purpose of better public relations). The environment requires caretaking *because* it serves human needs.

Humanist environmentalists argue that the betrayal of humanist values, not humanism per se, fosters ecological degradation. Not prerogatives of reason, but rather unchecked abuses of power sit at the root of our ecological predicament. As C. S. Lewis remarked, "what we call Man's power over Nature turns out to be a power exercised by some men over other men with Nature as its instrument."[29] Many humanist environmentalists maintain that the destructive attempt to master nature arises from preexisting social inequalities and structural domination. Murray Bookchin, who occupies the social anarchist wing of humanist environmentalism, wrote: "Men did not think of dominating nature until they had already begun to dominate the young, women, and, eventually, each other. And it is not until we eliminate domination in all its [social] forms, as we shall see, that we will really create a rational, ecological society."[30] To achieve an ecologically sustainable and reinvigorated world, Bookchin argues, humanity must put an end to oppressive social, economic, and political institutions. Hierarchical human relations, Bookchin insists, beget destructive ecological relations.

Anthropocentrists charge that biocentrists, through their antihumanist crusade, depreciate the value of human life. In doing so, biocentrists undermine the social conditions on which ecological care might be better sustained. They argue that biocentrists promote, or at a minimum do nothing to discourage, the political and economic inequalities and injustices that inevitably translate into ecological degradation. Biocentrists, in response, insist that humanists treat nature as a warehouse. Their effort to exploit natural resources efficiently reinforces deep-seated biases of the human prerogative to master nature. These biases inevitably re-infect the social realm. In the end, biocentrists argue, we are saddled with destructive forms of domination in both our social relations and our relations to the natural world.

The debate over population control illustrates these differences. Bio-centrists tend to view human overpopulation as an example of anthropocentric hubris. They deem the proliferation of human beings on the earth an illegitimate usurpation of the planet's habitats and resources. Effectively, we are robbing other species of their homes and livelihoods. In turn, overcrowding brings out the worst in the human species, as the struggle for scarce resources produces inequity and violence. Population control and the immediate preservation and expansion of wildlife habitats are the proffered solutions. Humanists, in contrast, tend to view human overpopulation as a social problem, stemming primarily from poverty, ignorance, and patriarchy. Mandated birth control is generally deemed a dangerous extension of state power that would be primarily exercised on the populations of the developing world, further increasing their oppression. Lessening or reversing population growth, humanists argue, will be a by-product of improving health care, educational opportunities, living conditions and, most importantly, women's economic and political status. In the meantime, declaring valuable natural resources off limits to starving populations is callous and hypocritical.

The conflict between biocentric and anthropocentric principles gives rise to other practical disagreements. Take the case of the preservation of biodiversity, and in particular the preservation of tropical rainforests. In making the argument that humanity's innate love of nature, its biophilia, might allow us to better care for the world, Edward O. Wilson has invoked Garrett Hardin's Cardinal Rule of Policy. He suggested that "[t]he only way to make a conservation ethic work is to ground it in ultimately selfish reasoning."[31] Wilson accordingly described biodiversity as "our most valuable but least appreciated resource."[32] He noted that one third of all prescribed medicine is based on natural substances and that the future of pharmaceutical medicine is linked to the preservation of rainforests, where the greatest biodiversity exists. Likewise, John Sawhill of The Nature Conservancy approvingly remarked that "some far-sighted pharmaceutical companies today have come to regard Mother Nature as the planet's most effective R&D operation."[33] The American Medical Association concurs. It passed a resolution in 1995 advocating the protection of biological diversity for the undiscovered cures a robust nature may hold. Current slash-and-burn practices in the tropics amount to torching the books of nature before we have a chance to read them.

From a strictly biocentric perspective, such arguments and advocacy are out of bounds. We are not justified in protecting nature simply because it serves our (medical, scientific, or economic) needs. Human utility is denied entry into the equation. If we choose to protect rainforests from slash-and-burn agriculture, the operative rationale cannot ground itself on the benefits accrued to humans by preserving pharmaceutical resources or keeping the "lungs of the planet" smoke free.

Certain biocentrists have attempted to make a virtue out of this necessity. They argue that the competing utilitarian claims of local, short-term gain and global, long-term loss from deforestation present an insoluble moral quandary. This quandary is made particularly prickly owing to the charges of imperialism that may be aimed at environmentalists of industrialized nations who involve themselves with the internal forestry affairs of developing countries. The benefits accrued from rainforest preservation, after all, will be globally shared while the economic costs are primarily borne south of the border.

Biocentric reasoning is offered as a means to resolve this dilemma. If the rainforest is deemed to have rights of its own, then it should be preserved regardless of anyone's short-term or long-term benefits or costs. Yet things are not that simple. As the authors of one such argument eventually admitted, the biocentric recognition of a "moral obligation" to preserve the natural environment for its own sake is "only a starting point" for a discussion in which the "difficult trade-offs of goods between competing groups of humans can be debated."[34] Once again we find ourselves saddled with utilitarian claims.

The problem, at base, is that the *rights* of ecosystems or specific (species of) animals and plants to be free from the harvesting hands of human beings are difficult if not impossible to uphold as the same sort of rights that human beings have not to be murdered, enslaved, or eaten. Deep ecological poet Gary Snyder has insisted that "[p]lants and animals are also people."[35] While Snyder might enjoy a good salad, however, and other deep ecologists do not shy away from consuming meat, no one suggests that we should be allowed to eat people. These sorts of inconsistencies have not been lost on humanist environmentalists who accuse deep ecologists of being unable to make biocentric theory coherent with its practice.

In the battle for the hearts and minds of the ecologically attuned but conceptually uncommitted, deep ecologists may gain ground by boldly heralding the rights of animals. Dave Foreman maintains that a grizzly bear in Yellowstone Park "has just as much right to life as any human has."[36] The majesty of these imposing creatures, Foreman is well aware, prepares us psychologically to uphold their "rights." In turn, a bear's right to life only minimally conflicts with our own, unless we come across an angry grizzly in the woods. Even then, maintaining equal rights for bear and human would permit our self-defense. Greenpeace's actions in defense of whales and seals and Defenders of Wildlife's efforts to protect wolves are grounded on similar principles and emotional attachments.

These sorts of arguments become less convincing when they are generalized. Do different standards apply to bugs and bacteria than to charismatic megafauna such as grizzlies, whales, seals, and wolves? What rights should we uphold for the mosquitoes that we are wont to swat in annoyance on muggy nights? Certain biocentrically oriented environmentalists argue that all species might benefit from the prevalent focus on megafauna, effectively riding on the coattails of charismatic species into a protected status. Defenders of Wildlife holds that charismatic animals are "so thoroughly ingrained in the human psyche as . . . symbol[s] of untamed nature" that winning support for them "fosters an improved public attitude toward nature in general."[37] As the largely successful campaign against wearing fur demonstrated, however, the respect for nature that trickles down to the less charismatic species from efforts to protect the more charismatic species may be quite minimal. Few of those who embraced the rights of seals and foxes to keep their skins felt the same way about cows. The first "Animal Bill of Rights," submitted to the 104th Congress by the 60,000 member Animal Legal Defense Fund, champions "[t]he right of animals to have their interests represented in court and safeguarded by the law of the land."[38] One doubts, however, that a legal crusade to end the householder's battle against the cockroach will ever find its way into the halls of justice.

To avoid inconsistency in the application of their principles, certain biocentrists have argued that one must rank various forms of life such that cows, chickens, mosquitoes, and cockroaches are accorded fewer rights or a lower grade of rights than grizzly bears, whales, seals, and wolves. Giving more or greater rights to wolves and whales over cows and cockroaches simply because the former supposedly exhibit a greater "richness of experience,"[39] however, begs the question of standards. What criteria are employed to determine experiential richness? The cockroach appears perfectly content with its life. Certainly its experience has served it well enough in terms of evolutionary success. These bugs have been around much longer than, and will likely outlive, most if not all of the vertebrate species currently existing. What we understand by rich experience, it turns out, is experience that we imagine to most closely resemble that enjoyed by human beings. Unacceptable anthropocentric biases leech back into any attempt to distribute rights according to standardized criteria.

Many biocentrists, such as Arne Naess, refuse to rank different forms of life according to schemes of "relative intrinsic value." Neither richness of experience, the presence of an eternal soul, the capacity for reason, self-consciousness, nor "higher" evolutionary development is considered relevant.[40] David Ehrenfeld has also argued that no credence should be given to such anthropocentric distinctions. "For those who reject the humanistic basis of modern life," Ehrenfeld wrote, "there is simply no way to tell whether one arbitrarily chosen part of Nature has more 'value' than another part, so like Noah we do not bother to make the effort."[41] Adopting this "Noah Principle" means that one should not distinguish between grizzly bears and cockroaches. All have an equal right to life; all deserve preservation.

This principle extends to every form of life. Thus, Ehrenfeld argues for the systematic preservation of the smallpox virus and other deadly pathogens. Social ecologists, such as Bookchin, find this position scandalous. When taken to its logical conclusion, moreover, this biocentric principle appears to dictate more than the controlled preservation of specimens of all species in Noah-like fashion. As simple "citizens" of the biosphere, we would have no claim to the prerogative of global zookeeper. Arguably, our noninterference with the uncontrolled proliferation of all species, including pathogenic bacteria and viruses, is required. Biocentrists seldom go this far. While whales are widely defended as friendly spirits who have a legitimate claim to share the earth with us, few biocentrists defend the rights of the AIDS virus.

Ironically, those biocentrists who maintain that individual animals should have rights to life and liberty equal to those of human beings may actually undermine ecological integrity. That is because the preservation of biodiversity may on occasion require the killing of invasive exotic species (such as the ecologically destructive and nonindigenous brown tree snakes of Guam, and the black and Pacific rats and feral pigs of Hawaii) or overpopulated predators that are decimating endangered species. Captive breeding programs aimed at restoring populations of endangered species might also be deemed unacceptable from a rights-based biocentrism. Yet abrogating the "rights" to life and liberty of individual animals may be necessary to secure ecosystemic health in certain situations. At times, allowing the commercial use or harvesting of species may also be the most practical means of ensuring their survival (assuming that the rates of harvest can be controlled and illegal poaching effectively interdicted). The enhanced economic value of these threatened species prompts local communities to protect them. Outlawing their legal economic exploitation may dòom them to extermination.

The ironies implicit in a thoroughgoing biocentrism mount. The biocentric acknowledgment of animals' equal rights to life and liberty may require more than our noninterference. Some have suggested that it entails humanitarian aid. Naess has observed that "if a rat is discovered in an inaccessible ventilator, it is clearly cause to warn the SPCA to come and end its suffering—by putting it out of its misery."[42] If we were to extend this principle, however, we might also be obligated to dispatch SPCA squads to the sewers to save rats from drowning, or ensure their painless demise, during severe rainstorms.

By deep ecological standards, sewer rats would merit our help just as, say, whales trapped under ice do. Both cases pose theoretical as well as practical problems. In 1988, newspaper readers and television viewers across the nation and around the world were apprised of the plight of three gray whales trapped beneath Arctic ice. In a much ballyhooed effort that took on an international scope, $5 million was spent trying to free the hapless cetaceans. Given that the U.S. Navy and Air Force used whales for target practice as late as the 1960s, the public interest, media attention, and governmental resources expended in the "rescue" of the three whales was remarkable.[43] While applauding this growth of environmental consciousness, however, one might question the ecological costs incurred (in terms of pollution and resource consumption) by having numerous ships, aircraft, and personnel from several nations involved in the rescue and its media coverage. Certainly these resources could have been better spent from an ecological perspective. The problem goes deeper. What of the fish and microorganisms that were robbed of a bountiful feast of whale meat as a result of our "humanitarian" intervention? And what of the maggots that would dine on drowned sewer rats? Must we not also compensate these scavengers for their losses? Of course, we could not compensate scavengers without taking yet other forms of life in the process. The practical and theoretical conundrums abound.

Albert Schweitzer, an iconic figure for many deep ecologists, insisted that the ethical person "tears no leaf from its tree, breaks off no flower, and is careful not to crush any insect as he walks." Schweitzer would remove bugs caught in pools of rainwater to safe ground, explaining that "when I help an insect out of his troubles all that I do is to attempt to remove some of the guilt contracted through [humanity's] crimes against animals."[44] To be consistent on this score, biocentrists would find it necessary to sweep the path before them lest they tread on some unsuspecting creature and wear gauze over their mouths lest they inadvertently inhale and thereby put an end to some minuscule life. Yet a handful of soil might contain 50,000 algae, twenty times as many fungi, and 200,000 times as many bacteria. The environmental dictum to "walk softly upon the earth" would have to be enforced quite literally.

Were we consistently to extend the biocentric ethic to include all forms of life, precluding the accusation of "zoocentrism," plant life would also merit equal protection and rights. A vegetarian would remain as culpable as a carnivore and a cannibal. As inevitable consumers of life, humans would become irredeemably guilty. This collective guilt, "biological 'original sin,' " as Bookchin has put it, would likely produce a misanthropic orientation.[45] One might worry that something has gone terribly wrong when biocentrism separates us from other species and fellow human beings with a veil of guilt, as is implicit in Dave Foreman's characterization of humanity as "a cancer on nature."[46]

Biocentric and Anthropocentric Practices

Biocentrists value wilderness areas "primarily as sanctuaries for wild ecosystems and unmanaged habitat for wild species." This is opposed to the viewpoint of humanists, caricatured as "the new yuppie Sierra Clubbers" who appreciate wilderness primarily for the "superlative scenery and playgrounds" that it provides.[47] In the case of the hiker, however, what is found to be of aesthetic or recreational value is a largely unadulterated state of nature. The extraction of aesthetic or recreational pleasure from nature does not necessarily injure or impair it.

Any intrusion into nature by large numbers of people is detrimental to wildlife and habitats We might rightly worry, with Leopold, that "all conservation of wildness is self-defeating, for to cherish we must see and fondle, and when enough have seen and fondled, there is no wilderness left to cherish."[48] In 1995, the top ten national parks attracted over sixty-five million visitors. Park use is increasing each year. Such "industrial tourism," to employ Edward Abbey's caustic term, has taken its toll on the flora and fauna of our national parks.[49] Still, limited numbers of humans deriving aesthetic or low-impact forms of recreational value from national parks and the natural world need not be viewed as an exploitive phenomenon. One can participate in nature without consuming nature.

Many environmentalists with biocentric leanings concur. They maintain that aspirations to complete ecological noninterference are counterproductive because they effectively separate humans from the natural world. A relatively benign integration in nature is advocated. Indeed, such integration is offered as the only means to ensure the preservation of wilderness. As is often the case, Leopold blazed the trail here. He wrote that "[w]hen we see land as a community to which we belong, we may begin to use it with love and respect. There is no other way for land to survive the impact of mechanized man, nor for us to reap from it the esthetic harvest it is capable, under science, of contributing to culture."[50] Affirming the need for humanity respectfully to "use" and aesthetically "harvest" nature is clearly distinguishable from a narrowly anthropocentric and crassly economic perspective. This latter perspective was given voice recently by a congressional representative who described the United States as "the lumber bin of the world," and by Don Young (R-Alaska), who as Chair of the House Resources Committee stated, "If you can't eat it, can't sleep under it, can't wear it or make something from it, it's not worth anything."[51] Yet the Leopoldian perspective is also distinguishable from a purely biocentric viewpoint that defends wilderness, and nature more broadly, against any human use.

The deep ecological perspective, according to Naess, promotes biospherical egalitarianism in principle. "The 'in principle' clause," he wrote, "is inserted because a realistic praxis necessitates some killing, exploitation, and suppression."[52] Devall and Sessions concur that biocentric equality is "an intuition [which] is true in principle, although in the process of living, all species use each other as food, shelter, etc. Mutual predation is a biological fact of life."[53] Earth First!ers generally adopt this perspective. Indeed, Foreman and other Earth First! leaders proudly displayed their heavy meat consumption as a sign of their participation in the food chain and their integration in the web of life.[54] There are, of course, many good reasons for a largely vegetarian diet. But the fact that humans have always been and remain predators of other animals is not in itself a bad thing from an ecological point of view. We must acknowledge that, despite its symbiotic and cooperative interactions, nature is often a violent affair of consuming and being consumed in turn. Ecological life persists, and diversifies, by way of predation between individual organisms and species. Human beings are inevitably part of this web of interdependent life and life-taking.

Rather than parsing environmental ethics to achieve a fully consistent logic, biocentrists frequently advocate the intuitive implementation of their sensibility. Christopher Manes wrote, "If anything, most radical environmentalists look at systematic philosophy as the problem, as an attempt to reduce the buzzing, howling, blossoming heterogeneity of the natural world to some abstract idea. If radical environmentalism has a watchword, it is probably its oft-repeated imperative 'Let your actions set the finer points of your philosophy.' "[55] Biocentric principles were never meant for direct and literal application. As Naess acknowledges, the confusion over the meaning of biocentrism ensues when "the rhetorics of a movement is treated like seminar exercises in university philosophy."[56] Talk of equal rights for all life is intended to serve a "mythic function," encouraging preservation of the ecosystem in its interactive diversity rather than the stifling of human life under the mandate of noninterference.[57]

After surveying the achievements of government in the area of environmental protection, Kirkpatrick Sale wrote disparagingly that "the only perspective Congress could ever muster was a narrow anthropocentric one. Only the Endangered Species Act had any pretense of having been passed on behalf of other than two-footed creatures (and even then benefits to 'the Nation and its people' came first), and all of the wilderness preservation legislation, even when it chose to deter commercial development, presumed that such areas were for human use and pleasure, even for eventual human medical scavenging."[58] Sale suggests that biocentrism should gain a political face and become a political force. Nature should secure actual representation in the nation's governmental institutions.

Arguably the job of Congress, like that of any democratic body, is to represent "the people." If this remains true, then the role of the environmental movement is not to reconstitute democratic institutions so that animals, plants, rocks, and rivers can gain political representation—however that might be achieved. Rather, the task is to transform citizens' attitudes, values, and practices such that the preservation of nature and the protection of the environment become popular priorities. Simultaneously, the task is to ensure that political representatives and political institutions adequately serve the interests of this enlightened citizenry.

Politically speaking, a pure biocentrism that spurns all reference to human utility is a nonstarter. Democratically speaking, the notion of political representatives no longer representing human constituents is unacceptable. As citizens develop ecological, and perhaps deep ecological orientations, however, their political representatives should, if democracy prevails, reflect these values. Speaking to his fellow environmentalists, David Brower has aptly observed that "[p]oliticians are like weather vanes. Our job is to make the wind blow."[59] Biocentric environmentalists primarily ensure that the wind carries the scent of wilderness. Humanist environmentalists primarily ensure that the wind is fit for human breathing and that the weather vanes turn in deference to public values rather than special, vested interests. The mutually supportive nature of these endeavors is grounded in the interdependent relations between human beings and the vast diversity of the earth's other species. There need be no conflict between democratic politics and a biocentric worldview, as long as citizens are willing and able to sustain their own political and ecological communities.

Championing biocentric activism, Foreman has insisted that "[t]he planet is either a collection of resources, or it's alive. That's the real choice."[60] Such categorical distinctions make for inspiring speeches, but they generally give way to a continuum of sensibilities when one examines concrete practices. As Bryan Norton has argued, the belief systems of environmentalists are much less monolithic than is often suggested. When it comes to actual policy positions—when theory gives way to practice—biocentrists often find their recommendations indistinguishable from those of "longsighted anthropocentrists."[61] While harboring deep ecological commitments, both Aldo Leopold and Rachel Carson ably employed utilitarian argumentation when it served their purposes. Not wholly unlike Gifford Pinchot, John Muir suggested that optimal conditions prevailed when "state woodlands are not allowed to lie idle [but] . . . are made to produce as much timber as is possible without spoiling them."[62] Muir's understanding of what level of harvesting spoiled forests was much more restrictive than Pinchot's. Nonetheless, Muir, like Leopold and Carson, often camouflaged, subsumed, or tempered deep ecological orientations within a humanist rhetoric when attempting to win over the public and politicians.

Bill Meadows, president of the Wilderness Society, observes that his organization regularly exploits "utilitarian arguments" to highlight the material and economic benefits of environmental care. These arguments, however, have always been made in tandem with efforts to promote "the spiritual and emotional values of wild places . . . that could inspire and move people to permanently change their attitudes and behavior toward the land."[63] Deep ecologists often find it necessary to adopt the language and values of anthropocentrism when they enter the public forum. With this in mind, Rodger Schlickeisen of Defenders of Wildlife suggested that sustainable development itself may serve as the ultimate Trojan horse.

It allows activists to bring biocentric intuitions into well-guarded anthropocentric establishments. "In our lifetimes, anthropocentric values will drive what we do on earth," Schlickeisen has acknowledged. The discourse of sustainable development promotes a program of "rational anthropocentrism" within which biocentric goals can be pursued.[64] Indeed, a future-focused, spatially expanded, rational anthropocentrism may be the most effective means to translate biocentric philosophy into action.

Most humanist environmentalists harbor deep ecological sentiments in varying degrees. They have, minimally, a reverence for nature that, though not usurping humanistic concerns, significantly bears upon their political and ethical deliberations. Their position is at its strongest when their claims reach beyond the need for the sustainable economic use of natural resources to include the aesthetic, recreational, and spiritual benefits received by humankind from a healthy, diverse biosphere. Since utility can be indefinitely expanded to include these other dimensions of human well-being, humanists can always argue from self-interest, albeit a long-term human interest broadly defined.

Likewise, biocentrists should not shy away from their humanist leanings or deny themselves anthropocentric arguments to secure their ecological goals. Arne Naess wrote that "human fulfillment seems to *demand* and *need* free nature. . . . Human nature may be such that with increased maturity a *human* need increases to protect the richness and diversity of life for *its own sake*. Consequently, what is useless in a narrow way may be useful in a wider sense, namely satisfying a human need."[65] With human needs and utility understood broadly enough, biocentric environmentalism and anthropocentric environmentalism begin to converge. To speak of the human *need* for unexploited wilderness and a rich, diverse biosphere is to bridge biocentric and anthropocentric orientations through the affirmation of ecological interdependence.

It is both "natural" and ecologically beneficial for us to be anthropocentric to some degree. Just as wolves typically conduct themselves in what might be called a *lupu*centric manner, eagles generally behave *aqui*centrically, and bees are fervent *api*centrists, we might conduct ourselves *anthropo*centrically without forsaking our embeddedness in or respect for the vast web of life. Even as "mere citizens" of the land, we have the right to secure our own interests. The challenge is to demonstrate that these interests largely converge with the caretaking of a vastly expanded biotic community. Holmes Rolston's dictum to "[t]hink of nature as a community first, a commodity second" well captures the conviction that ecological priorities must be integrated with rather than supersede economic concerns.[66] That conviction is inherent to a coevolutionary perspective.

Religion, Spirituality, and Environmental Ethics

In *A Sand County Almanac*, Leopold wrote that "[o]bligations have no meaning without conscience, and the problem we face is the extension of the social conscience from people to land. No important change in ethics was ever accomplished without an internal change in our intellectual emphasis, loyalties, affections, and convictions. The proof that conservation has not yet touched these foundations of conduct lies in the fact that philosophy and religion have not yet heard of it."[67] A little over a decade after Leopold's classic book appeared, Rachel Carson's *Silent Spring* altered the intellectual loyalties and convictions of thousands, sowing the seeds of an environmental ethic in the American public. Carson explained how the age-old attempt to gain "control of nature" was self-destructive because it failed to understand the intricate workings of "the whole fabric of life."[68] Gaining a sense of the relations of interdependence within the fabric of life is the first step toward the formulation of an environmental ethic.

Beginning in the late 1960s, philosophers and religious thinkers also began to explore the ethical import of ecology. In the March 1967 edition of *Science*, a medieval historian by the name of Lynn White, Jr., published a controversial article that saddled Christianity with the burden of much of the world's ecological woes.[69] White held that Christianity's destruction of pagan animism substantially contributed to the present ecological crisis. Biblical injunctions exacerbated the problem. Genesis 1:28, God's first commandment to humankind, reads: "Go forth and multiply and fill the earth and subdue it: and have dominion over the fish of the sea, and over the fowl of the air, and over every living thing that moves upon the earth." The quest for such dominion over life, White observed, is not environmentally benign. Since the roots of the ecological crisis were religious, White suggested, the remedy must also be religious. A reorientation of our deepest commitments was proposed. Reviving the nature philosophy of St. Francis, White believed, would make a good start.

Sparked by White's article, religious thinkers began to ponder, preach, and publish about the relationship between environmental degradation and faith or sacred scripture. Authors and poets such as Wallace Stegner and Wendell Berry were, by the 1960s and early 1970s, creatively addressing the connection between the preservation of wilderness, environmental caretaking, and spirituality. Gregory Bateson, in developing a psychological approach to ecological matters, maintained that religious beliefs have direct consequences for human relations to nature. "If you put God outside," Bateson observed of the modern, Western tradition, "and set him vis-à-vis his creation and if you have the idea that you are created in his image, you will logically and naturally see yourself as outside and against the things around you. And as you arrogate all mind to yourself, you will see the world around you as mindless and therefore not entitled to moral or ethical consideration. The environment will seem to be yours to exploit."[70] Debates concerning such assertions grew. Environmentalism was stimulating reflection about the very meaning of earthly life in ways that provoked and challenged traditional religious thought.

Within two decades of the publication of *A Sand County Almanac*, philosophy and religion had certainly "heard" of the environment. By 1979, the National Council of Churches adopted a policy statement on "the ethic of ecological justice." The Christian dictum to do unto others what you would have others do unto you, as well as the imperative to love your neighbor as yourself, were transformed to reflect ecological values. The definition of *neighbor* was expanded "to encompass all humans in past, present and future generations, as well as the rest of creation."[71] Environmental interdepedence had gained an ethical and religious voice.

In 1989, Pope John Paul II issued a document entitled "Peace with God the Creator, Peace with All of Creation." John Paul held that world peace was threatened by "the lack of necessary respect for nature," and the "disordered exploitation of her resources." He concluded that the ecological crisis had "assumed such proportions as to be everyone's responsibility."[72] Likewise, Protestant, Roman Catholic, and Jewish leaders of the National Religious Partnership for the Environment declared in 1993 that "the cause of environmental integrity must occupy a position of priority for people of faith."[73] On Earth Day 1994, the partnership mailed out environmental "education and action" kits to 53,800 congregations across the nation.[74] The mailings have become an annual event. Paul Gorman, executive director of the partnership, suggested that the key message to transmit to the religious faithful was that nature and society existed interdependently and that humans had the responsibility of preserving this relationship. "We are neither at the center of the universe nor astride it," Gorman wrote, "but we are embedded in it, as loving stewards."[75] Members of the Evangelical Environmental Network have also taken on the task of environmental protection and have criticized legislators who "believe in the creator but don't give a hang about his works." Co-founder Calvin DeWitt identifies the Endangered Species Act as the "Noah's Ark of our day." Network members have established 1000 "Noah Congregations" and "Creation Awareness Centers" around the country to carry on the work of environmental stewardship.[76]

The idea of responsible stewardship lies at the roots of American environmentalism. It is also a fundamental religious value. Perhaps it should not surprise us, then, that there is a strong empirical relationship between religiosity and environmentalism.[77] A Sierra Club activist stated, "My environmental activism comes from a much deeper place than concern for trees and birds. It's a spiritual, religious conviction that the natural order is much more than we feel and see and smell." Such religious connections to the environment are widespread. Sherman Morrison, administrative director of the American Coalition on Religion and Ecology, observed that "[t]wenty-five years ago, no one was talking about religion and ecology." Today, Morrison remarked, "this is definitely a movement."[78]

Today more than ever, organized religion and environmental protection prove mutually supportive. Yet organized religion continues to play a relatively small, though far from insignificant, role in the environmental movement. Of greater import in defining contemporary environmentalism is what might simply be called a spiritual sensibility.

The relationship of spiritual sensibility to the love of nature is patent. Love of nature frequently springs from and contributes to a belief in the sacredness of life. This belief need not be grounded in sectarian faith. Many environmentalists, particularly the more socially and politically active ones, do not consider themselves to be religious. Yet they are quick to affirm a spiritual connection to nature. Edward Abbey wrote that "[t]he love of wilderness is more than a hunger for what is always beyond reach; it is also an expression of loyalty to the earth, the earth which bore us and sustains us, the only home we shall ever know, the only paradise we ever need—if only we had the eyes to see."[79] Abbey is not a religious thinker in any traditional sense, and he openly disparages belief in a heavenly afterlife. Yet his orientation to wilderness, his "loyalty to the earth," is best described as a spiritual connection. In the absence of such a connection, many believe, environmentalism would founder. Californian Senator Tom Hayden, who teaches eco-theology in college classes, has argued that "[o]nly when we believe the sacred is present in the living Earth will we revere our world again. . . . I am convinced that we cannot resolve the environmental crisis without rediscovering its lost spiritual significance . . . The next wave of environmentalism must include a passionate, spiritual alternative."[80] Hayden gives voice to a widespread conviction.

Typically, this spiritual connection to wilderness and to nature is characterized in the language of interdependence. Humans are portrayed as intrinsic parts of a sacred web of life. A new "worldview" grounded in this ethical and spiritual appreciation of interrelatedness, eco-philosophers contend, is foundational to environmental care.[81] For Greens in particular, spirituality is defined as "a way of being in the world that acknowledges and celebrates our connectedness to the Earth, to each other, and to all life."[82] Those individuals who are not consciously integrated into the sacred web of life but remain isolated in a "human monoculture" put their spiritual and psychic health at risk. The field of *ecopsychology* has developed to explore this threat. Ecopsychology is grounded in the belief that isolation from and violence perpetrated against the ecological integrity of the planet produces psychic and spiritual distress and pathologies.[83]

Asked how certain she was about her environmental convictions, an activist with the Greens responded: "With facts I'm ready to change my opinion. But the deep part of it, the connection to all life, isn't open to debate. It's experiential. It's not something you can decide about." Asked on what his activism was ultimately based, a member of Rainforest Action Network responded that it was a "spiritual" orientation, one grounded in the American Indian belief that "everything is circular and all life is connected." Another environmentalist reaffirmed this sensibility, stating, "I am environmentally conscious because having been trained as a Buddhist I understand that everything is interconnected on this earth. Life forces are connected. What goes around comes around. What we do to the earth will come back to us." Likewise, an activist with Friends of the Earth stated, "I feel that we have a responsibility to the earth. We are of the earth. We are its expression of itself. Therefore we have a responsibility to nature as well as to our families. It is religious, spiritual—I feel spiritually connected to everything." Interdependence, one might say, is not only a scientific fact for environmentalists. It is a spiritual truth.

Thomas Berry, a Catholic priest, illuminates the nature of this spiritual connection to life by suggesting, in Leopoldian and Muirian fashion, that we should "move beyond democracy to biocracy, to the participation of the larger life community in our human decision-making processes."[84] Biocracy corresponds to what Morris Berman has called "participant consciousness."[85] It is demonstrated in Arne Naess's belief that the Kantian maxim never to use another person only as a means should be expanded to "never use any living being only as a means."[86] This ethicospiritual connection to life—at some level and to some degree—is endemic to coevolutionary environmentalism.

Collective action generally requires the existence of what is called a "solidary group" with which members can identify.[87] The sensibility of ecological interdependence that fourth-wave environmentalism fosters takes the biosphere as the fundamental solidary group. Spiritual and ethical solidarity with the earth's biotic community is what Warwick Fox has characterized as a *transpersonal identification* with life.[88] As spiritual and ethical members of an interdependent, ecological community, environmentalists do not identify themselves as masters of nature. They see themselves as participants, unique parts of a greater whole whose fates are linked. That is the ethical and spiritual import of coevolutionary thought.

Henry David Thoreau wrote that "[l]ife consists with wildness. The most alive is the wildest. Not yet subdued to man, its presence refreshes him."[89] Why, we might ask, does wildness refresh us? Wildness is that which has not yet been subdued by humankind and does not yet bear its imprint. By escaping physical and conceptual capture, wildness remains radically other, refreshingly nonhuman. As Thoreau observed, "The highest that we can attain to is . . . a discovery that there are more things in heaven and earth than are dreamed of in our philosophy."[90] In the same vein, Walt Whitman, Thoreau's romantic contemporary, poeticized:

> What is better than to tell the best,
> It is always to leave the best untold.[91]

This judgment follows, Whitman concluded, because the "best of the earth cannot be told anyhow." John Muir had this same insight. His environmental awakening occurred upon an encounter with wild orchids. Muir chanced upon the orchids in a Canadian swamp. They literally brought him to tears because he knew these radiant flowers had grown and blossomed without relevance or relation to human beings. Their astounding beauty was beyond us.

Some semblance of this ethical and spiritual attitude, environmentalists insist, must be cultivated if the fertile diversity of the earth is to be preserved. The love of nature is a love of that which is beyond our control and often beyond our comprehension. One Sierra Club activist insisted that it was our duty to "[l]eave big enough chunks [of wild nature] that they're not just museum pieces, rather that they're self-sustaining, wild enough to get lost in, get killed in." At times, the desire environmentalists exhibit for a natural world that escapes human control might appear to verge on the masochistic. An advocate for the restoration of grizzly bears to Yellowstone wrote:

> My own sense is that the Yellowstone population of *Ursus arctos horribilis* is probably as important as any endangered species, any subspecies, and population of animals on earth, insofar as it helps the planet's most arrogant people (at this point in history, anyway) to recall the acrid taste of an ancient ecological fear. No one wants to share [the] awful fate [of those killed by grizzlies]. But we profit from having a place in which that fate is still possible. A walk in these woods, we know, entails some small chance of being killed and eaten by a larger, more terrible being. We need that knowledge; we need that dimension of humility. There could be no better reminder for us of humanity's true role—a middling one, since all ecological roles are middling—in the eat-and-be-eaten mandala of living things. . . . When grizzly disappears from its Yellowstone island, an ocean of forgetting will drown us all.[92]

Ecology—the science of biological interdependence—is by this account an inherently spiritualizing exercise. It humbles us. It demonstrates humanity's "middling" role as a small part of a much greater whole. It fosters what Dave Foreman has called a "generosity of spirit." Spiritual generosity prompts us to share the planet with other forms of life. It induces us to celebrate the power and even facilitate the prominence of the nonhuman.[93]

Critics of environmentalism agree that a spiritual impetus is endemic to the movement. The accusation of religiosity is often used to undermine support for environmental policies and programs. In America, constitutional law dictates that religion, a private affair, may not unduly intrude into our public lives. Exploiting this popular ideal of the separation of church and state, anti-environmentalists criticize programs to save endangered species or protect biodiversity on the grounds that these efforts remain grounded in "religious beliefs." Since the preservation of species diversity is based only on "faith," one anti-environmentalist insisted, it should have no claim to public funding.

Environmentalists rely on the natural and social sciences as much if not more than spirituality to justify the protection of biodiversity. In any case, the spiritual justification for wildlife protection need not be based on faith, at least not in the sense of dogmatic belief. Roger G. Kennedy, director of the National Park Service, has suggested that environmentalists' spiritual motivations remain distinctly religious but are not for that reason dogmatic. He wrote about the role of wilderness:

> Let us consider wilderness as a religious concept. Religion is the recognition of the limits of human competence in the presence of the unknowable and the uncontrollable before which all humans stand in awe. So is wilderness. . . . Wilderness is that which we do *not* command. Wilderness is that which lies beyond our anxious self-assertion as humans . . . Wilderness is a place, but also is a mystery, a profound mystery. It is more than a gene pool—it is a fund of fathomless truths. . . . Wilderness areas are not big zoos; *we* are in the zoos, wilderness areas are outside the zoos. . . . Wilderness is a sort of physical, geographical Sabbath. In wilderness we can find surcease from the consequences of our bad management elsewhere, of what we have done to the world and to ourselves during 'the rest of the week.'[94]

Kennedy suggests that understanding ourselves as parts of a greater whole accomplishes two significant tasks. It integrates us into a larger community of life, and it underlines the importance of recognizing our limitations. Both are ecological virtues.

Environmental spirituality connotes a cherishing of nature, a veneration of that which escapes complete human comprehension and control. It is a sense of awe in the face of mystery. That is why religious or spiritual orientations—but not necessarily faith and certainly not dogmatic value systems—are conducive to environmental care. Studies suggest that members of all religious denominations are equally likely to adopt environmental values if they do not insist on "certainty" and "rigidity" in values and if they embrace a "benign story of God" that emphasizes the goodness of life. Fundamentalist believers, therefore, are less likely to have environmental values than members of other religious creeds.[95] Dogmatic religion and environmental values do not easily mesh. Spirituality and environmentalism typically go hand in hand.

The religiously oriented align their sense of belonging to a sacred web of life by way of a "benign story of God." Nonreligious environmentalists opt for a more naturalistic understanding of their role within the web of life, often defined in evolutionary terms. In both cases, an appreciation of humanity's part in a greater and still largely mysterious ecological whole is cultivated. An orientation to participation in rather than domination over the community of life provides the cornerstone of an environmental ethic. Jack Ward Thomas, former chief of the Forest Service, has offered a good description of how the development of this ethic might be characterized:

> We tend to revere technological inventors and interveners as heroes, as modern woodsmen penetrating the frontiers of human knowledge. I think we need a new kind of hero, one whose mission is not to breach limits but to understand them and to show us how to abide by those that are necessary and just—a hero capable of restraining what he *can* do in favor of what he *ought* to do for the good of the entire community. Some scientists, most corporate executives, and all members in good standing of the economic-technological orthodoxy will characterize this idea as a travesty, a capitulation of the questing human spirit. I call it growing up. As a child matures, he learns he is but one rightful member of a human community that sets limits on the satisfaction of his wishes. He then learns, I hope, what Aldo Leopold sketched as the "land ethic"—that his community extends beyond the human and includes other forms of life. And he also needs to learn that his known community opens around him into mysteries both beautiful and sacred, mysteries to which he belongs, mysteries which do not belong to him.[96]

To belong to the mystery of life, Thomas suggests, is to understand oneself as an interdependent part of an evolving whole. That is not to say that the earth depends upon us to the same extent that we depend upon the earth. On the contrary, our spiritual health largely rests with our acknowledged need of a biosphere that does not need us. Only such a world—a world that we may partake of but that we did not create and do not control—is still capable of fostering wonder. Wonder at this world stimulates the humble stewardship of its community of life.

The Ethics of Interdependence

Biocentric environmentalists seek to preserve biodiversity because they view themselves as participants in and caretakers of the sacred web of life. Anthropocentric environmentalists seek to preserve biodiversity, to employ E. O. Wilson's phrase, in order to safeguard our "most valuable but least appreciated resource." Whatever the mix of deep ecological and humanist motivations, environmental ethics remain grounded in the affirmation of interdependence. The sensibility of ecological interdependence does not obviate the conflicts between biocentrists and anthropocentrists. It does give environmentalists common ground on which their advocacy and activism may be better sustained.

Barry Commoner stated that the "first law of ecology" is that "everything is connected to everything else."[97] Arne Naess wrote that "[t]he study of ecology indicates an approach, a methodology which can be suggested by the simple maxim 'all things hang together'."[98] If all things hang together, or to employ John Muir's phrase, if "everything is hitched," then every human action has distinct environmental effects. These effects may be immediate, direct, and patent, or distant in time and mediated through various ecological relationships that manifest themselves only in retrospect or after extensive investigation. Ecological ethics, it follows, is defined by manifold, complex, extended relationships.

If all things hang together in the web of life, then each human action may have widespread ecological ramifications. Every endeavor may be seen, from an ecological perspective, as either part of an environmental problem or part of an environmental solution. But are all issues really environmental issues? Is something as seemingly divorced from ecological affairs as, say, urban crime, truly an environmental concern?

The answer, for many environmentalists, is a resounding yes.[99] While the connections may not be obvious at first glance, on reflection they become almost endless. "Ecological literacy," environmental educator David Orr reminds us, boils down to the ability to ask and answer the question, "What then?"[100] When crime gets reported by the media, it stimulates fear of crime. What then? Fear of crime stimulates flight to the suburbs. Thus, urban crime exacerbates suburban sprawl. Sprawl usurps remaining agricultural lands, woodlands, and other natural areas. That clearly is an environmental issue. Urban crime also discourages the use of public transport, thus increasing automobile driving, the consumption of fossil fuels, pollution, and global warming. Crime also heightens the consumption of other goods and services. It stimulates increased packaging to deter tampering and shoplifting, increased street and house lighting, the installation of burglar systems, increased law enforcement, and the building of prisons. The resources employed for these goods and services might have been put to use in environmental and social programs. These are some of the environmental effects of urban crime. It appears that there may also be environmental causes. Pollution, one study indicates, may contribute to violent crime. Toxic chemicals, it appears, disrupt brain chemistry of the neurological mechanisms that normally inhibit antisocial behavior.[101]

With these sort of connections in mind, a scholar of conservation biology and a founder and member of a number of conservation organizations observed that "[e]very personal act—of production, consumption, travel, communication, recreation, disposal, voting—is an ecological act. And every ecological act should be a conscious one."[102] The statement displays the affirmation of ecological interdependence characteristic of coevolutionary environmentalism. It is an endorsement of the quest for environmental integrity. Since everything is hitched, each of our actions must be engaged with an eye to its beneficial and harmful ecological effects and side effects.

Carl Anthony, president of Earth Island Institute, has insisted that "[t]he environment is not limited to what's out there—it's what's everywhere."[103] A staff member with the Environmental Defense Fund displayed a similar orientation. His own activism was initiated by a strong desire to preserve isolated tracts of wilderness. As he "matured," however, his sense of what it meant to care for the environment changed. He learned that "the environment is everything around us." His personal sense of obligation expanded accordingly to include caretaking of an integrated social and biological system. Since relationships of complex interdependence define the whole, the well-being of the natural parts can only be adequately protected in tandem with the well-being of the social parts. The reverse, of course, is also true.

The problem with such affirmations of interdependence is that they may become too inclusive. In casting their nets so broadly, environmentalists risk catching too much—and being unable to haul anything in. If, because of its vast relations of interdependence, the environment is everywhere and everything, then environmental protection would subsume all other social, political, economic, religious and scientific concerns. But nothing so broad could have any useful meaning, let alone practical effect. Critics have dubbed such catholic orientations to environmental protection "everythingism."[104] Everythingism undercuts the power of activism by making environmental protection too inclusive. To speak indiscriminately of the interdependence of everything may inhibit one from doing anything. Everythingism causes us to lose focus.

Every human act does, in some manner or form, affect the state of the earth. Clearly, however, good judgment is in order to determine, as precisely as possible, how ecologically intrusive or restrained, destructive or reparative, particular human actions are as isolated events and in their aggregate forms and future repercussions. All species and habitats are not equally fragile and do not have the same powers of recuperation. Every life form is not equally important to the health of its local habitat, every local habitat is not equally important to the viability of its eco-system, and every ecosystem is not equally important to the health of the earthly biosphere. Every human practice and artifice is not equally threatening or beneficial to society or to its natural surroundings. Judg-ments, therefore, must be firmly grounded in sound science as well as moral inquiry. To avoid "everythingism," environmentalists need not abandon the fundamental ecological insight that everything is connected. They simply have to pursue detailed understandings of *how* all the parts are connected. Ecological knowledge is required.

We never know with certainty which parts of the interdependent web of life are crucial to its overall well-being. As Paul Ehrlich suggests, the biosphere is like an airplane in flight.[105] Humankind is a tinkerer who methodically extracts the craft's rivets one by one. Each rivet is gleefully pocketed with an eye to its resale value. The airplane keeps flying, apparently none the worse, except perhaps for some extra creak-ing and shaking. At a certain point, however, one rivet too many gets scavenged and the craft loses a vital part. The airplane and its tinkerer plunge to their destruction. Each of the earth's species, the argument goes, is like one of the plane's rivets. In reducing biodiversity piecemeal, we are risking the catastrophic collapse of entire ecosystems.

Mark Shaffer of the Wilderness Society has employed a similar im-age to illustrate the dangers of species loss. He wrote: "If we think of the living world as a machine, then we have neither a full set of blue-prints (we don't know all the species) nor the operator's manual (we don't know all the interactions). We understand the general concepts of this machine. We are fairly sure there are some backup systems and even some noncritical accessories. But we can no more predict the conse-quences of continued species loss than we can a random stripping of a car's parts—except, of course, the ultimate consequence."[106] To say that everything is connected is not to deny that there exist "noncritical ac-cessories" and "backup systems" in nature. In the face of our ignorance, however, caution is the better part of wisdom.

All economic enterprises and technological endeavors have eco-logical repercussions. Yet human beings are curious creatures with an unquenchable thirst for discovery. We are incorrigible tinkerers. Clearly, we cannot walk so softly on the earth as to leave no footprints whatsoever. When we ignore the laws of ecology, however, we abandon the principles of what Aldo Leopold called "intelligent tinkering." Leopold wrote that "[t]he last word in ignorance is the man who says of an animal or plant: 'What good is it?' . . . If the biota, in the course of eons, has built something we like but do not understand, then who but a fool would discard seemingly useless parts. To keep every cog and wheel is the first precaution of intelligent tinkering."[107] Given humanity's dependence on biospheric health, unintelligent tinkering will inevitably come back to haunt us. The pragmatic watchword for environmentalists, born of ecological insight, is "Whenever in doubt, preserve and restore."

The many "laws" of ecology that environmentalists ask us to heed ultimately reduce themselves to three. Each has its attendant corollary. The first law of ecology is that interdependence is ubiquitous, both in the natural world and in society. Hence, the full impact of the countless linkages among the multiple parts will necessarily remain largely unknown to us. The second law, which follows from the first, is that every action has its ecological effects or costs. Hence, our actions inevitably will have unknown and often negative repercussions. The third law is that the security, stability, and beauty of the natural environment derive from its diversity and complexity. Preserving the health of ecosystems requires that limits be placed on human activities that pose threats to its integrity.[108] The first law is an acknowledgment of our ignorance. The second law is an acknowledgment of our power. The third law suggests how and why our power must be restrained in the face of our ignorance. These laws express not only the scientific but also the ethical import of ecological interdependence.

The problem is not that human beings affect their environment—all organisms do that. The problem is not even that human beings try to control their environment, for many organisms do that as well. Beavers change entire topographies to suit their purposes, flooding vast tracts of land to create their aquatic habitats. Human manipulations, however, are of far greater magnitude. Thus, limitations are in order.

James Lovelock has argued that the aggregate effect of attempts by countless species of plants and animals to maintain themselves in nurturing environments is the life-supporting condition of the planet as a whole. Life creates and maintains the conditions for life. In proposing this "Gaia" hypothesis, named after the ancient Greek goddess of the earth, Lovelock effectively outlined the effects of complex interdependence on a global scale.[109] Unfortunately, the human species does not play a symbiotic role on the planet. Its unrestrained pursuit of mastery over nature couples immense power with tremendous ignorance. By carelessly plundering the environment for technological dominion and economic gain, human beings jeopardize the continued existence and potential evolution of the myriad of other life forms that are engaged in vastly more humble, environmentally benign projects of interactive living.

For these reasons, the "precautionary principle" defines the quest for environmental integrity just as it defines the quest for environmental sustainability. Each time we disrupt natural processes and relationships we are testing the strength of a resilient but not invulnerable system. A spider web will rebound to its former shape even after repeated proddings. It does not collapse when a mooring is severed as would another structure once its foundation was undermined. Yet touch a spider web anywhere and the vibrations are felt everywhere. Cut a strand anywhere and the whole is weakened. Cut enough strands, or just a few key strands, and the whole web collapses. The web of life exhibits a similar strength and a similar vulnerability.

To affirm ecological interdependence is to acknowledge that we can never be certain which strands of the web of life are crucial to its overall resilience. If the strands that are severed served as anchors, the so-called "keystone" species upon which the viability of entire ecosystems depends, the entire web may be jeopardized. We should therefore sever as few strands as possible, engage in at least as much reparative work as we do damage, and seek to increase our knowledge of the complex dynamics of the web itself. Our environmental judgments will have to be context specific and informed by the best science available. When in doubt as to the long-term effects of our proddings and pluckings, as we inevitably will be, we should err on the side of caution.

Oscar Wilde once said that the tragedy of not getting what you want in this life is only overshadowed by the tragedy of getting it. Nowhere is this truer than in our efforts to master nature. Humans are the most powerful, adaptable, and successful species to inhabit the planet. Yet precisely for that reason, they jeopardize the earth's capacity to sustain life. As Morris Berman observed, "If you fight the ecology of a system, you lose—especially when you 'win.' "[110] The ethics and pragmatics of ecology, environmentalists are at pains to point out, neatly correspond. Both are grounded in an affirmation of complex interdependence.

The ethical, spiritual, and religious sensibilities that actuate most environmentalists today, far from constituting a dogmatic faith, are grounded in an appreciation of the mystery of life. The love of nature is fed by a humbling knowledge, or intuition, of the principles of ecological interdependence. These principles find their scattered origins in the ancient past, in the primeval understanding that humans are but a small part of the grand community of life. Only in modern times, as industrial civilization began to undermine the capacity of the earth to sustain its diversity of life and the conditions for evolution, has this understanding become both scientifically and ethically formalized.

The affirmation of interdependence was first formulated as an environmental ethic in the writings of Aldo Leopold.[111] Leopold recognized that such an ethic is necessarily "tentative." As the fluid project of "social evolution," it remains an ethics-in-the-making.[112] The sense of responsibility that humans bear within their ecologically interdependent communities is dynamic. Over the years it has gained an ever more secure foothold in hearts and minds. Today it actuates the environmental movement as a whole and is increasingly disseminated to the public that this movement proselytizes.

Environmentalism for
a New Millennium

For hundreds of years, Americans have claimed the right to use and abuse their land for pleasure and profit. Whether aided or opposed by government, individuals and corporations have vehemently and at times militantly asserted the prerogative to exploit natural resources. In the late 1970s, a collective assertion of such rights arose in the western states. It came to be known as the Sagebrush Rebellion. Fostered chiefly by energy companies and wealthy public lands ranchers, but significantly supported by grassroots westerners, the Sagebrush Rebels set themselves the goal of having most if not all federal lands, including national parks and wilderness areas, turned over to the states. The idea was that these lands would subsequently be made more easily available for private exploitation. In support of the rebellion, a number of western states laid claim to federal lands in court. The Sagebrush Rebels pitted themselves against the national government in its role as the caretaker of public lands. To a lesser extent, they also pitted themselves against environmentalists who advocated the further regulation and better caretaking of public lands. Identifying himself as a Sagebrush Rebel, Ronald Reagan set in motion the quick sale of federal public lands to private individuals and corporations.

The Sagebrush Rebellion petered out rather quickly. This was largely due to the fact that rebels found so many allies in the White House and the Interior Department. The federal government was giving ranchers and resource extraction industries open access to its lands at bargain prices, and in many cases with federal subsidies attached. Paying market prices for land and maintaining it at private expense no longer seemed like such a good idea when it could be easily accessed and cheaply exploited with public support. Revolutionary change became unnecessary and uneconomical. By 1983, Reagan abandoned his effort to privatize federal public lands. The Sagebrush Rebellion faded. In the late 1980s, however, a new Anti-environmental movement arose, resurrecting many sagebrush rebels and gaining substantial numbers of new adherents. This time, it would not go away.

The Anti-environmental Backlash: Wise Use and Property Rights

In 1988, a conference was held in Reno, Nevada, that galvanized anti-environmental sentiments that had been brewing for a decade. The Multiple Use Strategy Conference, sponsored by the Center for the Defense of Free Enterprise, was attended by 250 representatives from natural resource–extraction corporations, leaders of various right-wing organizations, and holdovers from the Sagebrush Rebellion. The attendees devised strategies to undo decades of environmental legislation and regulation.

The conference produced *The Wise Use Agenda: The Citizen's Policy Guide to Environmental Resource Issues*. The *Agenda* was edited and published by Alan Gottlieb, a conservative fund-raiser and the founder, in 1984, of the Center for the Defense of Free Enterprise. The *Agenda* was delivered to the new Bush administration as a "task force" evaluation on the environment. It demanded the opening up of public lands and wilderness areas to increased mining, drilling, logging, commercial development, and motorized recreational use. It also demanded the right for industrialists and developers to sue the government when environmental regulation threatened or harmed their economic enterprises. Finally, it demanded delisting of all "nonadaptive species" and "species lacking the biological vigor to spread in range" from the Endangered Species Act.[1] The *Agenda* was a declaration of war against environmentalism. It was the birth announcement of the Wise Use movement.

Closely related to the Wise Use movement, and bearing a parallel lifeline, is the Property Rights movement. Beginning in the 1970s, a number of property owners became engaged in litigation against federal agencies charged with environmental protection. The plaintiffs argued that the use of their property had been unjustly denied or restricted or that their property values had been unduly lowered owing to environmental legislation and regulation. Typically the litigants invoked the Fifth Amendment of the Constitution, which reads, "No person shall be . . . deprived of life, liberty, or property without due process of law; nor shall private property be taken for public use, without just compensation." Property rights litigation grew in the 1980s. The meaning of the "takings" clause of the Fifth Amendment became a topic of much debate, particularly after 1985, when Richard Epstein, a professor at the University of Chicago law school, published *Takings: Private Property and the Power of Eminent Domain*.[2] Epstein's book argued that a broad definition of the takings clause required public financial compensation to be paid for any reduction in use or value of private property due to governmental legislative or regulatory action.

In the past, the Supreme Court has ruled that all private property is held under the implied obligation that the owner's use of it shall not be injurious to the community and that government may prevent property owners from using their property in ways that are injurious to others. Courts also ruled that private property may be taken for public use, such as the building of a highway, under eminent domain proceedings providing that fair market prices are paid to the property owners. These decisions vindicated Justice Oliver Wendell Holmes's judgment in 1922 (*Pennsylvania Coal Co. v. Mahon*) that regulation should be recognized as a taking if it "goes too far" but that government "could hardly go on" if compensation were given for every adverse effect government action had on some form of private property or enterprise.[3] When the Food and Drug Administration approves or disapproves a drug for public use or when the Federal Reserve raises or lowers interest rates, various stocks and bonds as well as certain real goods and properties rise in market worth while others fall. The government has never been obliged to compensate the losing stock market players or corporations in these cases. Court decisions generally went against those who sued government for the "injury" done to their property by governmental actions as long as these actions remained within the government's traditional domain.

In 1988, President Reagan issued Executive Order 12630, which required federal agencies to conduct reviews of all actions that "may affect the use or value of private property." The order went hand in glove with Reagan's earlier appointments of judges to federal courts (such as the U.S. Claims Court and the Supreme Court) who were predisposed to rule in favor of private property rights. In turn, a number of states enacted takings legislation of their own. By 1995, eighteen states had some form of a takings law on the books.[4] With Reagan appointees in place, takings legislation advancing piecemeal, and a Property Rights movement coalescing, environmentalists worried that the very concept of governmental protection of the environment was in jeopardy. If forced to assess the potential reduction in property values ensuing from every piece of environmental legislation or regulation, federal agencies would be mired in endless studies. If forced to compensate every property owner for the costs of abiding by environmental legislation and regulation, the federal government would be loath to enact any new laws or regulations or even enforce those already promulgated.

The Wise Use and Property Rights movements have two essential objectives: the reduction or elimination of restraints on owners' uses of private property and the reduction or elimination of restraints on corporations' and private individuals' uses of public properties. Both movements exploit the general public's distrust of government and bureaucracy. They argue against any governmental role (that is to say, public role) in land use decisions. For many members of the Wise Use and Property Rights movements, an unbridled faith in market mechanisms prohibits the endorsement of most any form of governmental protection of the environment.

The Wise Use movement is strongest in the western states, where most public lands are found. It is chiefly concerned with maintaining industrial and agricultural access to public lands and keeping the costs of private access to public lands below market prices. It primarily serves large timber, mining, and ranching interests, though the movement has a substantial grassroots membership of loggers and ranchers. There are also increasing numbers of hunters and motorized outdoor recreationalists in its ranks who support the "multiple use" of federal lands.

The Property Rights movement is strongest in the eastern states, and in certain western suburbs. It is primarily focused on the takings issue and appeals primarily to commercial developers and other landowners faced with environmental restrictions to the use of property. Given that almost three quarters of all privately owned land in America is owned by less than 5 percent of the population, with many of the largest owners being the timber and mining corporations, it is understandable why the Wise Use and Property Rights movements, though originally separate, have begun to merge.

While relatively few voices in America intone the anti-environmental message, those that do are particularly loud and many are well placed.

The Wise Use and Property Rights movements have been able to create a perception of greater power and numbers by astute mobilization and clever media work.[5] Their cause has found its way into most major newspapers and journals and has even graced the cover of *Time* magazine.[6]

Wise Use and Property Rights organizers claim to represent a widespread grassroots movement. This claim is hotly disputed by environmentalists. Jay Hair of The National Wildlife Federation has insisted that Wise Use groups are really "a wise disguise for a well-financed, industry-backed campaign that preys upon the economic woes and fears of U.S. citizens."[7] "These organizations are not grassroots," Hair has charged. "They are Astroturf laid down with big corporate money."[8] George Frampton, Jr., president of the Wilderness Society, described the Wise Use movement as "an assault by commercial interests trying to preserve their traditional freedom to plunder the West without restriction—indeed, with taxpayer subsidies. They claim they lead a grass-roots movement, but they are in fact speaking for industry and their grass is watered by corporate money."[9] Christopher Bosso likewise has observed that the Wise Use movement is not the popular uprising its organizers pretend it to be. It may amount to little more than "a few hundred hardcore activists backed by an interlocking phalanx of professional organizers and conservative foundations, promoted by allies in industry, libertarian think tanks, and conservative media personalities like Rush Limbaugh."[10] This is wishful thinking. The coherence of the Wise Use movement is, to be sure, well orchestrated by its umbrella organizations. There is little doubt, moreover, that the grassroots of the Wise Use movement are well watered by big business. Yet grassroots support does exist, and it is a hardy variety that likely will not wither, even in the absence of corporate support.

The anti-environmental backlash has significantly increased the cooperative efforts of environmentalists. It has also underlined their shortcomings. Foremost on the list is the failure of environmental groups to branch out into rural areas and gain grassroots support. This failure is demonstrated by the low memberships of people from the midwestern, western, and southern states in environmental organizations.[11] Many of the mainstream groups moved to Washington, D.C., in the 1960s and 1970s with their lawyers, fund raisers, and lobbyists in tow. They neglected to maintain a broad geographic and socioeconomic base of support. In the 1980s and 1990s, to their dismay, environmentalists discovered that Wise Use activists were often doing a better job of gaining the allegiance of rural and land-based peoples.

Many land-based peoples believe that environmentalists are asking them to sacrifice their way of life so that urban and suburban vacationers can satisfy their appetites for wilderness recreation. As a group of south-western locals declared in a 1990 letter to the Group of Ten, "In their zeal to guard, conserve and watch what is not theirs, [environmentalists] have participated in the historical process of taking the land from the people."[12] To win the hearts and minds of rural peoples, environmentalists must demonstrate that economic and ecological sustainability go hand in hand. The long-term interests of land-based peoples must come to the fore.

Ecological care, fourth-wave environmentalists are learning, must be grounded in local support for sustainable development. Otherwise, environmental protection is too easily demonized by anti-environmentalists as just another incidence of governmental interference, bureaucratic imposition, or the self-serving moralism of Yuppie recreationists.[13] As one commentator aptly observed, "The environmental community cannot hope to be successful in rural communities until it becomes part of those communities, working to integrate environmental values with high-quality, sustainable economic opportunities."[14] Even traditionally hard-line wilderness advocates now demonstrate this understanding. Earth First!ers have acknowledged that the "moral flaunting" engaged by many environmentalists alienates land-based peoples. "We must make it clear that our fight is against those in distant places who profit inordinately from the extraction of [resources in] our wild lands, and not against the life-styles of this generation of workers. As long as the battle can be painted as tree-hugger versus decent, hard working people, those calling the shots will manipulate workers into a frenzy of hate."[15] The task environmentalists have set themselves, in other words, is to beat the Wise Users at their own game. They must become better advocates for rural communities.

To achieve this goal, environmentalists must demonstrate, on the ground, that sustainable development is a viable option. They have to prove that sustainability is not simply, as Wise Users have maintained, "a buzzword that Mrs. Brundtland made up" or a "code word" used by elitist environmentalists to "tell people whose lives are directly affected [by the land] what they can do." One Wise Use member observed that "[i]t's up to localities [to engage in conservation]. It depends on local cultures, values. Centralized control will not work. You can't micromanage from Washington D.C. . . . The people who have been living with a certain ecosystem for fifty years are the ones to protect it. They may have questions for experts, but they're the ones to manage it." Fourth-wave environmentalists largely concur with this statement and have reoriented their agendas accordingly. But building trust is a painfully slow affair. It is the task of demonstrating locally that sustainable development works.

Mary Hanley of the Wilderness Society has acknowledged that "[a] lot of people involved in the Wise Use movement are decent people who care about the land but have been misinformed. You can't make sweeping attacks because they are people we need to work with, like farmers and ranchers. But they feel threatened. We need to broaden our constituency and reach out to them." Hanley went on to note that the Wilderness Society has been doing "community stability work" since the late 1980s. "We work with timber communities in the West where we used to be hung in effigy. We work with Chambers of Commerce promoting tourism. We have a workbook to use with communities on how to keep their economies healthy beyond a dependence on resource exploitation."[16] These programs are not easy to develop. They require skills and ground support that the national organizations have in short supply. Despite the obstacles, developing such programs is a priority for many environmental groups. National environmental organizations are beginning to understand that their long-term success depends not only on generating revenues from government agencies, philanthropic foundations, business corporations, and urban "checkbook" members. It depends on widespread citizen involvement.

Jim Baca, former New Mexico Land Commissioner and director of the Federal Bureau of Land Management, has maintained that the existence of Wise Use groups such as People for the West "may very well be the best thing to happen to land-based environmentalists since James Watt . . . depending on our response."[17] Baca was reflecting on Ronald Reagan's appointment of James Watt as Interior Secretary. Environmentalists mobilized in response to Watt's radical anti-environmentalist positions. The "Watt boom," as it came to be known, brought a million new recruits to the environmental movement.[18] The response to the current threat that Baca demands of environmentalists today is a redoubled effort to create sustainable local economies from the bottom up. Only this response will gain the endorsement of land-based peoples. Ironically, then, a reoriented environmental movement may flourish owing to the activities of its foes. Anti-environmentalism may serve as a productive foil for the environmental movement, keeping it active and honest by way of its vigilant opposition. The threat of impending death, Dr. Johnson wryly observed, does a wonderful job of concentrating the mind. The presence of a fearsome antagonist might serve the environmental movement equally well. Anti-environmentalists, quite despite themselves, may do environmentalism its greatest service.

Education, Public Values, and Public Practices

Though enjoying a great deal of media coverage and substantial political success at both state and federal levels, the Wise Use and Property Rights movements remain dwarfed by the environmental movement in terms of membership and public support. Even in rural areas, where it is strongest, the anti-environmental backlash may not have the public support it needs to sustain itself. In many respects, the Wise Use and Property Rights movements serve a sociopolitical function analogous to that of radical environmentalists: they are fringe movements that may shift the ideological spectrum but in most respects remain peripheral.[19] The anti-environmental backlash has not significantly affected the values or behavior of the general public.

To date, environmental problems have never topped the public's list of the most important problems facing the nation. Occasionally, the plight of the environment does not even make the list. The problems of crime and drug abuse, and the fate of the economy, consistently score higher.[20] Within the general public, environmentalism is a second-tier commitment. While environmental concerns have become widespread, the general public demonstrates neither the level of concern nor the level of commitment of environmental leaders and activists.[21] Nonetheless, survey researchers have documented a clear "environmental consensus" in America. Environmental values, though shallow, are pervasive. Anti-environmental values, on the other hand, remain marginalized. No anti-environmental alternative has gained a coherent and consistent voice.[22] Environmentalism appears to have become a mainstream American value that is largely impervious to radical challenges.

As many as four out of five Americans identify themselves as environmentalists. As many as one in three accepts the label of "strong environmentalist." One in seven claims to be presently active in the environmental movement, more than one in five financially contributes to environmental organizations, and one in ten is a member of an environ-

mental group. In the 1990s, pollsters and survey researchers have confirmed and reconfirmed earlier trends: the public, regardless of partisan
affiliation, is strongly supportive of environmental protection. More specifically, between two out of three and four out of five Americans hold
that

- environmental problems adversely affect their health, and environmental problems will increasingly affect the health of their children and
  grandchildren;
- government is not doing enough to protect the environment;
- more federal spending should be devoted to environmental protection;
- stricter laws are needed to protect the environment in general, with
  substantial majorities advocating greater protection specifically for water, air, wild and natural areas, wetlands, and endangered species;
- legislation to protect the environment is warranted even when it in-
  terferes with the autonomy of businesses;
- legislation to protect the environment is warranted even when it in-
  terferes with citizens' rights to make their own decisions;
- landowners do not have the right to do whatever they want with their
  land, and restrictions on land use are justified to protect plants and
  animals;
- a company's environmental reputation affects their patronage of it;
- they are willing to pay higher prices to protect the environment and,
  if forced to choose, would opt for a healthy environment over economic growth.

When asked how concerned they were about environmental problems,
38 percent of Americans said a "great deal," with another 51 percent
saying a "fair amount." Only 11 percent said they were "not very much"
or "not at all" concerned. Only 15 percent of Americans feel environmental laws are too extensive.[23]

Of twenty-four other countries surveyed, only four (Nigeria, Philippines, Portugal, and Canada) exhibited higher levels of environmental
concern in their citizens than the United States.[24] The strength and per-
vasiveness of American environmentalism are likely to grow in the future. Two out of three Americans are more concerned about the environment today than they were in previous years.[25] And one of the few
demographic variables strongly and consistently correlated to environmental concern is age, with younger generations showing greater levels

of concern.[26] Indeed, 85 percent of Americans under thirty years of age consider themselves to be environmentalists.[27] Hence, public support for environmental protection is projected to increase over time, as younger cohorts with stronger environmental values come of age.[28] In sum, environmental values and concerns in America, notwithstanding periodic fluctuations, have grown for three decades to levels that are presently among the highest in the world. Environmental values and concerns have been effectively—and by the looks of it permanently—incorporated into mainstream American culture.

The mainstreaming of environmental values is a success story in education. The education of the public is the most common task taken on by environmental organizations. Most organizations spend the lion's share of their resources on educational projects.

This educational focus is abetted by federal tax laws that allow donations to "charitable" organizations to be declared as tax deductions. This tax-exempt status is given only to environmental organizations whose mandates are primarily educational. It is denied to advocacy groups that attempt to influence legislation or elections. To maximize their (tax deductible) support, therefore, many environmental groups restrict themselves to educational activities. Those who engage in lobbying and election campaigning must forgo the beneficial tax-exempt status.[29]

Though maximizing revenues is a priority for most environmental groups, their focus on education is not simply a question of maintaining a tax-exempt status. Environmentalists are in the business of changing public attitudes and values; they are engaged in "consciousness raising." A National Wildlife Federation activist observed that environmental organizations focus on "education in the sense of getting to the hearts and minds of people." Indeed, this is what many activists find most satisfying in their work. A Greenpeace staff member emphasized that the chief benefit of her work was interacting "with people one on one or in groups and getting at least one person to change their mind." The main task at hand for environmentalists is to inform the "environmentally illiterate" and to "shape the debate" among the informed but still uncommitted. Only an informed citizenry can understand that its long-term interests depend on the maintenance of a healthy environment. Only an informed citizenry can effectively ensure that government serves these enlightened interests. Environmental education is thus considered "both a prerequisite and tool for sustainable development."[30]

The public's deficiencies in environmental knowledge are evident. Only one out of four people surveyed, for example, disagreed with the false statement that "[t]he greenhouse effect is caused by a hole in the Earth's atmosphere." Despite the extensive media coverage received by the issues of ozone depletion and global warming, the general public still confuses them. The level of knowledge in other environmental areas is not much better.[31] The "single greatest failure" of the environmental movement, observers suggest, is its limited success in educating the public.[32]

Progress has been made. Gaylord Nelson has observed that grade school children today know more about the environment than college students did in the early 1970s. He stated that

> [p]rior to Earth Day 1970, there was no environmental education being taught in any grade school or high school in America that I know of. Now, thousands of schools do. . . . There were no environmental institutes on campuses in the United States that I know of except the one on the University of Wisconsin campus previous to 1970, now every college that I speak to of any size and many of the small ones have environmental departments, environmental courses, environmental institutes. All brand new. . . . [Then] there were no environmental law courses being taught in any law school in the United States that I know of and now every major law school has environmental law courses.[33]

In 1990, Wisconsin became the first state in the nation to mandate environmental education in the entire school system, grades K through 12. Other states followed suit, albeit slowly. Environmental education of some sort is now a formal requirement for high school graduation in at least thirty states. At the same time, the promotion of business interests in schools, with its attendant biases and lack of focus on sustainability, is often more pervasive.[34] Moreover, Wise Use groups and other anti-environmentalists have been increasing their efforts to undermine formal environmental education. In some states, such as Arizona, environmental education has been defunded and removed from the classroom. Nonetheless, 95 percent of Americans continue to approve of environmental education.[35]

Many environmental organizations take on their educational mandate directly with children and youth. They train teachers in workshops, produce education videos, sponsor the development of schoolyard wildlife habitats, and prepare educational programs and teaching kits, such as Friends of the Earth's Green Schools Project or Zero Population Growth's Kid's and Teen PACKs (Population Awareness Campaign Kits). Environmental organizations also devote sections of their newsletters to young readers or, like the Cousteau Society and the National Wildlife Federation, publish separate magazines for children. The National Wildlife Federation's children's publications reach well over a million homes.[36] This outreach to youth is worthwhile. Most activists attribute their ecological values and convictions to their education or upbringing. Reiterating this view, Rodger Schlickeisen of Defenders of Wildlife stated, "If you judge the future by the past, it's hard not to be worried about the environment. . . . My personal source of optimism has to do with the kids. They are much more educated and knowledgeable about the environment than ever before. When they come into power they are likely to advance the environmental agenda. My job is to try to hold on as long as possible until children come into power who can be successful."[37] Schlickeisen expressed the widely held belief that the greatest hope for the environmental movement lies in the long-term payoff of education.

When environmental organizations educate and inform, they are effectively giving their members what they most want. The desire for information is the primary reason people give for joining environmental organizations.[38] This is true even when one controls for the effects of age, gender, education, and environmental and political values. In turn, the most satisfying aspect of involvement in environmental organizations for a significant majority of activists is their own personal education. Many activists terminate their memberships in particular groups when information and communication are not sufficiently forthcoming.

Academics and scientists occasionally suggest that environmental activists have an anti-intellectual bias. They accuse environmental organizations, with some justification, of playing the media game of sound bites and strategic rhetoric at the expense of science. Yet environmentalists on the whole are, if anything, too infatuated with the power of knowledge. Many remain naively hopeful about its effects. A member of Friends of the Earth expressed a common belief that "the more educated people get the more concern they'll give the environmental side." At times, environmentalists' faith in education seems utterly quixotic. A member of the National Wildlife Federation stated, "I don't think that it's possible that people could be in control of all the facts and then vote to harm the environment." One Sierra Club member went so far as to say that only "the intellectual elite" harbor sufficient environmental concerns and that the chief problem at hand was the ignorance of the masses.

Studies over two decades have repeatedly confirmed that education is highly correlated with environmental concern. While income and gender appear to be largely unrelated to environmental values, and while youth, liberal ideology, and urban residence are positively correlated, many studies suggest that the strongest predictor of environmental concern and behavior is education.[39] At the same time, higher education is not always environmentally beneficial. Much college and university education today chiefly prepares students to become more powerful resource consumers.

Even environmental education has its limits. Environmental learning—gaining knowledge and acquiring information about the environment—often becomes a means of coping with fears and anxieties about environmental destruction rather than informing or inducing action. For a large segment of the population of one study, learning about the environment "did not lead them to change their behavior vis-à-vis the environment. Rather, it led them to seek more information and knowledge, and to a certain extent such further learning became even a substitute for social action." The desire to "know what they will have to face in the future," not the desire actively to oppose environmental degradation, was the primary motivation for pursuing environmental knowledge.[40] In short, environmental education does not necessarily produce environmentally responsible behavior. Information does not always translate into environmental concern, and environmental concern does not always translate into commitment and action.

Those who assess the environmental movement from a historical and comparative perspective, judging its development with reference to the fates of other social movements, mark its notable accomplishments. Dunlap and Mertig have noted that environmentalism has "remained a viable sociopolitical force for more than two decades" and on this basis alone must be deemed "a resounding success." Yet these observers caution that "history will judge [the environmental movement] in terms of its success in halting environmental deterioration rather than in simply avoiding its own demise."[41] Indeed, history might judge the environmental movement a failure despite its sociopolitical longevity and despite its undeniable success at propagating environmental values. That is because the environmental movement's notable success in educating the public has not produced a comparable change in the public's behavior.

The translation of widely proliferated environmental values into widespread practices remains a task largely undone. The public wants to know what to expect in an environmentally precarious world and is largely convinced of the need for environmental protection. At the same time, the public remains largely unwilling to apply itself to the tasks at hand. Thus, despite the expression of relatively high levels of environmental awareness and concern, Americans consistently display relatively low levels of environmentally responsible behavior, such as recycling, reducing consumption, conserving resources, supporting green businesses, eliminating waste and pollution, or engaging in various forms of environmental protection by way of individual or collective efforts.[42] The co-optation of environmentalism in the 1980s tragically underlined this lapse. Although environmental values were mainstreamed, behavior changed very little. This pattern has carried on to the present day. The public pursues environmental knowledge and is steeped in environmental values but remains reluctant to engage in environmentally responsible behavior. Even when knowledge is translated into conviction, the further translation of conviction into action is unsteady at best.

A recent poll found that two out of three Americans believe that an animal's right to live free of suffering is as important as a human's right to live free of suffering. Nonetheless, 98 percent of those polled ate meat, poultry, or fish, with two out of three respondents eating meat, poultry, or fish on a frequent basis.[43] Given the "inhumane" conditions under which most meat and poultry and some fish are raised and killed for human consumption, an undeniable contradiction between values and practices exists. One might object that championing animal rights and maintaining ecological values are two very different things. Indeed, these orientations are different and may even conflict.[44] Still, inconsistencies between the words and deeds of environmentally oriented people are rife.

Take the problem of overconsumption. In distinction to population growth and technological development, overconsumption has traditionally been the "neglected god" of the trinity of issues that environmentalists address.[45] Fourth-wave environmentalists have brought the problem of overconsumption sharply into focus. Defusing the "time bomb of overconsumption" has become a priority.[46] There is widespread acknowledgment today by citizens of industrialized nations that they remain responsible for most to the world's environmental problems and that they disproportionately contribute to the unsustainable consumption of the world's resources.[47] The environmental movement can justifiably take credit for this growth of self-critical awareness.

Unfortunately, increased awareness has not led to markedly altered practices. Citizens of industrialized countries continue to use ten to nineteen times more energy, timber, iron, steel, paper, chemicals, and aluminum and three to six times more fresh water, grain, fish, cement, fertilizers, and meat than citizens of developing countries on a per capita basis. Despite constituting only one fifth of the world's population, they consume over half of all the aforementioned resources (with the exception of fresh water, grain, and fish).[48] All told, the richest 25 percent of the world's population consumes 85 percent of the world's resources. They also produce 90 percent of its wastes. Assuming that relative levels of consumption and waste remain unchanged, the near sixty million people born in rich countries through the 1990s will have caused more pollution that the 900 million people born elsewhere.[49] Indeed, one study suggests that if the richest 25 percent of the world's people reduced their level of consumption by 25 percent, the global level of pollution would decline by twice as much as it would were the poorest 75 percent of the world's people to disappear from the planet altogether.[50]

Owing to the environmental degradation that it visits upon the earth, largely because of its wasteful consumer culture, the United States is considered the world's most "overpopulated" nation.[51] Indeed, the average U.S. citizen consumes forty to fifty times the resources of a citizen of a developing country and produces twice as much trash as the average European. Per capita emissions of carbon dioxide from fossil fuels are over twice as high in the United States as in Japan and over twenty-five times higher than in India.[52] Were everyone to grow their food as Americans do and eat similarly, the world's current oil production would be required for agricultural use alone and all known oil reserves would be exhausted in little more than a decade.[53]

Levels of overconsumption in the United States show no sign of decreasing despite the rise of environmental concern. Per capita consumption of virgin wood, for instance, has risen 30 percent since 1970.[54] Between 1969 and 1995, the number of cars in the United States increased six times faster than the population and twice as fast as the number of drivers.[55] Meanwhile, the average miles driven per gallon of gasoline actually decreased in 1996, as consumers bought larger, gas-guzzling vehicles. The increase in per capita consumption of green space through suburban sprawl is equally alarming. Between 1970 and 1990, metropolitan Cleveland's population declined by 8 percent, yet its urban land area increased by one third; metropolitan Chicago grew in population by 4 percent while its developed land area increased by 46 percent; and the population of Los Angeles grew by 45 percent, while its land area expanded by 300 percent.[56] For all these reasons, the United States generally ranks well below countries such as Germany, Japan, The Netherlands, Norway, and Sweden in its environmental policy performance over the last two decades.[57] The rise of environmentalism, one might suggest, has made Americans much more aware of what they do or neglect to do to protect the environment. It has not changed many habits.

American environmentalists have little good news to report in their efforts to translate public values into public practice. Over half the population agrees with the statement, "I do what is right for the environment, even when it costs more money or takes up more time." Only one out of every six respondents openly disagrees with this statement.[58] What is the practical upshot of such strong environmental concern? Efforts at recycling have grown, to be sure. Curbside recycling nationwide has increased more than eleven fold in the last eight years. Yet less than half the population recycles regularly. In almost every other area, environmentally responsible behavior has stagnated or is in decline. Indeed, environmentally responsible behavior is engaged in regularly by fewer than one in five individuals.[59] On the political front—certainly one of the most crucial arenas of struggle for environmentalists—the situation is no better. Over 70 percent of polled respondents in the general public say that they have never voted for or against a political candidate based on his or her environmental views or record.[60]

However much information is disseminated by the environmental movement, and however successful the movement is at inculcating environmental values, the conversion of these values into practice remains disappointingly weak and sporadic. When environmentally responsible behavior involves "personal sacrifice" or specific costs, people tend to shirk responsibility in the hope that others will bear the burden. They choose to be environmental free riders.[61] Studies confirm that "[t]here is indeed little causal relationship between environmental value orientations, awareness, concern, information and knowledge acquisition on the one hand, and behavior on the other."[62] Stated values and stable practice, it turns out, seldom converge. In environmental affairs, a "real disparity between words and deeds" exists. The conclusion drawn is that "many people have learned the language of environmentalism without developing a simultaneous behavioral commitment."[63]

Empirical research indicates that mainstream environmental behavior, such as recycling, taking public transport, and keeping abreast of environmental affairs, is stimulated by specific experiences. It is primarily a function of one's encounters with nature, one's first-hand experience of environmental degradation, and to a more limited extent, one's level of education. Low-level activism, which along with mainstream behavior includes such activities as signing petitions, informing others about environmental issues, and engaging in efforts to protect the local environment, is primarily a function of one's nature experiences and one's previous participation in environmental action. High-level activism, which along with the demands of mainstream behavior and low-level activism includes engaging in public opposition to environmentally destructive projects, is almost entirely a function of one's previous participation in environmental action.[64]

For these reasons, membership in environmental organizations remains a relatively good predictor of environmentally responsible behavior. Although the provision of information is the main incentive for joining environmental groups, those who join *solely* for information, the "information purists," constitute only 14 percent of all members. The majority of members, 58 percent, join and maintain their membership not only for educational purposes but also to further the political and practical aims of the organization. Thus, volunteer environmental activists and staff members or executives of environmental organizations consistently demonstrate the highest levels of environmentally responsible behavior in society. Nonactivist environmentalists who belong to environmental groups are next in line, followed by self-declared environmentalists with no group affiliation. Self-declared nonenvironmentalists exhibit the lowest levels of environmentally responsible behavior.[65]

Four general strategies are likely to facilitate environmentalists' efforts to foster environmental values and effect the translation of these values into practices. First, environmentalists need to encourage and promote regular nature experiences. Second, they need to educate and mobilize those who are subject to environmental degradation in their own communities. This requires both detailed reporting of environmental degradation and environmental injustice at local, national, and global levels and the provision of practical means to address the problems at hand. Third, environmental organizations need to keep activists involved. Nothing stimulates heightened activism like continued activism. Importantly, these measures must be extended beyond a select membership. They cannot be limited to those who can afford to donate substantial sums of money to environmental organizations or take expensive eco-tours in far-off lands. Nature experiences and inducements to activism must be locally grounded. They must be incorporated into the lives and livelihoods of average citizens. Fourth, environmental organizations have to lead by example. To the extent that environmental organizations fail to serve as paragons of environmental virtues, they further undermine the bridge between attitudes and action that the public is already reluctant to cross. Environmental groups must better practice what they preach. A number of the national organizations have made a good faith effort. They have built or retrofitted highly energy-efficient offices. They use chlorine-free and recycled paper for their publications or use alternative (nontree) sources of paper such as hemp or kenaf. They generally promote conservation measures in their daily practices. But much more could be done. Their magazines remain rife with advertising that promotes a wasteful and polluting consumer culture. Their massive direct mail campaigns are a double insult to the environment, destroying forests and producing vast amounts of needless trash. Finally, their efforts to achieve a socially diverse and integrated membership and staff fall far short of the mark.

Environmental education remains key to all four strategies. To be effective at stimulating and sustaining environmentally responsible behavior, however, environmental education must not only pass along information and generate knowledge about environmental issues. It must also develop the skills people need to effect change, including investigative, problem-solving, and action-taking skills. It must induce a sense of individual and collective responsibility and empowerment. And by facilitating extensive interaction with the natural world, it must instill a love of nature.[66]

It is not education, but education of a certain kind, that is effective. Environmental educator David Orr has observed, "Were we to confront our creaturehood squarely, how would we propose to educate? The answer, I think, is implied in the root of the word *education, educe,* which means 'to draw out.' What needs to be drawn out is our affinity for life. That affinity needs opportunities to grow and flourish, it needs to be validated, it needs to be instructed and disciplined, and it needs to be harnessed to the goal of building humane and sustainable societies."[67] The transmission of information, even when coupled with the transformation of values, does not serve environmentalism particularly well. Citizens must also deliberate among themselves and with their political leaders about the need for and means to achieve sustainable development.[68] They must cultivate an affinity for life. And they must act. Effective environmental education helps people respond to the social, political, and ecological demands of protecting the environment.[69]

A broad base of citizens educated in the ways of generational, social, and ecological interdependence makes the collective protection of the environment a realistic challenge. For those involved in environmental education, it follows, a "schooling for interdependence" is the chief task at hand. The aim is to enable people "to grasp and live out their interdependence with others and the natural world."[70] Researchers confirm that to be successful, environmental education must instill a sense of the "complexity" that natural and social systems display owing to the manifold interdependent relations that compose them.[71] Environmental educator C. A. Bowers has observed that in successful school programs, "[i]nterdependence not only serves as the root metaphor for understanding how energy is exchanged within the connected web of ecosystems but also provides students a radical alternative to the individually-centered view of moral responsibility. . . . [The] curriculum provides both the experience of being an interdependent member of a human/biotic community and an understanding that interdependence is the basic relationship that connects past, present, and future generations."[72] An education in interdependence facilitates the translation of environmental values into practice because it actively integrates the individual into social, political, and ecological communities.

Such learning must be stimulated well beyond our institutions of formal education. Environmental education of the general public, viewed expansively as the creation and sharing of knowledge, the cultivation of values, the instilling of commitments, and the collaborative solving of problems, remains the chief task and greatest hope for environmentalists today.

Conclusion

When as many as four out of five Americans consider themselves to be environmentalists, yet very few engage in environmentally responsible behavior, one might worry that the environmental movement has lost a crucial battle—the battle over its very meaning. When a movement creates an attractive collective identity, the selective provision of that identity may escape its control. Theorists of social movements observe the irony that a movement's "very success results in the creation of a public good available to nearly everyone. To adopt the collective identity carries with it no obligation to participate in the activities of the movement. When a movement finds itself in this situation, it may already be dying."[73] The environmental movement is not moribund. Its longevity is partly ensured by the fierce internal battles that continue to be fought over its direction and purpose. Yet the co-optive mainstreaming of the environmental movement clearly threatens its vitality. American environmentalism may have much of its life squeezed out of it by its inactive supporters. To avoid this fate, environmentalists will have to offer the public an education that meshes ecological knowledge with committed practice.

Better translating environmental values into environmentally responsible behavior is not an easy task. Undoubtedly, however, history will judge environmentalists based on their success in carrying out this task. Of course, the legislative and legal gains of the environmental movement over the last three decades have not been negligible. Perhaps these political and legal victories mandating environmental protection, rather than the deficiencies in the public's voluntary efforts, should be the chief measure of the movement's success. Even in the case of legal and legislative action, however, the accomplishments are more encouraging on the books than on the ground. For every species that has been removed from the Endangered Species list since 1973 because of its successful recovery, a hundred new species have been added owing to their newly threatened status.[74] Currently, there are over 1,450 species of plants and animals listed as endangered or threatened. It is a testimony to the strength of the environmental movement that it has been able to pressure the government and its agencies to expand the listing of species in need of protection. The problem is that the list is growing not primarily because people care more about already-threatened species. It is growing because more and more species are desperately in need of care. To make matters worse, the Fish and Wildlife Service acknowledges that fewer than 10 percent of all the endangered and threatened species for which it is responsible are actually improving in status. Nearly 40 percent are declining, while about a quarter remain stable.[75] Environmentalism is clearly faring better than the environment.

Barry Commoner stated that "the environmental movement is old enough . . . to be held accountable for its successes and failures. Having made a serious claim on public attention and on the nation's resources, the movement's supporters cannot now evade the troublesome, potentially embarrassing question: What has been accomplished?"[76] Commoner tallies up many of the victories and defeats the movement has celebrated and suffered. In the end, his judgment is unfavorable and pessimistic. Rick Sutherland, president of the Sierra Club Legal Defense Fund, likewise lamented: "My primary emotion when recalling the past twenty years of environmental law is one of profound disappointment."[77] Reflecting on the twentieth anniversary of Earth Day, Denis Hayes, its original organizer, assessed that "by any number of criteria that you can apply to the sustainability of the planet, we are in vastly worse shape than we were in 1970."[78] David Brower, after nearly sixty years at the front line of environmental activism, offered a similar evaluation, observing that "[a]s environmentalists, all we have been able to do is to slow down the rate at which things have been getting worse."[79] Reflecting on Brower's gloomy assessment, Earth Island Institute concurred that slowing the rate at which things are getting worse was all that it has been able to accomplish over the last two decades, and all that older groups such as the Sierra Club have managed to do over the last two centuries.[80] Policy analysts evaluating the effectiveness of governmental actions across the globe arrive at equally chary conclusions.[81]

Gloominess runs deep within the environmental movement. Some of the gloom is offset by the tendency of organizations to advertise their successes and trumpet their victories in the effort to increase memberships and donations. Convincing the public that your organization constitutes an effective force for change is key to sustaining support. For all the self-promotion environmentalists engage in, however, doomsaying remains pervasive. For some, an environmentalist is simply a pessimist with a conscience. As Rick Bass states: "I am an environmentalist. I find myself trying to find ways to apologize to the future."[82]

In a book that celebrates "ecological optimism," Gregg Easterbrook faults Rachel Carson for being "wrong" in predicting a silent spring. Easterbrook waxes panglossian about the present heyday of environmental care, and cautions against repeating Carson's mistake by "slamming society's fist down on the panic button." More moderate responses to environmental problems are in order. Yet Easterbrook himself acknowledges that "[i]f chemical-use trends had continued forever exactly as they were in the 1960s, then the severe ecological harm Carson foretold would have come to pass."[83] Carson pushed the panic button, as did thousands of Americans who read her book. Pushing that button was necessary to avoid catastrophe, given the powerful forces in the chemical industry that fought against Carson's message and mitigated its effect. Like Carson, present-day environmentalists frequently put themselves in the position of working assiduously to falsify their own predictions. The more successful they are, the more wrong they will be.

Doomsaying should not be *de rigueur* for environmentalists. Rachel Carson gave an accurate account of the then-current trends. These trends would not likely have been reversed, or so quickly reversed, if Carson's dire predictions were not made and heeded. Environmentalists adhere to René Dubos's dictum, "Trend is not destiny." That is what sparks activism. While pessimism is often self-defeating, Easterbrook's brand of eco-optimism is equally dangerous. Its heady estimation of the corrective capacities of natural systems and the saving powers of technology are formulas for complacency. Carson's active realism remains a worthy ideal for environmentalists.

An editorial written for the twentieth anniversary of Greenpeace reads: "It would be nice to look back on our second 20 years and see that we'd raised as much hell and come as far as we did in the first 20. It would be even nicer, in the year 2012, to know the world had come far enough that it no longer needed a Greenpeace at all."[84] Unfortunately, we are likely to need organizations like Greenpeace a good deal longer. The struggle to maintain biodiversity and preserve wilderness shows few signs of success or even progress. Habitat destruction and species extinction are accelerating as human populations grow and economic development expands. Every 2 seconds, the world's population increases by more that five people and the world's remaining wildlands decrease by almost three acres.[85] The Endangered Species Coalition reports that one quarter of the world's species could be lost over the next half century and that currently as many as 100 species are becoming extinct daily.[86] Though specific threats vary over time and space, the global decline of habitat and species is accelerating.

The Audubon Society was formed in the late 1800s with a mandate of saving the birds of Florida's Everglades. At the time, the birds were being killed for feathers to adorn hats. Milliners are no longer much of a threat to Florida's wild life. After a century of protective efforts, however, the wading bird population of the Everglades has dwindled to 5 percent of its original size.[87] Now agricultural, commercial, and residential development constitutes the chief danger as wildlife habitat dwindles. Half of the Everglades has already been lost. In the state of Florida as a whole, hundreds of acres of wildlife habitat are lost to development each day.[88] Specific threats to biodiversity may change over time, but the overall danger remains. Indeed, it grows. In both developed and developing countries, environmentalists will face an uphill battle for decades, if not centuries, in their efforts to preserve biodiversity. It often seems a losing battle. The fabric of life becomes more threadbare with each day of human handling.

The war against pollution and toxics, many environmentalists hope, might be more easily won. Despite encouraging signs, victory is far from assured. Certainly it is far from imminent. Technological advances and regulation have yielded significant improvements in air and water pollution abatement in recent years, particularly in industrialized nations. Yet technological and regulatory efforts to reduce or detoxify emissions are often more than offset by the increasing number of factories, energy plants, and automobiles spewing pollutants. The problem of pollution control efforts being negated by economic growth applies across a wide range of contaminants and waste.[89] Between 1975 and 1989, for instance, the 3M Corporation claims that its 3P (Pollution Prevention Pays) program prevented seventy-two million pounds of pollutants from being released every year. The company neglects to mention that growth in production actually increased its total output of pollutants during that period.[90] In the last twenty years, automobiles have been designed to achieve, on average, double their original fuel efficiency and emit between 75 percent and 90 percent fewer pollutants. The number of vehicles on the road, however, has also doubled, as has the number of miles driven. The overall result is a small, if any, net gain. To make matters worse, fuel efficiency has begun to decline in recent years, as cheap gasoline prompts the manufacture and sales of bigger cars and trucks.

The signs of the times are not encouraging. Per capita waste production also continues to grow, despite efforts in recycling. Two out of five U.S. lakes, rivers, and estuaries remain too polluted to allow fishing or swimming at certain times of the year.[91] Only 2 percent of the synthetic chemicals that Americans are exposed to have been adequately tested to determine their effects on human health.[92] About half of all Americans drink water containing cancer-causing chemicals. Breast milk from American mothers, owing to its high level of contamination from pesticides and industrial poisons, such as PCBs, has on occasion failed to meet FDA standards for food suitable for human consumption.[93] Ninety-five percent of Americans have measurable levels of pesticides in their bodies.[94] Meanwhile, insecticide use has grown more than tenfold since 1945, notwithstanding the fact that crop losses to insects have roughly doubled in that same period.[95] At a global scale, the situation is worse. Assessing the progress made since the 1992 Earth Summit, Wilfried Kreisel, executive director of the World Health Organization, observed in a 1997 study that air pollution has increased and water quality has worsened throughout the world. The study concludes that one out of every five children in the least developed countries dies of environmental causes before reaching his or her fifth birthday.[96]

Were significant environmental victories won in the fight against pollution and toxics, one might presume that the environmental organizations chiefly engaged in these struggles would soon disband. The characteristics of fourth-wave environmentalism make this unlikely. As the grassroots toxics movement demonstrates, campaigns against health-impairing pollution are part of a more comprehensive struggle for democratic rights and social justice. These sorts of struggles tend to be enduring ones. As an environmental justice spokesperson wrote, "Our movement is chiefly based on health concerns combined with basic issues of justice and human rights. Our adversaries will not be able to address these issues without fundamental changes in the way decisions are made."[97] The struggle for communities to gain control over the environmental hazards they face will not be quickly won. Like all democratic struggles, its battles will be hard fought and its victories will stand in need of constant vigilance.

There is another reason why environmentalism is likely to persist. Its successes are always tentative. Asked if and when he finds environmental activism rewarding, an Audubon member responded: "Yes, when I work for some particularly valuable area and we get it protected and properly managed, and over the years being able to go back and see that it's all still there. It's a continuing battle, however, because it all can be lost at once. There's a certain amount of anxiety that never goes away." What is saved today may be lost tomorrow. Environmental victories are inherently provisional.

Environmental defeats, in contrast, are often final, at least at human time scales and always in the case of species extinction. What took years to protect, preserve, or restore may be destroyed in a flash, never to regenerate. Environmental dangers are seldom dispatched, they are only mitigated. Environmental problems are seldom obviated, they are only deferred. Kirkpatrick Sale observed that environmentalists fight an "ecocidal hydra."[98] Every successful effort to protect the environment remains haunted by the possibility a future defeat.

If the environmental movement is not in jeopardy of expiring, that is because some feature of the environment always is. Faced with this fact, activists do not view environmental protection as a final goal to be reached but as a struggle indefinitely waged. Aldo Leopold wrote that "[w]e shall never achieve harmony with land, any more than we shall achieve absolute justice or liberty for people. In these higher aspirations the important thing is not to achieve, but to strive. It is only in mechanical enterprises that we can expect that early or complete fruition of effort which we call 'success.' "[99] Never blessed with a complete or secure victory, environmentalists gird themselves for continuous battle. Part of the environmentalist's struggle, it follows, is waged against the enervating effects of defeatism.

Despite their propensity for despair, environmentalists remain upbeat enough to implement change. They have widely disseminated their values and have begun, however tentatively, the difficult task of translating these values into common practices. For all that, fourth-wave environmentalists do not have *the* answer to the problem of environmental degradation. There is no single answer. As Paul Wapner wrote, "[E]nvironmental issues are not puzzles in search of solutions but rather perennial challenges that successive generations must persistently confront anew. These challenges involve constantly searching for more sensitive and sustainable ways of interacting with the natural environment."[100] The challenge of coevolution is perpetual.

In an update of Leopold, Kai Lee has observed that "sustainable development is not . . . a condition likely to be attained on earth as we know it. Rather, it is more like freedom and justice, a direction in which we strive."[101] The same might be said of coevolution. It is a direction, not a destination. The challenge of coevolution entails the pursuit of intergenerational, social, and ecological justice. The achievement of these forms of justice will necessarily remain partial. Yet this limitation need not prove debilitating. As Rabbi Tarphon observed two millennia ago, "It is not for us to complete the task, but neither are we free to refrain from getting it started."[102]

In the arena of environmental protection, one is inevitably burdened with the anxiety that more could always be done. Caution is in order here. Sometimes more is too much. There are extremes to be avoided. Environmental radicals who demand that human beings walk so softly on the earth that they leave no footprints of any sort—industrial, technological, agricultural, architectural, or ecological—effectively deny the value of human culture itself. It is a hypocritical denial. Great art, literature, and science—even collectively organized environmental protection—would become impossible in such a world. The Earth First! slogan

Back to the Pleistocene! makes an inspiring chant at rallies. It would never find itself in print were it actually heeded. As cultural beings, it is only natural for human beings to leave some footprints on the earth. As foresightful, aesthetic, and moral beings, it is only natural for us to restrain our stompings such that a high quality of life for ourselves, for our progeny, for our local and global neighbors, and for the vast diversity of other species that share the earth with us is made possible.

Certainly it is also natural for individuals to pursue their self-interest. The task for environmentalists, in this context, is simply to expand the meaning of self-interest—generationally, sociogeographically, and ecologically. Schooling the public in our interdependence across time, space, and species is the challenge environmentalists face as they enter a new millennium. It is the idealistic challenge of pursuing sustainable development within a coevolutionary framework. This ideal bears within it an undeniable realism. The fundamental reality of the human species is that we exist as participants in an intricate web of life that spans a finite planet. Ours is a shared fate.

The goal of the environmental movement, Victor Scheffer wrote, "is to somehow strike a balance between idealism and realism; to preserve the diversity and wondrous beauty of our world while recognizing that billions [of people] must steadily draw upon its substance for survival."[103] To be an environmentalist today is to walk a thin line between economic realism and ecological idealism, between pragmatic efforts of reform and uncompromising ethical principles, between concern for distant progeny and the demands of the present, between local caretaking and global oversight, between biocentric intuitions and anthropocentric habits, between an enervating doomsaying and a complacency-inducing optimism, between celebrating the wonders and enduring strengths of nature's evolutionary legacy and despairing at our vast and accelerating destruction of it. Accepting the challenge of coevolution means skillfully walking this line. Warning against both the folly of doing nothing and the danger of blind activism, one member of several local and national environmental groups observed that "environmental protection is always a balancing act." After all is said and done, that is the final word.

Appendix

*Notes on the
Methodology and Interviews*

Employing the methodology of Max Weber, the famous sociologist and economist, I have grounded my study of the environmental movement on observed "correlations" of events, institutions, attitudes, and values. Weber called these correlations "ideal types." An ideal type is not a moral category but rather a conceptual construct that amplifies and accentuates particular characteristics within historical events, institutions, or systems of thought in order to represent them more elegantly. Ideal types put diverse events, institutions, attitudes, and values into their "most consistent and logical forms."[1] Social life is not a neat and tidy affair that slides cleanly into our conceptual containers. Hence, achieving consistency and logic in our theories, Weber noted, entails doing a certain amount of "violence to historical reality."[2] This interpretive violence is regrettable but necessary if our historical narratives and social analyses are to yield meaningful overviews.

Viewed negatively, ideal types are Procrustean beds that make social reality conform to our categories by lopping off the more unwieldy, and perhaps more interesting, features. Viewed positively, ideal types are the conceptual lenses we require to create focused images of a diffuse reality. They allow us to grasp at meaning and project trends. Without these lenses, the complex interplay of cultural and material forces that produce social history would remain an unintelligible blur.

My brief characterization of the first three waves of environmentalism and my extended account of the fourth wave are submitted as ideal types. I have made an effort to mitigate the interpretive violence of the study by keeping it firmly tethered to the ground. With this in mind, three vantage points were employed to "triangulate" my research.

First, the publications of environmental organizations were studied over a period of six years, from 1993 to 1998. I examined the direct mail solicitations, pamphlets, letters to members, newsletters, and magazines of over thirty different environmental groups. These were local, state, regional, and national environmental organizations, with the latter more heavily represented because local, state, and regional groups tend to publish less than the nationals, if they publish at all. Such publications obviously speak with a bias. Organization periodicals have something to sell—ideas, commitments, proposals, practices. They actively engaged in self-promotion. Nonetheless, these publications were a window to leadership and volunteer values and activities, policy orientations, practical achievements, and rhetorical tactics. A list of the consulted periodicals appears in the bibliography.

As a second vantage point for research, I consulted numerous scholarly sources. These included broad-based survey data and polls of environmental movement members and the general public. Surveys and polls were analyzed to investigate trends and verify conclusions drawn from the interviews and organizational publications. I also relied extensively on secondary literature written by historians, observers, analysts, theorists, and critics of the environmental movement. This provided a crucial source of information and served to anchor the data from the interviews and organizational publications within a wider, and often more critical, framework. The bibliography provides a selected listing of these sources.

Third, in-depth interviews were conducted with seventy-six individuals in 1995. Interviewees were randomly selected from membership lists provided by organizations or identified by word of mouth. The interviewees included executives and other paid staff members of environmental groups, volunteer activists, nonactivists who were nonetheless members of environmental groups, and (self-declared) environmentalists who did not belong to any environmental groups. All told, the interviewees belonged to fifty-three different environmental organizations, including twenty-nine national organizations, six regional organizations, eight state organizations, and ten local organizations. The majority of individuals belonged to more than one organization. I also interviewed a number of (self-declared) nonenvironmentalists as well as active members of eleven different "anti-environmental" organizations of the Wise Use and Property Rights movements.

Most broad-based survey studies, historical accounts, and conceptual theorizations of the environmental movement do not employ the intensive interviewing necessary to put attitudes and behavior in context. They allow access to *what* is believed, but not to *how* it is believed. Lester Milbrath's important study of environmentalists' attitudes and values, for instance, is grounded solely on survey questionnaires. Milbrath himself specifically identified one of its shortcomings to be the lack of in-depth interviews that would have allowed a more thorough exploration.[3] The intensive interviews conducted for this study were undertaken to gain a deeper, contextual understanding of environmentalists' values, commitments, and practices. The interviews were semistructured in design. The intent, following Jennifer Hochschild's approach, was that they be "open enough to allow for unanticipated value judgments and unorthodox world views, but structured enough to permit comparisons among respondents and obedience to the discipline of a 'more detached and abstract understanding.' "[4]

Basic demographic data are available for seventy-one interviewees, fifty-two of which were environmentalists (including twenty-one volunteer activists, seven activists who were staff members or executives of environmental organizations, eight nonactivists who were members of environmental organizations, and sixteen nonactivists who did not belong to environmental organizations). The interviewees also included twelve nonenvironmentalists and seven activists with the Wise Use and Property Rights movements.

The small sample size of interviewees did not allow for much statistical inference (see chapter 6, note 65). Below are some simple demographic frequencies. I have noted certain characteristics of various groups that may be of interest.

The age of the interviewees ranged from the early twenties to the late seventies, with the plurality in their forties. All but two interviewees were U.S. citizens. Fifty-one percent were male, though 70 percent of volunteer environmental activists and 56 percent of paid environmental activists were male. Ninety percent of the interviewees were white.

Over 60 percent of the interviewees were married. Forty-five percent of the interviewees had no children. Over 80 percent of volunteer environmental activists were married, and over 50 percent of these volunteers had no children. None of the paid environmental activists had

Foundings of Environmental Organizations, 1860–1990

1860 1880 1900 1920 1940 1960 1980 2000

+ American Society for the Prevention of Cruelty to Animals (1866)
    + Boone and Crockett Club (1885)
    + New York Audubon Society (1886)
      + Sierra Club (1892)
        + American Scenic and Historic Preservation Society (1895)
          + National Audubon Society (1905)
            + American Game Protective Association (1911)
              + National Parks and Conservation Association (1919)
                + Izaak Walton League (1922)
                  + The Wilderness Society (1935)
                    + National Wildlife Federation (1936)
                      + Ducks Unlimited (1937)
                        + Defenders of Wildlife (1947)
                          + The Nature Conservancy (1951)
                            + World Wildlife Fund (1961)
                              + Environmental Defense Fund (1967)
                                + Zero Population Growth (1968)
                                  + Friends of the Earth (1969)
                                    + National Resources Defense Council (1970)

+ Environmental Action (1970)
+ Sierra Club Legal Defense Fund (1971)
+ Clean Water Action (1971)
+ Greenpeace (1971)
+ Cultural Survival (1972)
+ Negative Population Growth (1972)
+ Cousteau Society (1973)
+ Environmental Policy Institute (1974)
+ Worldwatch Institute (1975)
+ Sea Shepherd Conservation Society (1977)
+ People for the Ethical Treatment of Animals (1980)
+ Earth First! (1980)
+ Citizens Clearinghouse for Hazardous Waste (1981)
+ Co-op America (1982)
+ Earth Island Institute (1982)
+ National Toxics Campaign (1984)
+ Rainforest Action Network (1985)
+ The Conservation Fund (1985)
+ Rainforest Alliance (1986)
+ Conservation International (1987)

children. Fifty-five percent of the interviewees identified themselves as Christian, 35 percent as nonreligious, and 10 percent as Jewish or belonging to other denominations. Forty-seven percent of the volunteer environmental activists and 85 percent of the paid environmental activists considered themselves nonreligious.

Twenty-six percent of the interviewees had annual incomes under $15,000, 37 percent earned between $15,000 and $30,000 per year, 27 percent earned between $30,000 and $50,000, and 10 percent earned over $50,000 per year. Sixty-three percent of the interviewees had obtained at least a bachelor's degree. However, 85 percent of volunteer environmental activists and all of the paid environmental activists had at least a bachelor's degree. Fifty-six percent of interviewees were registered Democrats, 24 percent were Republican, 10 percent were Independent, 3 percent had other partisan affiliations, and 7 percent claimed no affiliation. Among the self-identified environmentalists, however, 71 percent were Democrats, with 14 percent being Republicans and 15 percent Independent or having other partisan affiliations. Six out of the seven Wise Use and Property Rights activists were Republicans; one was Independent.

Notes

Preface

1. Gregg Easterbrook, *A Moment on the Earth: The Coming Age of Environmental Optimism* (New York: Viking, 1995), p. 53.

2. The Earth Works Group, *50 Simple Things You Can Do to Save the Earth* (Berkeley: Earthworks Press, 1989). See also Jeremy Rifkin, ed., *The Green Lifestyle Handbook: 1001 Ways You Can Heal the Earth* (New York: Henry Holt, 1990).

Introduction

1. See Russell J. Dalton, *The Green Rainbow: Environmental Groups in Western Europe* (New Haven, Conn.: Yale University Press, 1994).

2. John Gowdy, *Coevolutionary Economics: The Economy, Society and the Environment* (Boston: Kluwer, 1994), pp. 22, 98.

3. *Sierra*, November–December 1994, p. 22.

4. David Pepper, *The Roots of Modern Environmentalism* (London: Croom Helm, 1984), p. 213.

5. In the spring of 1998, for example, the Sierra Club was subject to a heated internal debate and held a referendum on whether the club should identify immigration as a partial cause of U.S. overpopulation and seek to limit it. By a 3-to-2 margin, Sierra Club members voted to reject any official position on immigration. Club president Adam Werbach threatened to resign if the action passed.

6. Martin W. Lewis, *Green Delusions* (Durham: Duke University Press, 1992), p. 41.

7. Donald Snow, *Inside the Environmental Movement: Meeting the Leadership Challenge* (Washington, D.C.: Island Press, 1992), p. 181.

8. G. Jon Roush, "Conservation's Hour," in *Voices from the Environmental Movement: Perspectives for a New Era*, ed. Donald Snow (Washington, D.C.: Island Press, 1992), p. 35.

9. *Wilderness*, Spring 1995, p. 28.

10. Wendell Berry, *The Unsettling of America: Culture and Agriculture* (San Francisco: Sierra Club, 1986), p. 47.

11. See Robert O. Keohane and Joseph S. Nye, *Power and Interdependence: World Politics in Transition* (Boston: Little, Brown, 1989).

12. Jim MacNeill, Pieter Winsemius, and Taizo Yakushiji, *Beyond Interdependence: The Meshing of the World's Economy and the Earth's Ecology* (New York: Oxford University Press, 1991), p. 4.

13. Donald Worster, *Nature's Economy: A History of Ecological Ideas* (Cambridge: Cambridge University Press, 1994), p. 429.

14. Michael Oppenheimer, "Context, Connection, and Opportunity in Environmental Problem Solving," *Environment* 37, 5 (June 1995), pp. 10, 12.

15. Deb Callahan, "Message from the President," *LCV Insider*, Fall 1997, p. 1.

16. John S. Dryzek, *The Politics of the Earth: Environmental Discourses* (Oxford: Oxford University Press, 1997), p. 123.

17. "On résiste à l'invasion des armées, on ne résiste pas à l'invasion des idées." Victor Hugo, *Histoire d'un Crime*, Oeuvres Completes. Paris: *Imprimerie Nationale, 1904–52* Series E, vol. 2, p. 187.

18. Fairfield Osborn, *Our Plundered Planet* (Boston: Little, Brown, 1948), p. 193.

19. *Audubon*, July–August 1995, p. 6.

20. Here I take issue with Mark Dowie's *Losing Ground: American Environmentalism at the Close of the Twentieth Century* (Cambridge: MIT Press, 1995).

21. Here I take issue with Bryan G. Norton's *Toward Unity among Environmentalists* (New York: Oxford University Press, 1991).

22. Here I take issue with Gregg Easterbrook's *A Moment on the Earth: The Coming Age of Environmental Optimism* (New York: Viking, 1995).

Chapter I

1. Peter Kalm's *Travels into North America* (1753), quoted in Robert McHenry and Charles Van Doren, eds., *A Documentary History of Conservation in America* (New York: Praeger Publishers, 1972), pp. 168–172.

2. William Bradford, quoted in Daniel G. Payne, *Voices in the Wilderness: American Nature Writing and Environmental Ethics* (Hanover, NH: University Press of New England, 1996), p. 9. See also Roderick Nash, *Wilderness and the American Mind*, revised edition (New Haven, Conn.: Yale University Press, 1973), p. 24.

3. Roderick Frazier Nash, *The Rights of Nature: A History of Environmental Ethics* (Madison: University of Wisconsin Press, 1989), p. 35.

4. Gifford Pinchot, *The Fight for Conservation* (New York: Doubleday, Page, 1910), p. 79.

5. Ibid., p. 48.

6. Nash, *Wilderness and the American Mind*, p. 149.

7. Pinchot, *Fight for Conservation*, pp. 27, 42.

8. Aldo Leopold, *A Sand County Almanac, with Essays on Conservation from Round River* (New York: Ballantine Books, 1966), p. 240.

9. Ibid., p. 262.

10. Donald Worster, *Nature's Economy* (Cambridge: Cambridge University Press, 1994), p. 284.

11. "Environment" was not included in the 1955 *New York Times Index*. A single citation, for "environmental science," can be found in the 1960 index. Neither *environmentalist* nor *environmentalism* made it into the 1971 *American Heritage Dictionary*, though both *conservation* ("the official preservation of natural resources, such as topsoil, forests, and waterways") and *conservationist* ("one who practices or advocates the preservation of natural resources") are included.

12. Samuel P. Hays, *Beauty, Health, and Permanence: Environmental Politics in the United States, 1955–1985* (Cambridge: Cambridge University Press, 1987), p. 13. See also Samuel P. Hays, *Conservation and the Gospel of Efficiency: The Progressive Conservation Movement* (Cambridge: Harvard University Press, 1958); and Robert C. Paehlke, *Environmentalism and the Future of Progressive Politics* (New Haven, Conn.: Yale University Press, 1989), pp. 146ff.

13. Bryan G. Norton, *Toward Unity among Environmentalists* (New York: Oxford University Press, 1991), p. 123.

14. Quoted in Peter Borrelli, "The Ecophilosophers," in *Crossroads: Environmental Priorities for the Future*, ed. Peter Borrelli (Washington, D.C.: Island Press, 1988), p. 69.

15. Paul R. Ehrlich, *The Population Bomb* (New York: Ballantine Books, 1968), p. 9.

16. Robert Gottlieb, *Forcing the Spring: The Transformation of the American Environmental Movement* (Washington, D.C.: Island Press, 1993), p. 257.

17. Senator Gaylord Nelson, "History of Earth Day," speech given at the University of Illinois, October 6, 1990, excerpt published in *The Wilderness Society*, Mark Dowie, *Losing Ground: American Environmentalism at the Close of the Twentieth Century* (Cambridge: MIT Press, 1995), p. 24.

18. Samuel P. Hays, "From Conservation to Environment: Environmental Politics in the United States since World War II," in *Environmental History: Critical Issues in Comparative Perspective*, ed. Kendall E. Bailes, (Lanham, Md.: University Press of America), p. 214.

19. Quoted in Philip Shabecoff, *A Fierce Green Fire: The American Environmental Movement* (New York: Hill and Wang, 1993), p. 115.

20. Quoted in Dowie, *Losing Ground*, p. 25.

21. In *Time* magazine, quoted in Dowie, *Losing Ground*, p. 32.

22. Barry Commoner, *The Closing Circle* (New York: Bantam, 1972).

23. Barry Commoner, "How Poverty Breeds Overpopulation (and Not the Other Way Around), *Ramparts* (1974), pp. 21–25, 58–59, excerpted in *Ecology: Key Concepts in Critical Theory*, ed. Carolyn Merchant (Atlantic Highlands, N.J.: Humanities Press, 1994), pp. 88–95; Barry Commoner, *Making Peace with the Planet* (New York: Pantheon Books, [1975] 1990).

24. Quoted in Garrett Hardin, *Living within Limits: Ecology, Economics, and Population Taboos* (New York: Oxford University Press, 1993), p. 37.

25. Donella H. Meadows, Dennis L. Meadows, Jorgen Randers, and William W. Behrens, III, *The Limits to Growth* (New York: Universe Books, 1974), pp. 24, 190.

26. Ibid., p. 196.

27. *Sierra*, March–April 1995, p. 59; *The Planet*, December 1997, p. 3.

28. Victor B. Scheffer, *The Shaping of Environmentalism in America* (Seattle: University of Washington Press, 1991), p. 113.

29. See Sherry Cable and Charles Cable, *Environmental Problems/Grassroots Solutions: The Politics of Grassroots Environmental Conflict* (New York: St. Martin's Press, 1995), pp. 79–84; Jacqueline Vaughn Switzer with Gary Bryner, *Environmental Politics: Domestic and Global Dimensions*, 2nd edition (New York: St. Martin's Press, 1998), p. 12.

30. Gottlieb, *Forcing the Spring*, p. 122.

31. Quoted in Kirkpatrick Sale, *The Green Revolution: The American Environmental Movement 1962–1992* (New York: Hill and Wang, 1993), p. 58.

32. Lois Marie Gibbs and Karen J. Stults, "On Grassroots Environmentalism," in *Crossroads: Environmental Priorities for the Future*, ed. Peter Borrelli (Washington, D.C.: Island Press, 1988), p. 242.

33. Dowie, *Losing Ground*, p. 133; *Everyone's Backyard*, Summer 1996, p. 3.

34. Joseph M. Petulla, *American Environmentalism: Values, Tactics, Priorities* (College Station: Texas A & M University Press, 1980), p. 228.

35. Arne Naess, "The Shallow and the Deep, Long-Range Ecology Movement: A Summary," *Inquiry* 16 (1973): 95–100, reprinted in *Ecology: Key Concepts in Critical Theory*, ed. Carolyn Merchant (Atlantic Highlands, N.J.: Humanities Press, 1994), p. 120.

36. "Movement on the Move," in *Ten Years of Triumph* (Falls Church, Va.: Citizens Clearinghouse for Hazardous Waste, 1993), p. 2.

37. David Helvarg, *The War against the Greens: The "Wise-Use" Movement, the New Right, and Anti-environmental Violence* (San Francisco: Sierra Club Books, 1994), p. 430.

38. Quoted in Sale, *Green Revolution*, p. 56.

39. Tom Athanasiou, *Divided Planet: The Ecology of Rich and Poor* (Boston: Little, Brown, 1996), p. 22.

40. Brian Tokar, *Earth for Sale: Reclaiming Ecology in the Age of Corporate Greenwash* (Boston: South End Press, 1997), p. xii.

41. Quoted in Stephen Fox, *John Muir and His Legacy: The American Conservation Movement* (Boston: Little, Brown, 1981), p. 182.

42. Quoted in Sale, *The Green Revolution*, p. 25.

43. Scheffer, *Shaping of Environmentalism in America*, pp. 19, 128.

44. Citizens Clearinghouse for Hazardous Waste, "Stop Dioxin Exposure Campaign," p. 5.

45. Manes, *Green Rage*, p. 56.

46. Wolfgang Sachs, "Global Ecology and the Shadow of 'Development,' " in *Deep Ecology for the 21st Century*, ed. George Sessions (Boston: Shambala, 1995), pp. 435–36.

47. Donald Snow, *Inside the Environmental Movement: Meeting the Leadership Challenge* (Washington, D.C.: Island Press, 1992), p. xxxiii; Gottlieb, *Forcing the Spring*, p. 208; Nicholas Freudenberg and Carol Steinsapir, "Not in Our Backyards: The Grassroots Environmental Movement," in *American Environmentalism: The U.S. Environmental Movement 1970–1990*, ed. Riley Dunlap and Angela Mertig (New York: Taylor and Francis, 1992), p. 29; Joni Seager, *Earth Follies: Coming to Feminist Terms with the Global Environmental Crisis* (New York: Routledge, 1993), p. 186.

48. Quoted in Timothy W. Luke, *Ecocritique: Contesting the Politics of Nature, Economy, and Culture* (Minneapolis: University of Minnesota, 1997), p. 36.

49. See Gottlieb, *Forcing the Spring*, p. 316.

50. See Shabecoff, *Fierce Green Fire*, p. 279; Dowie, *Losing Ground*.

51. Sale, *Green Revolution*, p. 8.

52. Rodger Schlickeisen, personal interview, Washington, D.C., September 8, 1995.

53. See Christopher Manes, *Green Rage: Radical Environmentalism and the Unmaking of Civilization* (Boston: Little, Brown, 1990), p. 72.

54. *Outside*, March 1994, pp. 67–72.

55. G. Jon Roush, "Introduction" and "Conservation's Hour," in *Voices from the Environmental Movement: Perspectives for a New Era*, ed. Donald Snow (Washington, D.C.: Island Press, 1992), pp. 6, 32.

56. *Nature Conservancy*, November–December 1995, p. 5.

57. *The Planet*, January–February 1998, p. 16.

58. *The Planet: Sierra Club Activist Resource*.

59. Snow, *Inside the Environmental Movement*, p. 98. Examine, for instance, the endnotes to Tokar's *Earth for Sale*.

60. See Laura R. Woliver, *From Outrage to Action: The Politics of Grass-Roots Dissent* (Urbana: University of Illinois Press, 1993), p. 73.

61. Jonathan B. Cook, "Managing Nonprofits of Different Sizes," in *Educating Managers on Nonprofit Organizations*, ed. Michael O'Neill and Dennis R. Young (New York: Praeger, 1988), p. 103, quoted in Snow, *Inside the Environmental Movement*, pp. 161–62.

62. Snow, *Inside the Environmental Movement*, p. 11.

63. Ibid, p. 112.

64. *Sierra*, May–June 1995, p. 83.

65. *Friends of the Earth*, 1996 annual report, Winter 1997, p. 13.

66. Snow, *Inside the Environmental Movement*, pp. xxiv, 23.

67. Dowie, *Losing Ground*, p. 5.

68. Christopher J. Bosso, "After the Movement: Environmental Activism in the 1990s," in *Environmental Policy in the 1990s*, 2nd edition, ed. Norman J. Vig and Michael E. Kraft, (Washington, D.C.: CQ Press, 1994), p. 38.

69. See Jim Maddy, "Changing the Balance," *Defenders*, Winter 1994/95, p. 33.

70. Christopher J. Bosso, "Seizing Back the Day: The Challenge to Environmental Activism in the 1990s." In *Environmental Policy in the 1990s*, 3rd edition, ed. Norman J. Vig and Michael E. Kraft, (Washington, D.C.: CQ Press, 1997), p. 62.

71. Fox, *John Muir and His Legacy*, p. 333.

Chapter 2

1. See *Earth Island Journal*, Spring 1996, pp. 6, 8; "Setting the Record Straight on Greenpeace, Tuna and Dolphins," Greenpeace mailing, October 1996; Penn Loh, "Creating an Environment of Blame" in *Groundwork* 6 (1996): 22–23, 55–56; *Everyone's Backyard*, Spring 1995, p. 25.

2. Brent Blackwelder, personal interview, Washington, D.C., September 8, 1995.

3. e b bortz, "Ecology and Consciousness," *Synthesis/Regeneration*, Winter 1998, p. 20.

4. Carolyn Merchant, "Introduction," in *Ecology: Key Concepts in Critical Theory*, ed. Carolyn Merchant (Atlantic Highlands, N.J.: Humanities Press, 1994), p. 20. See also Carolyn Merchant, *Earthcare* (New York: Routledge, 1995).

5. *Nature Conservancy Reporter*, Fall 1996, p. 2.

6. The Wilderness Society, *Annual Report 1994*, p. 2.

7. Donald Snow, *Inside the Environmental Movement: Meeting the Leadership Challenge* (Washington, D.C.: Island Press, 1992), p. 159.

8. Michael McCloskey, "Twenty Years of Change in the Environmental Movement: An Insider's View," in *American Environmentalism: The U.S. Environmental Movement 1970–1990*, ed. Riley Dunlap and Angel Mertig (New York: Taylor and Francis, 1992), p. 85.

9. Ibid.

10. Jane Perkins, Friends of the Earth president, newsletter of June 6, 1994.

11. McCloskey, "Twenty Years of Change," p. 85.

12. *Outside*, October 1996, p. 50.

13. *Earth First!*, February—March 1997, p. 22.

14. Roger Schlickeisen, personal interview, Washington, D.C., September 8, 1995.

15. Quoted in Dick Russell, "The Monkeywrenchers," in *Crossroads: Environmental Priorities for the Future*, ed. Peter Borrelli (Washington, D.C.: Island Press, 1988), p. 30.

16. Blackwelder interview.

17. Quoted in Christopher Manes, *Green Rage: Radical Environmentalism and the Unmaking of Civilization* (Boston: Little, Brown, 1990), p. 18.

18. Mary Hanley, personal interview, Washington, D.C., September 8, 1995.

19. Donald Worster, *Nature's Economy* (Cambridge: Cambridge University Press, 1994), pp. 18, 139, 157.

20. Quoted in Manes, *Green Rage*, p. 18.

21. Murray Bookchin and Dave Foreman, *Defending the Earth* (Boston: South End Press, 1991), p. 39. See also Dave Foreman, *Confessions of an Eco-Warrior* (New York: Crown Trade Paperbacks, 1991), p. 139.

22. Gary McFarlane and Darryl Echt, "Cult of Non-violence," *Earth First!*, November–December 1997, pp. 3, 17.

23. Quoted in Roderick Frazier Nash, *The Rights of Nature: A History of Environmental Ethics* (Madison: University of Wisconsin Press, 1989), p. 191.

24. *Outside*, November 1995, pp. 74–75.

25. Quoted in *Outside*, August 1996, p. 28; Brian Tokar, *Earth for Sale: Reclaiming Ecology in the Age of Corporate Greenwash* (Boston: South End Press, 1997), p. 155.

26. *Earth First!*, May–June 1997, p. 9.

27. Martin W. Lewis, *Green Delusions* (Durham: Duke University Press, 1992), pp. 42, 250.

28. Robert Young, " 'Monkeywrenching' and the Processes of Democracy," in *Ecology and Democracy*, ed. Freya Mathews (London: Frank Cass, 1996), pp. 199–214.

29. John Sawhill, President, *Nature Conservancy*, March/April 1995, p. 5.

30. *Nature Conservancy Reporter*, Summer 1996, p. 2.

31. Fred Powledge, "A Time of Change—and Promise," in *Gale Environmental Almanac*, ed. Russ Hoyle (Detroit: Gale Research, 1993), pp. 173–74.

32. *Sierra*, May/June 1996, p. 14.

33. Ducks Unlimited membership appeal, direct mail flyer, April 1996.

34. *Sierra*, September/October 1996, p. 47.

35. Dunlap and Mertig, *American Environmentalism*, p. 7.

36. Jerome Frank and Earl Nash, "Commitment to Peace Work," *American Journal of Orthopsychiatry* 35 (1965): 115; A. F. C. Beales, *The History of Peace: A Short Account of the Organized Movements for International Peace* (London: G. Bell and Sons, 1931), p. 45.

37. Robert C. Paehlke, *Environmentalism and the Future of Progressive Politics* (New Haven: Yale University Press, 1989), p. 213.

38. Zero Population Growth letter to members, November 15, 1994.

39. Environmental Careers Organization, *Beyond the Green: Redefining and Diversifying the Environmental Movement* (Boston: Environmental Careers Organization, 1992).

40. Interview in *The Workbook* (Southwest Research and Information Center), Spring 1992, pp. 20–21.

41. Bryan G. Norton, *Toward Unity among Environmentalists* (New York: Oxford University Press, 1991), p. 202.

42. The seminal article is Paul R. Ehrlich and Peter Raven, "Butterflies and Plants: A Study in Co-evolution," *Evolution*, 18 (1965): 586–608.

43. Fritjof Capra, "Systems Theory and the New Paradigm," in *Ecology: Key Concepts in Critical Theory*, ed. Carolyn Merchant (Atlantic Highlands, N.J.: Humanities Press, 1994), p. 335. See also Fritjof Capra and Charlene Spretnak, *Green Politics* (New York: Dutton, 1984), p. xix.

44. Fritjof Capra, *The Turning Point* (New York: Simon and Schuster, 1982), p. 16.

45. Fritjof Capra, *The Web of Life: A New Scientific Understanding of Living Systems* (New York: Doubleday, 1996), p. 298.

46. Ibid., p. 301.

47. Gregory Bateson, *Steps to an Ecology of Mind* (New York: Ballantine, 1972), p. 451.

48. R. Lewin, "In Ecology, Change Brings Stability," *Science* 234 (1986): 1071–73; Stuart Pimm, *The Balance of Nature?* (Chicago: University of Chicago Press, 1991).

49. *Calypso Log*, April 1994, p. 20.

50. Donald Windsor, "Endangered Interrelationships," *Wildearth*, Winter 1995/96, pp. 78–83.

51. *Nature Conservancy*, January/February 1996, p. 21. See also Donald Worster, "Nature and the Disorder of History," in *Reinventing Nature*, ed. Michael E. Soulé and Gary Lease (Washington, D.C.: Island Press, 1995). Already

in the 1960s, writers such as Rachel Carson recognized that "[t]he balance of nature is not a *status quo*; it is fluid, ever shifting, in a constant state of adjustment." Rachel Carson, *Silent Spring* (Boston: Houghton Mifflin, 1962), p. 246.

52. *Florida Naturalist*, Spring 1998, p. 8.

53. Dave Foreman, "Wilderness: From Scenery to Nature," in *Wildearth*, Winter 1995/96, p. 11.

54. T. H. Watkins, *The Wilderness Year 1996* (Washington, DC: Wilderness Society, 1997), pp. 10, 34. See also Josh Gordon and Jane Coppock, "Ecosystem Management and Economic Development," in *Thinking Ecologically: The Next Generation of Environmental Policy*, ed. Marian R. Chertow and Daniel C. Esty (New Haven: Yale University Press, 1997), pp. 37–48.

55. John Muir, *My First Summer in the Sierra* (Boston: Houghton Mifflin, 1911), p. 157.

56. *Sierra*, January/February 1996, p. 21.

57. Gregg Easterbrook, *A Moment on the Earth: The Coming Age of Environmental Optimism* (New York: Viking, 1995), p. 98.

58. *Focus*, World Wildlife Fund, January/February 1996, p. 2.

59. *Nature Conservancy*, March/April 1996, p. 5.

60. National Wildlife Federation, *1994 Annual Report*, p. 2.

61. Peter Vitousek, Paul Ehrlich, Anne Ehrlich, and Pamela Matson, "Human Appropriation of the Products of Photosynthesis," *BioScience* 36 (1986): 368–73.

62. Lester Brown, Christopher Flavin, and Sandra Postel, *Saving the Planet: How to Shape an Environmentally Sustainable Global Economy* (New York: Norton, 1991), p. 116.

63. Michael Soulé, "Thresholds for Survival: Maintaining Fitness and Evolutionary Potential," in *Conservation Biology: An Evolutionary-Ecological Perspective*, ed. Michael Soulé and Bruce Wilcox (Sunderland, Mass.: Sinauer, 1980), pp. 166, 168.

64. David Brower, *Let the Mountains Talk, Let the Rivers Run: A Call to Those Who Would Save the Earth* (New York: HarperCollins, 1995), p. 99.

65. *Wildearth*, Winter 1995/96, p. i.

66. Richard Norgaard, "Coevolution of Economy, Society and Environment," in *Real-life Economics*, ed. Paul Ekins and Manfred Max-Neef (London: Routledge, 1992), pp. 781–82, 786.

67. *National Wildlife*, February–March 1995, p. 35.

68. Aldo Leopold, *A Sand County Almanac, with Essays on Conservation from Round River* (New York: Ballantine Books, 1966), p. 263.

69. *Audubon*, January/February, 1996, p. 101. See also *National Wildlife*, February–March 1996, pp. 42–46.

70. See Norgaard, "Coevolution," pp. 76–86.

71. World Commission on Environment and Development, *Our Common Future* (Oxford: Oxford University Press, 1987), pp. 43, 48.

72. Donald Mann, president of Negative Population Growth, quoted in Garrett Hardin, *Living within Limits: Ecology, Economics, and Population Taboos* (New York: Oxford University Press, 1993), p. 206.

73. Worster, "Shaky Ground of Sustainability," pp. 417–18, 424.

74. Wolfgang Sachs, "Global Ecology and the Shadow of 'Development,' " in *Deep Ecology for the 21st Century*, ed. George Sessions (Boston: Shambala, 1995), p. 434.

75. John Bellamy Foster, *The Vulnerable Planet: A Short Economic History of the Environment* (New York: Monthly Review Press, 1994), p. 131.

76. Quoted in Hardin, *Living within Limits*, p. 190.

77. *Sustainable Development: A New Consensus*, final report of the President's Council on Sustainable Development. Internet, 1995. http://www.whitehouse.gov/PCSD.

78. William Ophuls and A. Stephen Boyan, Jr., *Ecology and the Politics of Scarcity Revisited: The Unraveling of the American Dream* (New York: Freeman, 1992), pp. 237–39.

79. T. H. Watkins, *Audubon*, September–October 1994, p. 44.

80. *An Environmental Agenda for the Future* (Washington, D.C.: Island Press, 1985), p. 7.

81. Worster, "Shaky Ground of Sustainability," p. 417.

82. Ismail Serageldein and Richard Barnett, eds., *Ethics and Spiritual Values: Promoting Environmentally Sustainable Development* (Washington, D.C.: The World Bank, 1996), pp. 10–11.

83. United Nations Development Program study, reported in the *New York Times*, June 5, 1990, p. A12.

84. Herman Daly, *Steady-State Economics* (San Francisco: Freeman, 1977), p. 17.

85. Herman E. Daly, "The Steady-state Economy: Toward a Political Economy of Biophysical Equilibrium and Moral Growth," in *Toward a Steady-State Economy* (San Francisco: W. H. Freeman, 1973), p. 167.

86. Paehlke, *Environmentalism*, p. 213.

87. Quoted in Norton, *Toward Unity among Environmentalists*, p. 115.

88. Ibid., p. 119.

89. Ophuls and Boyan, *Politics of Scarcity Revisited*, p. 15.

90. See Joel Jay Kassiola, *The Death of Industrial Civilization: The Limits to Economic Growth and the Repoliticization of Advanced Industrial Society* (Albany: SUNY Press, 1990), p. 47.

91. See, for instance, "Measuring True Progress," *Green Politics*, Spring 1996, p. 2.

92. David Helvarg, *The War against the Greens: The "Wise-Use" Movement, the New Right, and Anti-environmental Violence* (San Francisco: Sierra Club Books, 1994), p. 431.

93. Norton, *Toward Unity among Environmentalists*, p. 253.

94. Ursula Mueller, "Swedish Greens: Our Basic Ideas," *Synthesis/Regeneration*, Spring 1998, p. 34.

95. Quoted in Charles Birch and John Cobb, *The Liberation of Life: From the Cell to the Community* (Cambridge: Cambridge University Press, 1981), p. 29.

96. Leopold, *Sand County Almanac*, p. 239.

97. The first of the Principles of Environmental Justice composed at the first National People of Color Environmental Leadership Summit reads: "Environmental justice affirms the sacredness of Mother Earth, ecological unity and the interdependence of all species, and the right to be free from ecological destruction." The principles appear as appendix C in Jim Schwab, *Deeper Shades of Green: The Rise of Blue-Collar and Minority Environmentalism in America* (San Francisco: Sierra Club Books, 1994), pp. 441–43.

98. David Harvey, *The Condition of Postmodernity: An Inquiry into the Origins of Cultural Change* (Cambridge: Blackwell Publishers, 1989), p. 218.

Chapter 3

1. Quoted in Al Gore, *Earth in the Balance: Ecology and the Human Spirit* (Boston: Houghton Mifflin, 1992), p. 263.

2. See Willett Kempton, James Boster, and Jennifer Hartley, *Environmental Values in American Culture* (Cambridge: MIT Press, 1995), p. 219. "Future focus" is one of the U.S. Greens' Ten Key Values. The others are ecological wisdom, social justice, grassroots democracy, nonviolence, decentralization, community-based economics, feminism, respect for diversity, and personal and global responsibility.

3. The original version of the phrase has been attributed to David Brower, *Let the Mountains Talk, Let the Rivers Run: A Call to Those Who Would Save the Earth* (New York: HarperCollins, 1995), p. 1.

4. Carl Pope, in *Sierra*, May–June 1994, pp. 14–15.

5. *Amicus Journal*, Winter 1996, p. 33.

6. Editorial by President John Sawhill, *Nature Conservancy*, September–October 1994, p. 5.

7. *Audubon*, September–October 1995, p. 6.

8. Aldo Leopold, *A Sand County Almanac, with Essays on Conservation from Round River* (New York: Ballantine Books, 1966), p. 117.

9. Edmund Burke, *Reflections on the Revolution in France* (Garden City: Doubleday, 1961), p. 110.

10. See *Friends of the Earth*, January–February 1996, p. 6; *Outside*, June 1996, p. 30; Robert Booth Fowler, *The Greening of Protestant Thought* (Chapel Hill: University of North Carolina Press, 1995). I discuss these issues in chapter 6.

11. Robert C. Paehlke, *Environmentalism and the Future of Progressive Politics* (New Haven: Yale University Press, 1989), p. 158.

12. Steve Trombulak, Reed Noss, and Jim Strittholt, "Obstacles to Implementing the Wildlands Project Vision," *Wild Earth*, Winter 1995/96, p. 86.

13. *Science News*, September 11, 1993, p. 169. See also D. Foreman, J. Davis, D. Johns, R. Noss, and M. Soulé, "The Wildlands Project Mission Statement," *Wild Earth*, special issue, 1992, pp. 3–4.

14. *The NPG Forum*, August 1995, p. 3.

15. For a similar thought experiment, see Lester W. Milbrath, *Envisioning a Sustainable Society* (Albany: SUNY Press, 1989), p. 2.

16. Christopher Manes, *Green Rage: Radical Environmentalism and the Unmaking of Civilization* (Boston: Little, Brown, 1990), p. 25.

17. Quoted in Roderick Frazier Nash, *The Rights of Nature: A History of Environmental Ethics* (Madison: University of Wisconsin Press, 1989), p. 66.

18. Rep. William Dannemeyer (R-California), statement given at the 1992 Wise Use Leadership Conference, Reno, Nevada. Reported in *Sierra*, November–December 1992, p. 61.

19. Stephen J. Gould, *Bully for Brontosaurus: Reflections in Natural History* (New York: W. W. Norton, 1991), pp. 16–17, 365.

20. Edward O. Wilson, *Biophilia* (Cambridge: Harvard University Press, 1984), p. 121.

21. *Focus*, World Wildlife Fund, May–June 1993, p. 7.

22. Quoted in *Focus*, World Wildlife Fund, March–April 1996, p. 6.

23. See Hilary F. French, "Learning from the Ozone Experience," in Lester R. Brown et al., *State of the World 1997* (New York: Norton, 1997), pp. 151–71; *Nucleus*, Winter 1995/96, pp. 1–3; "Can We Save Our Skins?" *Friends of the Earth*, July–August 1996, pp. 8–11.

24. Manes, *Green Rage*, p. 241.

25. *Synthesis/Regeneration: A Magazine of Green Social Thought*, Winter 1996, pp. 33–36; Richard L. Brodsky and Riclard L. Russman, "A Constitutional Initiative," *Defenders*, Fall 1996, pp. 37–38; Winona LaDuke, "Looking Ahead," *Earth Island Journal*, Spring 1998, p. 42.

26. William A. Gamson, *Talking Politics* (Cambridge: Cambridge University Press, 1992), pp. 7, 114.

27. Ibid., p. 85.

28. Matthias Finger, "From Knowledge to Action? Exploring the Relationships between Environmental Experiences, Learning, and Behavior," *Journal of Social Issues* 50, 3 (Fall 1994): 159.

29. Patrick Watson, *Dimensions*, 5 (March 1990): 3.

30. Paul Harrison, *The Third Revolution: Population, Environment, and a Sustainable World* (New York: Penguin, 1993), p. 305.

31. Gore, *Earth in the Balance*, p. 371.

32. Bryan G. Norton, *Toward Unity among Environmentalists* (New York: Oxford University Press, 1991), p. 121.

33. Kempton et al., *Environmental Values in American Culture*, pp. 101–2; see also p. 128.

34. David Durenberger, "A Dissenting Voice," *EPA Journal* (March–April 1991), in *Taking Sides: Clashing Views on Controversial Environmental Issues*, 5th edition, ed. Theodore Goldfarb (Guilford, Conn.: Dushkin Publishing, 1993), p. 99.

35. Brian Tokar, *Earth for Sale: Reclaiming Ecology in the Age of Corporate Greenwash* (Boston: South End Press, 1997), pp. xiv–xv.

36. Michael Robbins, editor, *Audubon*, September–October 1994, p. 4.

37. Zero Population Growth letter to members, February 14, 1995. See also *ZPG Reporter*, March–April 1995, p. 7.

38. Jacques Yves Cousteau, *Cousteau Society* letter to members, May 1995; *Calypso Log*, April 1996, p. 18.

39. Soil Conservation Service, *Summary Report 1992 National Resources Inventory* (Washington, D.C.: U.S. Department of Agriculture, 1994); David Pimentel et al., "Environmental and Economic Costs of Soil Erosion and Conservation Benefits," *Science* 267 (February 14, 1995): 1117–23; *Population-Environment Balance*, November 1997, p. 2.

40. See World Commission on Environment and Development, *Our Common Future* (Oxford: Oxford University Press, 1987), p. 46.

41. See Sharachchandra M. Lele, "Sustainable Development: A Critical Review," *World Development* 19, 6 (June 1991): 607–21.

42. *NPG Forum*, March 1997, p. 1.

43. Gaylord Nelson, personal interview, Washington, D.C., September 8, 1995.

44. *Audubon*, September–October 1995, p. 112.

45. *Wilderness*, Fall 1995, p. 33.

46. John Barry, "Sustainability, Political Judgement and Citizenship: Connecting Green Politics and Democracy," in *Democracy and Green Political Thought*, ed. Brian Doherty and Marius de Geus (London: Routledge, 1996), pp. 128–29.

47. Mark Sagoff, *The Economy of the Earth: Philosophy, Law and the Environment* (Cambridge: Cambridge University Press, 1988), pp. 62–64.

48. Dave Foreman, *Confessions of an Eco-Warrior* (New York: Crown Trade Paperbacks, 1991), p. xi. Emphasis added.

49. Edith Brown Weiss, *In Fairness to Future Generations* (Tokyo: United Nations University, 1989), p. 38.

50. Garrett Hardin first made the assertion, "We can never do merely one thing," in 1963. See Garrett Hardin, *Living within Limits: Ecology, Economics, and Population Taboos* (New York: Oxford University Press, 1993), p. 199.

51. Lester Milbrath, "Environmental Understanding: A New Concern for Political Socialization," in *Political Socialization: Citizenship Education, and Democracy*, ed. Orit Ichilov (New York: Teachers College Press, 1990), p. 292.

52. Allan Schnaiberg, *The Environment: From Surplus to Scarcity* (New York: Oxford University Press, 1980), p. 375.

53. Frank Egler, quoted in Manes, *Green Rage*, p. 71.

54. Nancy Newhall, quoted in Brower, *Let the Mountains Talk*, p. 105.

55. Richard B. Norgaard, *Development Betrayed: The End of Progress and a Coevolutionary Revisioning of the Future* (London: Routledge, 1994), esp. pp. 11–22, 46–47. Norgaard usefully emphasizes the complexities not only of ecological systems but also of social systems embedded in natural environments.

56. Norman Myers, *Ultimate Security: The Environmental Basis of Political Stability* (Washington, D.C.: Island Press, 1993), pp. 204–5.

57. See Theo Colborn, Diane Dumanoski, and John Peterson Myers, *Our Stolen Future* (New York: Dutton Books, 1996).

58. See Chris Bright, "Tracking the Ecology of Climate Change," in Lester R. Brown et al., *State of the World 1997* (New York: Norton, 1997), pp. 78–94.

59. Wendell Berry, *Gift of Good Land*: Further Essays Cultural and Agricultural (San Francisco: North Point Press, 1981), pp. ix, 116.

60. World Wildlife Fund, *1992 Annual Report*, p. 1.

61. Richard Elliot Benedick, "Equity and Ethics in a Global Climate Convention," in *Taking Sides: Clashing Views on Controversial Environmental Issues*, 5th edition, ed. Theodore Goldfarb (Guilford, Conn.: Dushkin Publishing, 1993), p. 314.

62. Quoted in Brower, *Let the Mountains Talk*, p. 95.

63. World Commission, *Our Common Future*, p. 53.

64. As formulated by a panel of thirty-two scientists, philosophers, lawyers, and environmental activists in January 1998 convened by the Science and Environmental Health Network at Wingspread, headquarters of the Johnson Foundation near Racine, Wisconsin. See also R. Costanza, "Three General Policies to Achieve Sustainability," in *Investing in Natural Capital: The Ecological Economics Approach to Sustainability*, ed. A. Jansson, M. Hammer, C. Folke, and R. Costanza (Washington, D.C.: Island Press, 1994), pp. 392–407; Christopher Flavin and Odil Tunali, *Climate of Hope: New Strategies for Stabilizing the World's Atmosphere*, Worldwatch Paper 130 (Washington, D.C.: Worldwatch Institute, 1996), p. 20; Stewart Hudson, "Principles of Basic Tax Reform," NWF monograph (Washington, D.C.: National Wildlife Federation, 1995), p. 7.

65. Robert Costanza, "Three General Policies to Achieve Sustainability," in A. Jansson et al, ed. *Investing in Natural Capital: The Ecological Economics Approach to Sustainability* (Washington, D.C.: Island Press, 1994), p. 399.

66. *Nucleus*, Spring 1996, p. 3; Summer 1997, pp. 1–3.

67. Quoted in Jim MacNeill, Pieter Winsemius, and Taizo Yakushiji, *Beyond Interdependence: The Meshing of the World's Economy and the Earth's Ecology* (New York: Oxford University Press, 1991), pp. 17–18.

68. Gore, *Earth in the Balance*, p. 170.

69. Robert L. Heilbroner, *An Inquiry into the Human Prospect: Looked at Again for the 1990s* (New York: Norton, 1991), p. 138.

70. Quoted in Robert McHenry and Charles Van Doren, eds., *A Documentary History of Conservation in America* (New York: Praeger, 1972), p. 172.

71. Natural Resources Defense Council, 1993, cited in *E The Environmental Magazine*, November–December 1996, p. 33.

72. John S. Dryzek, *Rational Ecology: Environment and Political Economy* (New York: Basil Blackwell, 1987), p. 56.

73. William Ophuls and A. Stephen Boyan, Jr., *Ecology and the Politics of Scarcity Revisited: The Unraveling of the American Dream* (New York: Freeman, 1992), p. 219; MacNeill et al., *Beyond Interdependence*, p. 46.

74. See Clive L. Spash, "Economics, Ethics, and Long-Term Environmental Damages," *Environmental Ethics* 15 (Summer 1993): 118, 127–28.

75. David Malin Roodman, *Paying the Piper: Subsidies, Politics, and the Environment*, Worldwatch Paper 133 (Washington, D.C.: Worldwatch Institute, 1995), p. 48.

76. E. F. Schumacher, *Small Is Beautiful: Economics as if People Mattered* (New York: Harper and Row, 1973), p. 20.

77. Herman E. Daly, "Farewell Lecture to the World Bank," *Focus* Carrying Capacity Network 4, 2 (1994): 9.

78. World Wildlife Fund, *1993 Annual Report*, p. 33.

79. Wilderness Society, *The Wilderness Year 1996* 1997, p. 4.

80. *Earth First!*, June–July 1997, p. 18.

81. Jacqueline Vaughn Switzer, *Green Backlash: The History and Politics of Environmental Opposition in the U.S.* (Boulder: Lynne Rienner, 1997), p. 58.

82. Natural Resources Defense Council, "Twenty-five Years Defending the Environment," 1995, p. 5; Martin W. Lewis, *Green Delusions* (Durham: Duke University Press, 1992), p. 21; *The Pelican*, Sierra Club Florida Chapter, Fall 1996, p. 8; Roodman, *Paying the Piper*, p. 19; Mark Shaffer, *Beyond the Endangered Species Act: Conservation in the 21st Century* (Washington, D.C.: Wilderness Society, 1992), p. 7; Friends of the Earth and the National Taxpayers Union Foundation, *The Green Scissors Report: Cutting Wasteful and Environmentally Harmful Spending and Subsidies* (1995), p. 8; Ted Williams, "The Unkindest Cuts," *Audubon*, January–February 1998, pp. 24–31.

83. Natural Resources Defense Council newsletter, March 30, 1995, p. 2.

84. *Wilderness*, Winter 1995, p. 3.

85. Roodman, *Paying the Piper*, pp. 17–18; *Green Scissors Report*, p. 7.

86. *An Environmental Agenda for the Future* (Washington, D.C.: Island Press, 1985), pp. 12–13; MacNeill et al., *Beyond Interdependence*, p. 37; John

T. Preston, "Technology Innovation and Environmental Progress," in Marian R. Chertow and Daniel C. Esty, ed. *Thinking Ecologically: The Next Generation of Environmental Policy* (New Haven, Conn.: Yale University Press, 1997), pp. 140–41.

87. "Stopping Sprawl," *The Planet*, April 1997, pp. 1–6.

88. Peter Montague, "Polluted Politics and Corporate Welfare," *Earth Island Journal*, Spring 1995, p. 29.

89. *Friends of the Earth*, January–February 1996, p. 9; Friends of the Earth membership letter, February 21, 1997.

90. Friends of the Earth, *Dirty Little Secrets* (1995).

91. Hardin, *Living within Limits*, p. 238.

92. See A. Jansson, M. Hammer, C. Folke and R. Costanza. ed. *Investing in Natural Capital: The Ecological Economics Approach to Sustainability.* (Washington, D.C.: Island Press, 1994) and Robert Costanza, ed., *Ecological Economics* (New York: Columbia University Press, 1991).

93. Stephan Schmidheiny, *Changing Course: A Global Business Perspective on Development and the Environment* (Cambridge: MIT Press, 1992), p. 17.

94. Responding to a member who objected to the assertion that sport utility vehicles (as well as pickups and minivans) are unnecessary gas-guzzlers, Sierra Club executive director Carl Pope explained that "If we charged appropriately for gasoline and gas-guzzlers . . . then people who genuinely need large vehicles could still buy them . . . and the rest of us would have an incentive to buy the right car for our needs." *Sierra*, May/June 1998, p. 14.

95. Quoted in Schmidheiny, *Changing Course*, p. 28.

96. Quoted ibid., p. xi.

97. Quoted ibid., p. 14.

98. P. Faeth, *Paying the Farm Bill* (Washington, D.C.: World Resources Institute, 1991), and D. Pimental, "Environmental and Social Implications of Waste in U.S. Agriculture," *Journal of Agricultural Ethics*, 1990, works cited in David W. Orr, *Earth in Mind: On Education, Environment, and the Human Prospect* (Washington, D.C.: Island Press, 1994), p. 173.

99. Alan Thein Durning, "Redesigning the Forest Economy," in *State of the World 1994*, Lester Brown et al. (New York: Norton, 1994), p. 34.

100. James J. MacKenzie, Roger C. Dower, and Don Chen, *The Going Rate: What It Really Costs to Drive* (Washington, D.C.: World Resources Institute, 1992); *Calypso Log*, October 1995, p. 17; *Friends of the Earth*, September–October 1996, p. 3; *The Planet*, April 1997, p. 6.

101. *Friends of the Earth*, Spring 1997, p. 16.

102. *The Washington Post*, January 21, 1990, p. A12, cited in Ophuls and Boyan, *Ecology and the Politics of Scarcity Revisited*, p. 138; E *The Environmental Magazine*, November–December 1996, p. 45.

103. *The Planet*, April 1997, p. 5.

104. Harvard School of Public Health report, cited in *The Planet*, January–February 1997, p. 11; Hal Kane, "Shifting to Sustainable Industries," in Lester R. Brown et al., *State of the World 1996* (New York: Norton, 1996), p. 156.

105. Charles W. Powers and Marian R. Chertow, "Industrial Ecology: Overcoming Policy Fragmentation," in Marian R. Chertow and Daniel C. Esty, *Thinking Ecologically: The Next Generation of Environmental Policy* (New Haven: Yale University Press, 1997), p. 23.

106. *Earth Island Journal*, Spring 1995, p. 15.

107. Gore, *Earth in the Balance*, p. 185.

108. Herman E. Daly and John B. Cobb, Jr., *For the Common Good: Redirecting the Economy toward Community, the Environment, and a Sustainable Future*, 2nd edition (Boston: Beacon Press, 1994), pp. 443–507.

109. *The Amicus Journal*, Summer 1996, p. 2.

110. *International Wildlife*, January–February 1996, p. 8.

111. *EDF Letter*, March 1995, p. 3, and August 1995, p. 2.

112. Steven Kelman, "Cost-Benefit Analysis: An Ethical Critique," in *Readings in Risk*, ed. Theodore S. Glickman and Michael Gough (Washington, D.C.: Resources for the Future, 1990), pp. 132–33.

113. *The Amicus Journal*, Spring 1994, p. 8.

114. Shaffer, *Beyond the Endangered Species Act*, p. 7; *Audubon*, January–February, 1996, p. 41.

115. Cited in Ronald Snodgrass, "The Endangered Species Act," in *Let the People Judge: Wise Use and the Private Property Rights Movement*, ed. John D. Echeverria and Raymond Booth Eby (Washington, D.C.: Island Press, 1995), p. 281.

116. See Charles Mann and Mark Plummer, "California vs. Gnatcatcher," *Audubon*, January–February 1995. pp. 39–48, 100–104.

117. Ophuls and Boyan, *Ecology and the Politics of Scarcity Revisited*, p. 226.

118. Jane N. Abramovitz, "Valuing Nature's Services," in Lester R. Brown et al., *State of the World 1997* (New York: W. W. Norton, 1997), p. 99. See also Gretchen C. Daily, ed., *Nature's Services: Societal Dependence on Natural Ecosystems* (Washington, D.C.: Island Press, 1997).

119. See John S. Dryzek, *The Politics of the Earth: Environmental Discourses* (Oxford: Oxford University Press, 1997) p. 72.

120. K. S. Shrader-Frechette, *Science Policy, Ethics, and Economic Methodology: Some Problems of Technology Assessment and Environmental-Impact Analysis* (Boston: Reidel Publishing, 1985), p. 307.

121. Ibid., p. 200. See also Leslie Paul Thiele, "Taking Risks: Ethics, the Environment, and Uncertainty (forthcoming).

122. 1990 *Wall Street Journal* poll, cited in *Amicus Journal*, Spring 1996, p. 51.

123. *Earth Island Journal*, Winter 1995/96, p. 3; *ZPG Reporter*, September–October 1996, p. 1.

124. *New York Times*, March 3, 1996, pp. A1, A14–15.

125. *Audubon*, July–August 1996, p. 70.

126. John J. Berger, "9 ways to save our national forests," *Sierra*, July–August 1997, pp. 38–39. See also *Audubon*, March–April 1998, p. 4.

127. *The Amicus Journal*, Fall 1994, p. 2.

128. Paul Hawken, *The Ecology of Commerce: A Declaration of Sustainability* (New York: HarperCollins, 1993), pp. 60–61. See also Michael Porter, "America's Green Strategy," in *Business and the Environment*, ed. Richard Welford and Richard Starkey (Washington, D.C.: Taylor and Francis, 1996), p. 34.

129. See Roger H. Bezdek, "Environment and Economy: What's the Bottom Line?" *Environment* 35,7 (September 1993): 7–11, 25–31; Robert Repetto, Dale S. Rothman, Paul Faeth, and Duncan Austin, *Has Environmental Protection Really Reduced Productivity Growth?* (Washington, D.C.: World Resources Institute, 1996); Robert Repetto, *Jobs, Competitiveness, and Environmental Regulation* (Washington, D.C.: World Resources Institute, 1995); Grant Ferrier, "Strategic Overview of the Environmental Industry," in *Let the People Judge: Wise Use and the Private Property Rights Movement*, ed. John D. Echeverria and Raymond Booth Eby (Washington, D.C.: Island Press, 1995), p. 208; Donald L. Connors, Michael Bliss, and Jack Archer, "The U.S. Environmental Industry," in *Let the People Judge: Wise Use and the Private Property Rights Movement*, ed. John D. Echeverria and Raymond Booth Eby (Washington, D.C.: Island Press, 1995), p. 225–26; study conducted by Eban Goodstein of Skidmore College and the Economic Policy Institute, cited in *The ZPG Reporter*, July–August 1995, p. 10; *Amicus Journal*, Winter 1995, p. 31; *Felling the Myth: The Role of Timber in the Economy of California's Sierra Nevada* (New York: Natural Resource Defense Council, 1995).

130. *Friends of the Earth*, September–October 1995, p. 8.

131. Mark Dowie, *Losing Ground: American Environmentalism at the Close of the Twentieth Century* (Cambridge: MIT Press, 1995), p. 85.

132. Linda Trocki, "Science, Technology, Environment, and Competitiveness in a North American Context," in *Let the People Judge: Wise Use and the Private Property Rights Movement* John D. Echeverria and Raymond Booth Eby, ed. (Washington, D.C.: Island Press, 1995), p. 246.

133. *ZPG Reporter*, September–October 1996, p. 1.

134. Kempton et al., *Environmental Values in American Culture*, p. 170.

135. Huey D. Johnson, "Environmental Quality as a National Purpose," in *Crossroads: Environmental Priorities for the Future*, ed. Peter Borrelli (Washington, D.C.: Island Press, 1988), p. 222; *Friends of the Earth*, 1996 Annual Report, Winter 1997, p. 5.

136. *National Wildlife*, December–January 1997, p. 12.

137. Gregg Easterbrook, *A Moment on the Earth: The Coming Age of Environmental Optimism* (New York: Viking, 1995), p. 472.

138. See Curtis Moore and Alan Miller, *Green Gold: Japan, Germany, the United States and the Race for Environmental Technology* (Boston: Beacon Press, 1994); *Friends of the Earth*, September–October 1995, pp. 7–11.

139. Connors et al., "U.S. Environmental Industry," pp. 225–226.

140. See Michael Porter, *The Competitive Advantage of Nations* (London: Macmillan, 1990); Michael Porter and Claas van der Linde, "Green and Competitive: Ending the Stalement," in *Business and the Environment*, ed. Richard Welford and Richard Starkey (Washington, D.C.: Taylor and Francis, 1996), pp. 61–77.

141. *Audubon*, September–October 1996, p. 27.

142. Frances Cairncross, *Green, Inc.: A Guide to Business and the Environment* (Washington, D.C.: Island Press, 1995), p. 60.

143. Porter and van der Linde, "Green and Competitive"; Roland Clift and Anita Longley, "Introduction to Clean Technology," in *Business and the Environment*, ed. Richard Welford and Richard Starkey (Washington, D.C.: Taylor and Francis, 1996), pp. 71–2, 109–28.

144. Bezdek, "Environment and Economy," p. 31.

145. Hawken, *Ecology of Commerce*, p. xiii.

146. Schmidheiny, *Changing Course*, p. 11.

147. Ibid.

148. *Nature Conservancy*, March–April 1996, p. 5.

149. Carl Pope, in *Sierra*, May–June 1994, p. 14.

150. Ophuls and Boyan, *Ecology and the Politics of Scarcity Revisited*, p. 235.

151. Kempton et al., *Environmental Values in American Culture*, pp. 99, 222.

152. John O'Neill, "Time, Narrative, and Environmental Politics," in *The Ecological Community*, ed. Roger Gottlieb (New York: Routledge, 1997), p. 32. See also John Passmore, *Responsibility for Nature* (London: Duckworths, 1974), p. 91.

Chapter 4

1. *The ZPG Reporter*, Annual Report 1994, July–August 1995, p. 2.

2. Lester W. Milbrath, *Learning to Think Environmentally (While There Is Still Time)* (Albany: SUNY Press, 1996), p. 120.

3. See Clive L. Spash, "Economics, Ethics, and Long-Term Environmental Damages," *Environmental Ethics* 15 (Summer 1993): 118, 127, 128.

4. World Commission on Environment and Development *Our Common Future*, (Oxford: Oxford University Press, 1987), p. 43.

5. *ZPG Reporter*, May–June 1996, p. 6.

6. Aron Sachs, "Upholding Human Rights and Environmental Justice," in Lester R. Brown et al., *State of the World 1996* (New York: Norton, 1996), p. 151.

7. National Wildlife Federation newsletter, August–September 1994, p. 28.

8. *Suwannnee-St. Johns Group Sierra Club Newsletter*, December 1995, p. 3.

9. Kirkpatrick Sale, *The Green Revolution: The American Environmental Movement 1962–1992* (New York: Hill and Wang, 1993), p. 85.

10. Ken Conca, Michael Alberty, and Geoffrey Dabelko, eds., *Green Planet Blues: Environmental Politics from Stockholm to Rio* (Boulder: Westview Press, 1995), p. 8.

11. Ibid, p. 7.

12. Norman Myers, *Ultimate Security: The Environmental Basis of Political Stability* (Washington, D.C.: Island Press, 1993), pp. 12–13, 231.

13. *The Planet, The Sierra Club* Activist Resource, November 1997, p. 3.

14. *Sierra*, January–February 1994, p. 17.

15. Michael Brown and John May, *The Greenpeace Story* (New York: Dorling Kindersley, 1991), p. 128.

16. Richard Rosecrance, *The Rise of the Trading State: Commerce and Conquest in the Modern World* (New York: Basic Books, 1986), p. 222.

17. *Noetic Sciences Review*, No. 20 (Winter 1991): 34. These same trends also evidence themselves among American youth (*American Freshman National Norms*, Los Angeles: Cooperative Institutional Research Program, 1990).

18. For a similar argument see Mark Sagoff, *The Economy of the Earth: Philosophy, Law and the Environment* (Cambridge: Cambridge University Press, 1988).

19. *Groundwork 7*, Summer 1998, p. 38.

20. See Paul Wapner, *Environmental Activism and World Civic Politics* (Albany: SUNY Press, 1996).

21. *Earth Island Journal*, Summer 1997, p. 25.

22. Wapner, *Environmental Activism*, p. 109.

23. World Commission, *Our Common Future*, p. 328.

24. Wapner, *Environmental Activism*, p. 2.

25. *The Wall Street Journal*, July 7, 1995, p. 1.

26. EDF Letter, 24: 3 (May 1993): 4.

27. See George Kateb's discussion of "political evil" in *The Inner Ocean: Individualism and Democratic Culture* (Ithaca: Cornell University Press, 1992), p. 213.

28. See, for example, Warwick Fox, *Toward a Transpersonal Ecology:*

Developing New Foundations for Environmentalism (Boston: Shambhala, 1990).

29. *New York Times*, March 27, 1990.

30. Al Gore, *Earth in the Balance: Ecology and the Human Spirit* (Boston: Houghton Mifflin, 1992), pp. 269–360.

31. Tom Athanasiou, *Divided Planet: The Ecology of Rich and Poor* (Boston: Little, Brown, 1996), p. 306.

32. Daniel Deudney, "The Case against Linking Environmental Degradation and National Security," *Millennium: Journal of International Studies* 19,3 (Winter 1990): 461–76. Reprinted and quoted in Conca et al., *Green Planet Blues*, pp. 263–64, 272.

33. See Thomas F. Homer-Dixon, "Environmental Scarcities and Violent Conflict: Evidence from Cases," *International Security* 19,1 (1994): 5–40.

34. See *Florida Population Forum*, January 1998, p. 1; Emanuel Sferios, "Immigration and the Environment: Is Eco-Fascism on the Rise?" Synthesis *Regeneration*, Spring 1998, p. 42.

35. Wapner, *Environmental Activism*, p. 70.

36. Adlai E. Stevenson, *The Papers of Adlai E. Stevenson, Vol. 8, Ambassador to the United Nations 1961–1965*, ed. Walter Johnson (Boston: Little, Brown and Company, 1979), p. 828.

37. A sense of a shared fate is perhaps a crucial feature of all social movements. See Carol McClurg Mueller, "Building Social Movement Theory," in *Frontiers in Social Movement Theory*, ed. Aldon Morris and Carol McClurg Mueller (New Haven: Yale University Press, 1992), pp. 3–25.

38. See, for example, Robert Gottlieb, *Forcing the Spring: The Transformation of the American Environmental Movement* (Washington, D.C.: Island Press, 1993); *Beyond the Green: Redefining and Diversifying the Environmental Movement* (Boston: Environmental Careers Organization, 1992); Robert Bullard, *Dumping in Dixie: Race, Class and Environmental Quality* (Boulder: Westview Press, 1990); Robert Bullard, ed., *Confronting Environmental Racism: Voices from the Grassroots* (Boston: South End Press, 1993); and Robert Bullard, ed., *Unequal Protection: Environmental Justice and Communities of Color* (New York: Random House, 1994).

39. Anil Agarwal and Sunity Narain, "Global Warming in an Unequal World: A Case of Environmental Colonialism," *Earth Island Journal*, Spring 1991, p. 40.

40. *International Wildlife*, November–December 1995, p. 4 (emphasis added). See also National Wildlife Federation, *1994 Annual Report*, p. 2.

41. The Wilderness Society, *Annual Report 1994*, n.p.

42. Charles T. Rubin, *The Green Crusade: Rethinking the Roots of Environmentalism* (New York: Free Press, 1994), pp. 24–25.

43. Garrett Hardin, "The Tragedy of the Commons," *Science* 162 (December 13, 1968): 1243–48.

44. Garrett Hardin, "The Tragedy of the Commons," in *Managing the Commons*, ed. (Garrett Hardin and John Baden, San Francisco: Freeman, 1977), p. 20.

45. Along with Hardin, other authoritarian/centralist positions were staked out by Robert L. Heilbroner, *An Inquiry into the Human Prospect* (New York: Norton, 1974), and, analytically if not prescriptively, by William Ophuls in "The Scarcity Society," *Harpers*, April 1974, and *Ecology and the Politics of Scarcity* (New York: Freeman, 1977).

46. World Commission, *Our Common Future*, p. 27.

47. Larry Lohmann, "Whose Common Future," *The Ecologist* 20,3 (May–June 1990): 82–84. On the question of the potentially pernicious uses of the terms *environmental security* and *ecological security*, see also Lothar Brock, "Security through Defending the Environment: An Illusion?" in *New Agendas for Peace Research: Conflict and Security Reexamined*, ed. Elise Boulding (Boulder: Lynne Rienner, 1992), pp. 79–102.

48. Susan Jane Buck Cox, "No Tragedy on the Commons," *Environmental Ethics* 7 (Spring 1985): 49–61; David Feeny, Fikret Berkes, Bonnie J. McCay, and James M. Acheson, "The Tragedy of the Commons: Twenty-two Years Later," *Human Ecology* 18,1 (1990): 1–19; Elinor Ostrom, *Governing the Commons: The Evolution of Institutions for Collective Action* (Cambridge: Cambridge University Press, 1990).

49. J. Martinez-Alier, "Environmental Policy and Distributional Conflicts," in *Ecological Economics*, ed. Robert Costanza (New York: Columbia University Press, 1991), pp. 118–36; A. Mayhew, "Dangers in Using the Idea of Property Rights: Modern Property Rights Theory and the Neoclassical Trap," *Journal of Economic Issues* 19 (1985): 959–66.

50. Richard J. Ellis and Fred Thompson, "Culture and the Environment in the Pacific Northwest," *American Political Science Review* 91 (December 1997): 888.

51. Sarah Anderson and John Cavanagh, *The Top 200: The Rise of Global Corporate Power*, cited in *Everyone's Backyard*, Winter 1996, p. 15; *Sierra*, May/June 1998, p. 7.

52. *Audubon*, May–June 1995, p. 6.

53. *Co-op America Quarterly*, Fall 1994, pp. 12–14.

54. Ibid.

55. *Friends of the Earth*, May–June 1996, p. 3.

56. *Earth Island Journal*, Fall 1997, p. 24.

57. *Public Citizen*, Annual Report, Spring 1996, p. 5.

58. International Forum on Globalization position statement, January 1995. See also Jerry Mander and Edward Goldsmith, eds., *The Case against the Global Economy—and for a Turn toward the Local* (San Francisco: Sierra Club Books, 1996).

59. *Earth Island Journal*, Fall 1997, p. 8.

60. *The Planet*, November 1997, pp. 3–4.

61. Rainforest Action Network, *Ten Years of Rainforest Action* (1995), p. 15.

62. Wendell Berry, *Sex, Economy, Freedom and Community* (New York: Pantheon Books, 1993), p. 19.

63. R. Buckminster Fuller, *Operating Manual for Spaceship Earth* (New York: Dutton, 1971), pp. 52, 132.

64. Daniel Press, *Democratic Dilemmas in the Age of Ecology: Trees and Toxics in the American West* (Durham: Duke University Press, 1994), p. 130.

65. Wendell Berry, *What Are People For?* (San Francisco: North Point Press, 1990), p. 166.

66. Berry, *Sex, Economy, Freedom and Community*, p. 24.

67. Garrett Hardin, *Living within Limits: Ecology, Economics, and Population Taboos* (New York: Oxford University Press, 1993), p. 278.

68. See Press, *Democratic Dilemmas in the Age of Ecology*, p. 130.

69. Wade Sikorski, "Building Wilderness," in *In the Nature of Things*, ed. Jane Bennett and William Chaloupka (Minneapolis: University of Minnesota Press, 1993), p. 28.

70. Peter Dykstra, *Greenpeace*, January–March 1991, p. 2.

71. *The ZPG Reporter*, January–February 1996, p. 2.

72. Zero Population Growth letter to members, April 1996.

73. *Wild Earth*, Winter 1995/96, pp. i, 37.

74. *Audubon*, July–August 1995, p. 99.

75. Berry, *Sex, Economy, Freedom and Community*, pp. 13–14.

76. *Popline*, July–August 1997, p. 2.

77. Bruce A. Williams and Albert R. Matheny, *Democracy, Dialogue, and Environmental Disputes: The Contested Languages of Social Regulation* (New Haven: Yale University Press, 1995), p. 171.

78. Kent Portney, *Siting Hazardous Waste Treatment Facilities*, cited in Williams and Matheny, *Democracy, Dialogue, and Environmental Disputes*, p. 172.

79. Quoted in Williams and Matheny, *Democracy, Dialogue, and Environmental Disputes*, pp. 183–84.

80. Sachs, "Upholding Human Rights," p. 144; Petra Kelly, "The Need for Eco-Justice," in *Green Planet Blues: Environmental Politics from Stockholm to Rio*, ed. Ken Conca, Michael Alberty, and Geoffrey Dabelko. (Boulder: Westview Press, 1995), p. 284; Mutombo Mpanya, "The Dumping of Toxic Waste in African Countries: A Case of Poverty and Racism," in *Race and the Incidence of Environmental Hazards*, ed. Bunyan Bryant and Paul Mohai (Boulder: Westview Press, 1992), pp. 204–14.

81. World Commission, *Our Common Future*, p. 350.

82. Quoted in William Greider, *Who Will Tell the People?* (New York: Simon and Schuster, 1992), p. 169–70.

83. Williams and Matheny, *Democracy, Dialogue, and Environmental Disputes,* p. 194.

84. Gregg Easterbrook, *A Moment on the Earth: The Coming Age of Environmental Optimism* (New York: Viking, 1995), p. 452.

85. Gifford Pinchot, *The Fight for Conservation* (New York: Doubleday, Page, 1910), pp. 26, 46, 81–82.

86. Donald Snow, *Inside the Environmental Movement: Meeting the Leadership Challenge* (Washington, D.C.: Island Press, 1992), p. 10.

87. Friends of the Earth Annual Report 1997, p. 2.

88. The relation between environmentalism and democracy is a growing concern for environmental theorists. See, for example, Freya Mathews, ed., *Ecology and Democracy* (London: Frank Cass, 1996), and Brian Doherty and Marius de Geus, eds., *Democracy and Green Political Thought* (London: Routledge, 1996).

89. Sachs, "Upholding Human Rights," p. 148.

90. Adolf G. Gundersen, *The Environmental Promise of Democratic Deliberation* (Madison: University of Wisconsin Press, 1995), pp. 5, 159–65. See also John Barry, "Sustainability, Political Judgment and Citizenship: Connecting Green Politics and Democracy," in *Democracy and Green Political Thought*, ed. Brian Doherty and Marius de Geus (London: Routledge, 1996), pp. 115–31.

91. *Earth Island Journal,* Spring 1995, p. 11.

92. *Everyone's Backyard,* Fall/Winter 1995, p. 29.

93. Snow, *Inside the Environmental Movement*, p. 160.

94. *Sierra,* July–August 1996, p. 52.

95. Freya Mathews, "Community and the Ecological Self," in *Ecology and Democracy*, ed. Freya Mathews (London: Frank Cass, 1996), pp. 66–100.

96. Richard J. Ellis and Fred Thompson, "Culture and the Environment in the Pacific Northwest," *American Political Science Review* 91 (December 1997): 891.

97. Ibid., pp. 891–92

98. Ibid.

99. Thomas Jefferson, letter to William Charles Jarvis, September 28, 1820. In Paul Leicester Ford, ed., *The Writings of Thomas Jefferson*, Vol. 10 (New York: Putnam's, 1899).

100. World Commission, *Our Common Future*, p. 65.

101. Daphne Wysham, "Restoring Democracy," *Greenpeace,* April–June 1993, p. 2.

102. *Calypso Log,* April 1994, p. 20.

103. Alberto Melucci, "The Symbolic Challenge of Contemporary Movements," *Social Research* 52 (1985): 801.

104. *Nature Conservancy,* March–April 1996, p. 5.

105. Charles Jordan and Donald Snow, "Diversification, Minorities, and the Mainstream Environmental, Movement," in *Voices from the Environmental*

Movement: Perspectives for a New Era, ed. Donald Snow (Washington, D.C.: Island Press, 1992), p. 88.

106. *International Wildlife*, November–December 1995, p. 5.

107. See Press, *Democratic Dilemmas*, p. 123.

108. For a good discussion of how green values may conflict with green, democratic agency, see Robert Goodin, *Green Political Theory* (Cambridge: Polity Press, 1992).

109. *Sierra*, May–June 1995, p. 13.

110. *The Planet*, December 1997, p. 7.

111. *Greenpeace*, January–March 1991, p. 2.

112. Mark Dowie, "Greens Outgunned," in *Earth Island Journal*, Spring 1995, p. 26.

113. *Sierra*, November–December 1996, p. 59.

114. See Ross Gelbspan, "The Heat Is On: The Warming of the World's Climate Sparks a Blaze of Denial," *Harpers Magazine*, December 1995, pp. 31–37.

115. *Public Citizen*, Annual Report, Spring 1996, p. 15.

116. *Public Citizen*, Fall 1996, pp. 2, 7.

117. See Stephen Fox, *John Muir and His Legacy: The American Conservation Movement* (Boston: Little, Brown, 1981), pp. 345–51.

118. Quoted in Fox, *John Muir and His Legacy*, p. 349.

119. *Time*, August 31, 1970, p. 42. Quoted in Christopher Manes, *Green Rage: Radical Environmentalism and the Unmaking of Civilization* (Boston: Little, Brown, 1990), p. 51.

120. Jordan and Snow, "Diversification, Minorities," p. 78.

121. Snow, *Inside the Environmental Movement*, p. 138.

122. See John D. Skrentny, "Concern for the Environment: A Cross National Perspective," *International Journal of Public Opinion Research* 5,4 (Winter 1993): 335–52; and Susan Howell and Shirley Laska, "The Changing Face of the Environmental Coalition," *Environment and Behavior* 24,1 (January 1992): 134–44.

123. Jacqueline Vaughn Switzer, *Environmental Politics: Domestic and Global Dimensions* (New York: St. Martin's Press, 1994), p. 30.

124. Richard Rodriguez, quoted in Mark Dowie, *Losing Ground: American Environmentalism at the Close of the Twentieth Century* (Cambridge: MIT Press, 1995), p. 220.

125. Quoted in Jordan and Snow, "Diversification, Minorities," p. 71.

126. The CEIP Fund, Inc., *The Minority Opportunity Study*, August, 1989. Cited in Jim Schwab, *Deeper Shades of Green: The Rise of Blue-Collar and Minority Environmentalism in America* (San Francisco: Sierra Club Books, 1994), p. 388.

127. Sale, *The Green Revolution*, p. 98.

128. Snow, *Inside the Environmental Movement*, p. 75.

129. *Green Politics*, Winter 1995, p. 3.

130. Switzer, *Environmental Politics*, p. 91.

131. William A. Gamson, *Talking Politics* (Cambridge: Cambridge University Press, 1992), p. 175.

132. Philip Shabecoff, *A Fierce Green Fire: The American Environmental Movement* (New York: Hill and Wang, 1993), p. 282.

133. Jordan and Snow, "Diversification, Minorities," p. 71.

134. Deeohn Ferris and David Hahn-Baker, "Environmentalists and Environmental Justice Policy," in *Environmental Justice: Issues, Policies, and Solutions*, ed. Bunyan Bryant (Washington, D.C..: Island Press, 1995), p. 71.

135. *Sierra*, May–June 1995, p. 83.

136. Aldo Leopold, *A Sand County Almanac, with Essays on Conservation from Round River* (New York: Ballantine Books, 1966), p. xvii.

137. Quoted in Stephan Schmidheiny, *Changing Course: A Global Business Perspective on Development and the Environment* (Cambridge: MIT Press, 1992), p. 135.

138. See Jordan and Snow, "Diversification, Minorities."

139. Dorceta Taylor, "Can the Environmental Movement Attract and Maintain the Support of Minorities?" in *Race and the Incidence of Environmental Hazards*, ed. Bunyan Bryant and Paul Mohai (Boulder: Westview Press, 1992), p. 39.

140. *National Wildlife*, April–May 1996, p. 61.

141. Matthias Finger, "From Knowledge to Action? Exploring the Relationships between Environmental Experiences, Learning, and Behavior," *Journal of Social Issues* 50,3 (Fall 1994): 153; Jody M. Hines, Harold R. Hungerford, and Audrey N. Tomera, "Analysis and Synthesis of Research on Responsible Environmental Behavior: A Meta-Analysis," *Journal of Environmental Education* 18,2 (Winter 1986/87): 1–8; Harold R. Hungerford and Trudi L. Volk, "Changing Learner Behavior through Environmental Education,"*Journal of Environmental Education* 21,3 (spring 1990): 8–21.

142. *Nature Conservancy Reporter*, Fall 1996, p. 2.

143. *An Environmental Agenda for the Future* (Washington, D.C.: Island Press, 1985), pp. 11, 23. The areas of concern are nuclear issues, human population growth, energy strategies, water resources, toxics and pollution control, wild living resources, private lands and agriculture, protected land systems, public lands, urban environment, and international responsibilities.

144. Dowie, *Losing Ground*, p. 44.

145. Snow, *Inside the Environmental Movement*, p. 63.

146. See Cheri Lucas Jennings and Bruce H. Jennings, "Green Fields/Brown Skin: Posting as a Sign of Recognition," in *In the Nature of Things*, ed. Jane Bennett and William Chaloupka (Minneapolis: University of Minnesota Press, 1993), p. 192; League of Conservation Voters, *National Environmental Scorecard*, October 1994, p. 10; Ivette Perfecto, "Pesticide Exposure of Farm Workers

and the International Connection," in *Race and the Incidence of Environmental Hazards*, ed. Bunyan Bryant and Paul Mohai (Boulder: Westview Press, 1992), pp. 177–203.

147. Allan Schnaiberg, *The Environment: From Surplus to Scarcity* (New York: Oxford University Press, 1980), p. 5.

148. Frederick H. Buttel, "Rethinking International Environmental Policy in the Late Twentieth Century," in *Environmental Justice: Issues, Policies, and Solutions*, ed. Bunyan Bryant (Washington, D.C.: Island Press, 1995), p. 206.

149. Deeohn Ferris and David Hahn-Baker, "Environmentalists and Environmental Justice Policy," in *Environmental Justice: Issues, Policies, and Solutions*, ed. Bunyan Bryant (Washington, D.C.: Island Press, 1995), p. 70.

150. Quoted in Jordan and Snow, "Diversification, Minorities," p. 99.

151. Quoted in Shabecoff, *Fierce Green Fire*, p. 263.

152. National Wildlife Federation, *1994 Annual Report*, pp. 16–17.

153. John Adams, "Letter from the Executive Director," in *Twenty-five Years Defending the Environment* (New York: Natural Resources Defense Council, 1995), p. 3.

154. Interview with Vernice Miller, director of the Environmental Justice Initiative, in *The ZPG Reporter*, March–April 1996, p. 5.

155. Brent Blackwelder, personal interview, Washington, D.C., September 8, 1995.

156. *Friends of the Earth*, 1994 Annual Report, pp. 4–5.

157. Bunyan Bryant, "Summary," in *Environmental Justice: Issues, Policies, and Solutions*, ed. Bunyan Bryant (Washington, D.C.: Island Press, 1995), p. 217.

158. Henry Vance Davis, "The Environmental Voting Record of the Congressional Black Caucus," in *Race and the Incidence of Environmental Hazards*, Bunyan Bryant and Paul Mohai (Boulder: Westview Press, 1992), pp. 59–60; Deeohn Ferris and David Hahn-Baker, "Environmentalists and Environmental Justice Policy," in *Environmental Justice: Issues, Policies, and Solutions*, ed. Bunyan Bryant (Washington, D.C.: Island Press, 1995), p. 71.

159. *Sierra*, September–October 1997, p. 27.

160. *General Social Survey 1994–95* (Chicago: National Opinion Research Center, 1995). Combining the 1993 and 1994 surveys, a cross-tabulation showed 227 out of 2,402 white respondents were members of environmental groups (9.45%) while 29 out of 345 black respondents were members of environmental groups (8.41%). Chi-square tests demonstrate that these differences are not significant at the .1 level.

161. David Helvarg, *The War against the Greens: The "Wise-Use" Movement, the New Right, and Anti-environmental Violence* (San Francisco: Sierra Club Books, 1994), pp. 430–31.

162. Robert Bullard, "Environmental Racism and the Environmental Jus-

tice Movement," in *Ecology: Key Concepts in Critical Theory*, ed. Carolyn Merchant (Atlantic Highlands, N.J.: Humanities Press, 1994), pp. 260, 263.

163. Adam Walinsky, quoted in Dowie, *Losing Ground*, p. xi.

164. Allan Schnaiberg and Kenneth Alan Gould, *Environment and Society: The Enduring Conflict* (New York: St. Martin's Press, 1994), p. 160.

165. Quoted in Jordan and Snow, "Diversification, Minorities," p. 89.

166. Paul R. Ehrlich, *The Population Bomb* (New York: Ballantine Books, 1968), p. 15.

167. Aaron Sachs, *Eco-Justice: Linking Human Rights and the Environment*, Worldwatch Paper 127 (Washington, D.C.: Worldwatch Institute, 1995), pp. 14–15.

168. See, for instance, Arun Agrawal, "Community in Conservation: Beyond Enchantment and Disenchantment," Gainesville, Florida: Conservation and Development Forum, 1997.

169. See Michael Wells and Katrina Brandon, *People and Parks: Linking Protected Area Management with Local Communities* (Washington, D.C.: The World Bank, World Wildlife Fund, and USAID, 1992); David Western and R. Michael Wright, eds., *Natural Connections: Perspectives in Community-Based Conservation* (Washington, D.C.: Island Press, 1994).

170. Sachs, *Eco-Justice*, pp. 11, 45.

171. Michael Renner, "Transforming Security," in Lester R. Brown et al., *State of the World 1997* (New York: Norton, 1997), p. 127; *Earth Island Journal*, Winter 1996/97, p. 17, and Spring 1997, p. 16.

172. Sachs, "Upholding Human Rights," p. 137.

173. Renner, "Transforming Security," p. 124; Jeremy Schmidt, "China's Coming Flood," *International Wildlife*, September–October 1996, pp. 35–43.

174. Stephen Minkin and James Boyce, "Net Losses," *Amicus Journal*, Fall 1994, pp. 39–40.

175. Riley E. Dunlap, George H. Gallup, and Alec M. Gallup, "Of Global Concern: Results of the Health of the Planet Survey," *Environment* 39,9 (November 1993): 36.

176. Leslie E. Anderson, *The Political Ecology of the Modern Peasant* (Baltimore: Johns Hopkins University Press, 1994), pp. 171, 173.

177. Ramachandra Guha, "Radical Environmentalism: A Third-World Critique," in *Ecology: Key Concepts in Critical Theory*, ed. Carolyn Merchant (Atlantic Highlands, N.J.: Humanities Press, 1994), p. 284.

178. World Wildlife Fund, *1993 Annual Report*, p. 17.

179. Quoted in Luc P. Deslarzes, "Strategic Elements of the WWF Environmental Education Programme for the 1990s," in *Environmental Education: An Approach to Sustainable Development*, ed. Hartmut Schneider (Paris: Organization for Economic Co-operation and Development, 1993), p. 116.

180. World Wildlife Fund, *1993 Annual Report*, p. 10.

181. *Focus*, World Wildlife Fund, September/October 1994, pp. 1–2.

182. The Nature Conservancy president's letter to members, January 13, 1995.

183. *Audubon*, March–April 1995, p. 23.

184. *Nature Conservancy*, May–June 1994, p. 5, and March–April 1997, p. 14. See also W. William Weeks, *Beyond the Ark: Tools for an Ecosystem Approach to Conservation* (Washington, D.C.: Island Press, 1997).

185. *Nature Conservancy*, July–August 1996, p. 20.

186. Coordinating Body for the Indigenous Peoples' Organizations of the Amazon Basin, "Two Agendas on Amazon Development," in *Green Planet Blues: Environmental Politics from Stockholm to Rio*, ed. Ken Conca, Michael Alberty, and Geoffrey Dabelko (Boulder: Westview Press, 1995), pp. 299–305.

187. Quoted in World Commission, *Our Common Future*, p. 61.

188. *International Wildlife*, May–June 1996, p. 7.

189. Rainforest Action Network, *1994 Annual Report*, p. 4.

190. World Wildlife Fund, *1993 Annual Report*, pp. 22, 34. See also Wapner, *Environmental Activism*, pp. 72–116.

191. *EDF Letter*, January 1997, p. 4.

192. See Kent H. Redford and Allyn M. Stearman, "Forest Dwelling Native Amazonians and the Conservation of Biodiversity: Interests in Common or in Collision?" *Conservation Biology* 7,2 (1993): 248–55.

193. *Sierra*, November–December 1996, p. 51.

194. *Defenders*, Spring 1996, p. 28.

195. Luis Torres, "The Dilemma of Choice for Land-Development Communities: The People for the West of Environmentalists?" *The Workbook* (Southwest Research and Information Center), Spring 1992, p. 16.

196. *Nature Conservancy*, May/June 1998, p. 19.

197. *National Wildlife*, December–January 1997, pp. 60–61.

198. *Greenpeace Magazine*, Summer 1998, pp. 10–12.

199. *The Planet*, January–February 1997, p. 5.

200. *Friends of the Earth*, 1996 Annual Report, Winter 1997, p. 11.

201. *Defenders*, Winter 1995/96, pp. 21–22.

202. *International Wildlife*, November–December 1995, p. 5.

203. John Sawhill, "President's Report," *Nature Conservancy*, January–February 1998, p. 8.

204. Sachs, *Eco-Justice*, p. 8.

205. Sachs, "Upholding Human Rights," p. 143.

206. *Wilderness*, Winter 1995, p. 10.

207. Linda L. Young, "Reforms Shift Regulations to State and Local Levels," *Pro Earth Times*, February 1997, p. 2.

208. 1997 World Population Data Sheet, Population Reference Group, cited in *Popline*, May–June 1997.

209. Tom Athanasiou, *Divided Planet: The Ecology of Rich and Poor* (Boston: Little, Brown, 1996), p. 304.

210. International NGO Forum, *Alternative Treaties* (Rio de Janeiro, June 1992), quoted in Malcolm Plant, "The Riddle of Sustainable Development and the Role of Environmental Education," *Environmental Education Research* 1 (October 1995): 261.

Chapter 5

1. Garrett Hardin, *The Limits of Altruism: An Ecologist's View of Survival* (Bloomington: Indiana University Press, 1977), pp. 27, 42.

2. Roger Gottlieb, "The Center Cannot Hold," in *The Ecological Community*, ed. Roger Gottlieb (New York: Routledge, 1997), p. ix.

3. Richard J. Ellis and Fred Thompson, "Culture and the Environment in the Pacific Northwest," *American Political Science Review* 91 (December 1997): 891.

4. See Willett Kempton, James Boster, and Jennifer Hartley, *Environmental Values in American Culture* (Cambridge: MIT Press, 1995), pp. 102, 111–15; *General Social Survey 1993–94* (Chicago: National Opinion Research Center, 1994); Associated Press poll, November 1995, *The Gainesville Sun*, December 3, 1995, p. 3A.

5. Matthias Finger, "From Knowledge to Action? Exploring the Relationships between Environmental Experiences, Learning, and Behavior," *Journal of Social Issues* 50,3 (Fall 1994): 155. See also Kempton et al., *Environmental Values in American Culture*.

6. David W. Orr, *Earth in Mind: On Education, Environment, and the Human Prospect* (Washington, D.C.: Island Press, 1994), p. 45.

7. See Donald Snow, *Inside the Environmental Movement: Meeting the Leadership Challenge* (Washington, D.C.: Island Press, 1992), p. 134. See also Lester Milbrath, "Environmental Understanding: A New Concern for Political Socialization," in *Political Socialization: Citizenship, Education and Democracy*, ed. Orit Ichilov (New York: Teachers College Press, 1990), p. 286.

8. Kempton et al., *Environmental Values in American Culture*, p. 115.

9. World Wildlife Fund, *1993 Annual Report*, p. 7.

10. Jan E. Dizard, "Going Wild: The Contested Terrain of Nature," in *In the Nature of Things*, ed. Jane Bennett and William Chaloupka (Minneapolis: University of Minnesota Press, 1993), p. 112.

11. See Shane Phelan, "Intimate Distance: The Dislocation of Nature in Modernity," in *In the Nature of Things*, ed. Jane Bennett and William Chaloupka (Minneapolis: University of Minnesota Press, 1993), p. 58.

12. Walter Truett Anderson, *To Govern Evolution* (Boston: Harcourt Brace Jovanovitch, 1987), p. 7.

13. Bill McKibben, *The End of Nature* (New York: Random House, 1989), pp. 210–17.

14. *Audubon*, March–April 1995, p. 63.

15. *The Planet*, January–February 1997, p. 7, and January–February 1998, p. 16.

16. Finger, "From Knowledge to Action?" p. 153.

17. Edward O. Wilson, *Biophilia* (Cambridge: Harvard University Press, 1984), p. 139. See also Stephen Kellert and Edward O. Wilson, eds., *Biophilia Hypothesis* (Washington, D.C.: Island Press, 1993).

18. Aldo Leopold, *A Sand County Almanac, with Essays on Conservation from Round River* (New York: Balantine Books, 1996), p. 134.

19. Brian K. Steverson, "Contextualism and Norton's Convergence Hypothesis," *Environmental Ethics* 17 (Summer 1995): 135.

20. See Holmes Rolston, III, *Environmental Ethics: Duties to and Values in the Natural World* (Philadelphia: Temple University Press, 1988), p. 114

21. Arne Naess, "A Defense of the Deep Ecology Movement," *Environmental Ethics* 6 (1984): 270.

22. Richard J. Ellis and Fred Thompson, "Culture and the Environment in the Pacific Northwest," *American Political Science Review* 91 (December 1997): 894.

23. Ross McCluney, ed., *Florida Population Forum*, December 1996, p. 2.

24. Bill Devall and George Sessions, *Deep Ecology* (Salt Lake City: Gibbs M. Smith, 1985), p. 67.

25. See Peter Singer, *Animal Liberation: A New Ethics for Our Treatment of Animals* (New York: Random House, 1976). Importantly, Singer refers only to sentient animal species, maintaining the capacity to feel pain as the prerequisite for ethical consideration. Ethics, for Singer, ceases to apply "somewhere between a shrimp and oyster."

26. Christopher Manes, *Green Rage: Radical Environmentalism and the Unmaking of Civilization* (Boston: Little, Brown, 1990), p. 141.

27. *National Wildlife*, June–July 1995, p. 2.

28. Rodger Schlickeisen, Defenders of Wildlife president, *Wildlife Advocate*, Winter 1995, p. 2.

29. C. S. Lewis, *The Abolition of Man* (London: Geoffrey Bles, 1946), p. 39, cited in William Leiss, *The Domination of Nature* (New York: George Braziller, 1972), p. 195. Leiss agrees with Lewis's assessment.

30. Murray Bookchin, *Remaking Society: Pathways to a Green Future* (Boston: South End Press, 1990), p. 44.

31. Wilson, *Biophilia*, p. 131.

32. E. O. Wilson, *The Diversity of Life* (Cambridge: Harvard University Press, 1992).

33. *Nature Conservancy*, May/June 1998, p. 5.

34. Eric Katz and Lauren Oechsli, "Moving beyond Anthropocentrism: Environmental Ethics, Development, and the Amazon," *Environmental Ethics* 15 (1993): 59.

35. Quoted in Roderick Frazier Nash, *The Rights of Nature: A History of Environmental Ethics* (Madison: University of Wisconsin Press, 1989), p. 3.

36. Dave Foreman, *Confessions of an Eco-Warrior* (New York: Crown Trade Paperbacks, 1991), p. 3.

37. *Defenders*, Winter 1996/97, p. 5.

38. Animal Legal Defense Fund petition, July 1996.

39. Charles Birch and John Cobb, Jr., *The Liberation of Life: From the Cell to the Community* (Cambridge: Cambridge University Press, 1981), p. 155. Birch and Cobb propose that we make a "rough estimate" of which species deserve what rights according to the richness of their experiences (p. 175).

40. Arne Naess, *Ecology, Community and Lifestyle: Outline of an Eco-sophy* (Cambridge: Cambridge University Press, 1989), p. 167.

41. David Ehrenfeld, *The Arrogance of Humanism* (New York: Oxford University Press, 1978), p. 208.

42. Naess, *Ecology, Community and Lifestyle*, p. 171.

43. See Paul Wapner, "Politics beyond the State: Environmental Activism and World Civic Politics," *World Politics* 47 (April 1995): 324.

44. Quoted in Nash, *Rights of Nature*, p. 61.

45. Bookchin, *Remaking Society*, p. 10.

46. See Al Gore, *Earth in the Balance: Ecology and the Human Spirit* (Boston: Houghton Mifflin, 1992), p. 217; and Michael Zimmerman, "Rethinking the Heidegger-Deep Ecology Relationship," *Environmental Ethics* 15 (Fall 1993): 205. When Earth First!ers chanted "Down with Human Beings!" around campfires at a gathering, Foreman understood this as "an honest expression" of a certain deep ecological perspective. Cited in Murray Bookchin and Dave Foreman, *Defending the Earth* (Boston: South End Press, 1991) p. 19.

47. George Sessions, quoted in L. M Benton, "Selling the Natural or Selling Out? Exploring Environmental Merchandising," *Environmental Ethics* 17,7 (Spring 1995): 3–26.

48. Leopold, *Sand County Almanac*, p. 108.

49. Edward Abbey, *Desert Solitaire: A Season in the Wilderness* (New York: Ballantine Books, 1968), p. 52.

50. Leopold, *Sand County Almanac*, pp. xviii–xix.

51. Young characterized environmentalists as "the most despicable group of individuals I've ever been around . . . [a] self-centered bunch . . . [of], waffle-stomping, Harvard graduating, intellectual bunch of idiots." Quoted in *Sierra*, May–June 1995, pp. 25–26; Wilderness Society memo, November 18, 1996.

52. Naess, *Ecology, Community and Lifestyle*, p. 28.

53. Devall and Sessions, *Deep Ecology*, p. 67.

54. See Susan Zakin, *Coyotes and Town Dogs: Earth First! and the Environmental Movement* (New York: Penguin Books, 1993).

55. Manes, *Green Rage*, p. 21.

56. Arne Naess, "A Defense of the Deep Ecology Movement," *Environmental Ethics* 6 (1984): 268.

57. Naess, *Ecology, Community and Lifestyle*, p. 165.

58. Kirkpatrick Sale, *The Green Revolution: The American Environmental Movement 1962–1992* (New York: Hill and Wang, 1993), pp. 30–31.

59. David Brower, *Let the Mountains Talk, Let the Rivers Run: A Call to Those Who Would Save the Earth* (New York: Harper Collins, 1995), p. 27.

60. Quoted in Dick Russell, "The Monkeywrenchers," in *Crossroads: Environmental Priorities for the Future* ed., Peter Borrelli (Washington D.C.: Island Press, 1988), p. 48.

61. Bryan G. Norton, *Toward Unity among Environmentalists* (New York: Oxford University Press, 1991), p. 68.

62. Quoted ibid., p. 35. See also Roderick Nash, *Wilderness and the American Mind*, revised edition (New Haven: Yale University Press, 1973), pp. 134–35.

63. Bill Meadows, "A Passion for Wilderness," *Wilderness America*, Summer 1998, p. 2.

64. Rodger Schlickeisen, personal interview, Washington, D.C., September 8, 1995.

65. Naess, *Ecology, Community and Lifestyle*, pp. 141, 177.

66. Rolston, *Environmental Ethics*, p. 310.

67. Leopold, *Sand County Almanac*, p. 246.

68. Rachel Carson, *Silent Spring* (Boston: Houghton Mifflin, 1962), p. 278.

69. Lynn White, Jr., "The Historical Roots of Our Ecological Crisis," *Science* 155 (March 10, 1967): 1203–7.

70. Gregory Bateson, *Steps to an Ecology of Mind* (New York: Ballantine Books, 1972), p. 462.

71. Quoted in Victor B. Scheffer, *The Shaping of Environmentalism in America* (Seattle: University of Washington Press, 1991), p. 14.

72. Quoted in Philip Shabecoff, *A Fierce Green Fire: The American Environmental Movement* (New York: Hill and Wang, 1993), p. 127.

73. Quoted in Gregg Easterbrook, *A Moment on the Earth: The Coming Age of Environmental Optimism* (New York: Viking, 1995), p. 133.

74. *Pro Earth Times*, February 1995, p. 3; *Amicus Journal*, Spring 1995, p. 12.

75. Quoted in James Carroll, "Faith in the Earth," *Boston Globe*, April 19, 1994.

76. *Friends of the Earth*, January–February 1996, p. 6; *Outside*, June 1996, p. 30.

77. Kempton et al., *Environmental Values in American Culture*, pp. 87–91.

78. *Amicus Journal*, Fall 1995, p. 32. See also Charlene Spretnak, *The Spiritual Dimensions of Green Politics* (Santa Fe: Bear and Co., 1986), esp. pp. 66–69; John E. Carroll, Paul Brockelman, and Mary Westfall, eds., *The Greening of Faith: God, the Environment, and the Good Life* (University of New Hamphire Press, 1996); *Sierra* November/December 1998, pp. 50–57.

79. Abbey, *Desert Solitaire*, p. 190.

80. Adapted from Tom Hayden, *The Lost Gospel of the Earth: A Call for Renewing Nature, Spirit and Politics* (Sierra Club Books, 1996), in *Earth Island Journal*, Winter 1996/97, pp. 38–39.

81. José Lutzenberger, "Science, Technology, Economics, Ethics, and Environment," in *Earth Summit Ethics: Toward a Reconstructive Postmodern Philosophy of Environmental Education*, ed. J. Baird Callicott and Fernando J. R. da Rocha (Albany: SUNY Press, 1996), pp. 43–44.

82. *Green Politics*, Summer 1996 supplement, p. 11; see also *Green Politics*, Summer 1997, p. 2.

83. See Theodore Roszak, *Ecopsychology: Restoring the Earth, Healing the Mind* (San Francisco: Sierra Club Books, 1995); and David Abram, *The Spell of the Sensuous: Perception and Language in a More-Than-Human World* (New York: Pantheon Books, 1996), pp. 22, 272.

84. Thomas Berry, *The Dream of the Earth* (San Francisco: Sierra Club Books, 1988), p. xiii.

85. Morris Berman, *The Reenchantment of the World* (Ithaca: Cornell University Press, 1981), p. 147.

86. Naess, *Ecology, Community and Lifestyle*, p. 174.

87. William A. Gamson, *Talking Politics* (Cambridge: Cambridge University Press, 1992), p. 84.

88. Warwick Fox, *Toward a Transpersonal Ecology: Developing New Foundations for Environmentalism* (Boston: Shambhala, 1990).

89. Henry David Thoreau, "Walking," in *Great Short Works of Henry David Thoreau* (New York: Harper and Row, 1982), p. 310.

90. Ibid., p. 320.

91. Walt Whitman, "A Song of the Rolling Earth," *Leaves of Grass* (New York: Holt, Rinehart and Winston, 1966), p. 189.

92. *Audubon*, March–April 1995, p. 91. It is estimated that in the early nineteenth century there were 100,000 grizzly bears in the lower forty-eight states. By the 1980s as few as 600 remained. See *An Environmental Agenda for the Future* (Washington, D.C.: Island Press, 1985), p. 83.

93. *Sierra*, January–February 1996, p. 21.

94. *Wilderness*, Spring 1995, pp. 28–29.

95. Andrew Greeley, "Religion and Attitudes toward the Environment," *Journal for the Scientific Study of Religion* 32:1 (March 1993): 19–28.

96. *Audubon*, September–October 1994, p. 44.

97. Barry Commoner's complementary axiom was "You can't do just one thing." Barry Commoner, *The Closing Circle* (New York: Bantam, 1972), p. 29.

98. Naess, *Ecology, Community and Lifestyle*, p. 36.

99. See, for instance, Paul Ruffins, "Save the Earth: Fight Crime," *Audubon*, January–February 1995, p. 120.

100. David W. Orr, *Ecological Literacy: Education and the Transition to a Postmodern World* (Albany: SUNY Press, 1992), p. 85.

101. The study was conducted by political scientist Roger Masters, reported in *World Press Review*, September 1997, p. 39. It is scheduled to appear in *Environmental Toxicology: Current Developments*.

102. Michael E. Soulé, *Defenders*, Fall 1994, p. 39.

103. Earth Island Institute, 1996–1997 Annual Report, in *Earth Island Journal*, Summer 1997, p. 3.

104. Charles T. Rubin, *The Green Crusade: Rethinking the Roots of Environmentalism* (New York: Free Press, 1994), p. 243.

105. Paul Ehrlich, *Extinction: The Causes and Consequences of the Disappearance of Species* (New York: Random House, 1981).

106. Mark Shaffer, *Beyond the Endangered Species Act: Conservation in the 21st Century* (Washington, D.C.: The Wilderness Society, 1992), p. 3.

107. Leopold, *Sand County Almanac*, p. 190.

108. See the 1976 Greenpeace Declaration of Interdependence, in *Radical Environmentalism: Philosophy and Tactics*, ed. Peter List (Belmont: Wadsworth Publishing, 1993), pp. 134–36.

109. James Lovelock, *The Ages of Gaia: A Biography of Our Living Planet* (New York: Norton, 1988).

110. Morris Berman, *The Reenchantment of the World* (Ithaca: Cornell University Press, 1981), p. 257.

111. One might discover earlier roots in Gilbert White's eighteenth-century work *Natural History of Selborne*. White's work may be understood as a critique of industrial degradation and social fragmentation and a paean to organic human life integrated into nature. See Donald Worster, *Nature's Economy* (Cambridge: Cambridge University Press, 1994), chap. 1.

112. Leopold, *Sand County Almanac*, p. 263.

Chapter 6

1. Quoted in David Helvarg, *The War against the Greens: The "Wise-Use" Movement, the New Right, and Anti-environmental Violence* (San Francisco: Sierra Club Books, 1994), pp. 77–78.

2. Richard Allen Epstein, *Takings: Private Property and the Power of Eminent Domain* (Cambridge: Harvard University Press, 1985).

3. Quoted in John D. Echeverria, "The Takings Issue," in *Let the People Judge: Wise Use and the Private Property Rights Movement*, ed. John D. Echeverria and Raymond Booth Eby (Washington, D.C.: Island Press, 1995), p. 145. See also Michael Raiden, "Confronting 'Private Property Rights Protection Act,'" *The Pelican*, December 1993, pp. 10–13; Glenn P. Sugameli, "Environmentalism: The Real Movement to Protect Property Rights," in Philip D. Brick and R. McGreggor Cawley, *A Wolf in the Garden: The Land Rights Movement and the New Environmental Debate* (Lanham, Md.: Rowman and Littlefield, 1996), pp. 59–72.

4. *National Wildlife,* December–January 1995/96, p. 50.

5. Helvarg, *War against the Greens*, pp. 8–9.

6. *Time*, October 23, 1995.

7. Quoted in Thomas Lewis, "Cloaked in a Wise Disguise," in *Let the People Judge: Wise Use and the Private Property Rights Movement*, ed. John D. Echeverria and Raymond Booth Eby (Washington, D.C.: Island Press, 1995), p. 14.

8. Quoted in Helvarg, *War against the Greens*, p. 120.

9. Quoted in Lewis, "Cloaked in a Wise Disguise," p. 17.

10. Christopher J. Bosso, "Seizing Back the Day: The Challenge to Environmental Activism in the 1990s," in *Environmental Policy in the 1990s*, 3rd edition, ed. Norman J. Vig and Michael E. Kraft (Washington, D.C.: CQ Press, 1997), p. 60.

11. Thomas A. Wikle, "Geographic Concentrations of Environmental Organization Members," *Environment* 37,9 (November 1995): 44. See also Victor B. Scheffer, *The Shaping of Environmentalism in America* (Seattle: University of Washington Press, 1991), pp. 115–16.

12. *Earth Island Journal,* Summer 1996, p. 24.

13. See Philip D. Brick and R McGreggor Cawley, *A Wolf in the Garden: The Land Rights Movement and the New Environmental Debate* (Lantham, Md.: Rowman and Littlefield, 1996).

14. Philip Brick, "Taking Back the Rural West," in *Let the People Judge: Wise Use and the Private Property Rights Movement*, ed. John D. Echeverria and Raymond Booth Eby (Washington, D.C.: Island Press, 1995), p. 65.

15. *Earth First!*, September–October 1997, p. 2.

16. Mary Hanley, personal interview, Washington, D.C., September 8, 1995.

17. Jim Baca, "People for the West! Challenges and Opportunities," in *Let the People Judge: Wise Use and the Private Property Rights Movement*, ed. John D. Echeverria and Raymond Booth Eby (Washington, D.C.: Island Press, 1995), p. 53.

18. David Brower, *Let the Mountains Talk, Let the Rivers Run: A Call to Those Who Would Save the Earth* (New York: HarperCollins, 1995), p. 190.

19. Helvarg, *War against the Greens*, pp. 290–91.

20. The environment is ranked as the most important problem facing the nation by 4 to 5 percent of respondents. Jacqueline Vaughn Switzer, *Green Backlash: The History and Politics of Environmental Opposition in the U.S.* (Boulder: Lynne Rienner, 1997), pp. 124, 288.

21. S. Robert Lichter misinterprets this fact to suggest that environmentalism itself is not a mainstream value. See S. Robert Lichter, "Liberal Greens, Mainstream Camoflage," *Wall Street Journal*, April 10, 1995, A10.

22. Willett Kempton, James Boster, and Jennifer Hartley, *Environmental Values in American Culture* (Cambridge: MIT Press, 1995), pp. 196–97.

23. Polls conducted by Gallup, ABC News/*Washington Post*, Peter Hart Research Associates, Times-Mirror, *Time*-CNN, the National Opinion Research Center, and the Wirthlin Group, as reported in Riley Dunlap, "Trends in Public Opinion Toward Environmental Issues: 1965–1990," in *American Environmentalism: The U.S. Environmental Movement 1970–1990*, ed. Riley Dunlap and Angel Mertig (New York: Taylor and Francis, 1992), pp. 89–116; Jacqueline Vaughn Switzer with Gary Bryner, *Environmental Politics: Domestic and Global Dimensions*, 2nd edition (New York: St. Martin's Press, 1998), pp. 18, 366; Mark Dowie, *Losing Ground: American Environmentalism at the Close of the Twentieth Century* (Cambridge: MIT Press, 1995), p. 4; Kempton et al., *Environmental Values in American Culture*, pp. 5, 49, 134–35, 216, 270; *LCV Insider*, Winter 1995, p. 2; *General Social Survey 1993–94*, (Chicago: National Opinion Research Center, 1994); Christopher J. Bosso, "Seizing Back the Day: The Challenge to Environmental Activism in the 1990s," in *Environmental Policy in the 1990s*, 3rd edition, ed. Norman J. Vig and Michael E. Kraft (Washington, D.C.: CQ Press, 1997), p. 56; Riley E. Dunlap, George H. Gallup, and Alec M. Gallup, "Of Global Concern: Results of the Health of the Planet Survey," *Environment* 39,9 (November 1993), p. 14; *National Wildlife*, April–May 1995, pp. 34–36, and October–November 1996, p. 58; *Amicus Journal*, Fall 1994, p. 25; *Sierra*, July–August 1994, p. 84a, and November–December 1995, p. 28; *Nature Conservancy*, September–October 1994, p. 7; *Everyone's Backyard*, Fall 1996, p. 11.

24. Dunlap et al., "Of Global Concern," p. 11.

25. Poll conducted by TNC, *Nature Conservancy*, September–October 1994, p. 7.

26. Kempton et al., *Environmental Values in American Culture*, p. 7.

27. 1993 Environmental Opinion Survey, League of Conservation Voters, *1994 Election Report*, p. 14.

28. Kempton et al., *Environmental Values in American Culture*, p. 226; Ronald Inglehart, "Public Support for Environmental Protection: Objective Problems and Subjective Values in 43 Societies," *PS: Political Science and Politics* 28 (March 1995): 65–66, 68; John Pierce, Mary Ann Steger, Brent Steel, and Nicholas Lovrich, *Citizens, Political Communication, and Interest Groups: Environmental Organizations in Canada and the United States* (West-

port, Conn.: Praeger, 1992), p. 43. In an effort to tap this trend, the Sierra Club chose 23-year-old Adam Werbach as its president in 1996. Werbach, who founded the now 30,000 member strong Sierra Student Coalition while in high school, was deemed young and "hip" enough to appeal to the MTV crowd. *The Planet*, July–August 1996, p. 8.

29. In 1966, when David Brower bought advertising space in *The New York Times, The Washington Post*, the *San Francisco Chronicle*, and the *Los Angeles Times* protesting the proposed damming of the Colorado River in the Grand Canyon, he included "coupons" for readers to send to members of Congress. Within twenty-four hours of the advertisement's appearance, the Internal Revenue Service revoked the Sierra Club's charitable tax status, having concluded that the advertising was primarily oriented to affect legislation. Contributions to the Club were no longer tax deductible, a change of status that cost the organization an estimated half a million dollars in revenues. It was largely for this reason that the Sierra Club Legal Defense Fund was created in 1971 as a separate, nonpolitical organization eligible for tax-deductible contributions. Likewise, in 1987 Greenpeace USA split into a charitable organization (Greenpeace Fund) donations to which are tax deductible and an advocacy organization (Greenpeace, Inc.) donations to which are not tax deductible. In 1976, Congress passed a law allowing "C3" organizations to spend up to 20 percent of their revenues on lobbying without losing their preferred tax status.

30. Hartmut Schneider, "Conclusion," in *Environmental Education: An Approach to Sustainable Development*, ed. Hartmut Schneider, (Paris: Organization for Economic Co-operation and Development, 1993), p. 249.

31. *General Social Survey 1993–94* (Chicago: National Opinion Research Center, 1994). The United States ranks eighth among twenty-one countries surveyed in levels of environmental and scientific knowledge. See also Tom W. Smith, "Environmental and Scientific Knowledge around the World," GSS Cross-National Report No. 16, University of Chicago, National Opinion Research Center, 1996.

32. Bryan G. Norton, *Toward Unity among Environmentalists* (New York: Oxford University Press, 1991), p. 250.

33. Gaylord Nelson, personal interview, Washington, D.C., September 8, 1995.

34. *Earth Island Journal*. Summer 1996, p. 17. See Bruce Selcraig, "Reading, 'Riting, and Ravaging: The Three Rs, Brought to You by Corporate America and the Far Right," *Sierra*, May/June 1998, pp. 60–65, 86–92.

35. *The ZPG Reporter*, July–August 1997, pp. 1, 6–7; Zero Population Growth letter to members, October 1997.

36. National Wildlife Federation, *1994 Annual Report*, p. 4.

37. Rodger Schlickeisen, personal interview, Washington, D.C., September 8, 1995.

38. Pierce et al., *Citizens, Political Communication, and Interest Groups*, pp. 106–7.

39. General education, not only environmental education, strongly correlates with environmental values. Recent studies reconfirming these findings include John D. Skrentny, "Concern for the Environment: A Cross National Perspective," *International Journal of Public Opinion Research* 5,4 (Winter 1993): 335–52; David Scott and Fern K. Willets, "Environmental Attitudes and Behavior," *Environment and Behavior* 26,2 (March 1994): 255–56; and Susan Howell and Shirley Laska, "The Changing Face of the Environmental Coalition," *Environment and Behavior* 24,1 (January 1992): 134–44.

40. Matthias Finger, "From Knowledge to Action? Exploring the Relationships Between Environmental Experiences, Learning, and Behavior," *Journal of Social Issues* 50,3 (Fall 1994): 147, 157–58.

41. Riley Dunlap and Angel Mertig, eds., *American Environmentalism: The U.S. Environmental Movement 1970–1990* (New York: Taylor and Francis, 1992), p. 8.

42. Kempton et al., *Environmental Values in American Culture*, p. 226.

43. Associated Press poll, November 1995, *The Gainesville Sun,* December 3, 1995, p. 3A.

44. See Gary E. Varner, "Can Animal Rights Activists Be Environmentalists?" in *Environmental Philosophy and Environmental Activism*, ed., Don E. Marietta, Jr., and Lester Embree (Lanham, Md.: Rowman and Littlefield, 1995), pp. 169–201.

45. Linda Starke, "Foreword," in Alan Thein Durning, *How Much Is Enough? The Consumer Society and the Future of the Earth* (New York: Norton, 1992), p. 11; *Action Alert*, February 1995, p. 1; Jacques Yves Cousteau, Cousteau Society letter to members, May 1995.

46. See John Adams, Executive Director, Natural Resource Defense Council letter, November 1995, p. 5; *Action Alert*, February 1995, p. 1; Jacques Yves Cousteau, Cousteau Society letter to members, May 1995.

47. Dunlap et al., "Of Global Concern," p. 15.

48. Durning, *How Much Is Enough?* pp. 49–50.

49. Paul Harrison, *The Third Revolution: Population, Environment and a Sustainable World* (New York: Penguin, 1993), pp. 256–57. See also Norman Myers, *Ultimate Security: The Environmental Basis of Political Stability* (Washington, D.C.: Island Press, 1993), pp. 161–62.

50. Mahathir Mohamad, "Statement to the U.N. Conference on Environment and Development," in *Green Planet Blues: Environmental Politics from Stockholm to Rio*, ed. Ken Conca, Michael Alberty, and Geoffrey Dabelko (Boulder: Westview Press, 1995), p. 287.

51. See Paul Ehrlich and Ann Ehrlich, *Healing the Planet* (Reading, Mass.: Addison-Wesley, 1991), pp. 8–9.

52. Christopher Flavin and Odil Tunali, *Climate of Hope: New Strategies for Stabilizing the World's Atmosphere,* Worldwatch Paper 130 (Washington, D.C.: Worldwatch Institute, 1996), p. 32.

53. Clive Ponting, *A Green History of the World* (New York: Penguin Books, 1991), p. 403.

54. *Sierra,* July–August 1996, p. 39.

55. Nationwide Personal Transportation Survey, *Gainesville Sun,* September 21, 1997, 5A.

56. John Turner and Jason Rylander, "Land Use: The Forgotten Agenda," in Marian R. Chertow and Daniel C. Esty, ed. *Thinking Ecologically: The Next Generation of Environmental Policy* (New Haven, Conn.: Yale University Press, 1997), p. 62.

57. John S. Dryzek, *The Politics of the Earth: Environmental Discourses* (Oxford: Oxford University Press, 1997), pp. 137–38.

58. *General Social Survey 1993–94* (Chicago: National Opinion Research Center, 1994).

59. Stephen Lester, "Recycling—A Modern Success Story," *Everyone's Backyard,* Winter 1996, p. 9; Frederick Allen and Gregg Sekscienski, "Greening at the Grassroots: What Polls Say about American's Environmental Commitment," *EPA Journal* 18,4 (September–October 1992): 52–53.

60. Nationwide poll conducted by Roper Starch Worldwide in March 1995, reported in *Friends of the Earth,* March–April 1996, p. 4.

61. Frederick W. Allen and Roy Popkin, "Environmental Polls: What They Tell Us," *EPA Journal* 14,6 (July–August 1988): 10. See also John Gillroy and Robert Shapiro, "The Polls: Environmental Protection," *Public Opinion Quarterly* 50 (1986): 270–79.

62. Finger, "From Knowledge to Action?" p. 158.

63. David Scott and Fern K. Willets, "Environmental Attitudes and Behavior," *Environment and Behavior* 26,2 (March 1994): 255.

64. Finger, "From Knowledge to Action?" pp. 141–60.

65. These are the results of a statistical analysis of interview data collected by the author from seventy-one individuals. Interviewees were asked to indicate in which environmental activities they regularly engaged. The listed activities included recycling, reducing consumption, reusing materials, composting, green shopping, limiting family size, taking public transportation, riding a bicycle, participating in local environmental protection, educating the public, voting on the basis of candidates' environmental record or platform, writing to congresspeople, attending hearings, lobbying, and protesting or demonstrating.

Wise Use/Property Rights activists actually scored quite high in their levels of participation, ahead of the least active category of environmentalist. However, this was due to their heavy involvement in advocacy work and political involvement. When these "political" activities were deleted from the list, leaving only explicitly environmental activities (i.e., the first nine activities listed above), Wise Use/Property Rights activists fell to last place.

Owing to the small sample size, the statistical tests conducted were limited and the findings must be considered tentative. Nonetheless, *T* tests (which measure differences in means) demonstrated that both volunteer environmental activists and staff members and executives of environmental organizations scored significantly higher in their levels of environmental behavior when measured against environmentalists with no group affiliation (*P* values of .0005 and .0465, respectively), against nonenvironmentalists (*P* values of .0001 and .0003, respectively), and against Wise Use movement members (*P* values of .0001 and .0001). Nonactivist environmentalists who belong to environmental groups scored significantly higher in their levels of environmental behavior when measured against Wise Use movement members, against nonenvironmentalists, and to a lesser extent, against environmentalists who did not belong to any groups (*P* values of .0001, .0004, and .0753, respectively). Environmentalists with no group affiliation scored significantly higher in their levels of environmental behavior when measured against Wise Use members and marginally so against nonenvironmentalists (*P* values of .0086 and .1007, respectively). Nonenvironmentalists scored higher, at a marginal level of significance, in their levels of environmental behavior when measured against members of Wise Use organizations (*P* value of .0993).

These findings do not prove that membership in an environmental group actually causes or increases environmentally responsible behavior. Those who already engage in such behavior may be predisposed to join environmental groups. See also Jody M. Hines, Harold R. Hungerford, and Audrey N. Tomera, "Analysis and Synthesis of Research on Responsible Environmental Behavior: A Meta-Analysis," *Journal of Environmental Education* 18,2 (Winter 1986/87): 4.

66. Archibald P. Sia, Harold R. Hungerford, and Audrey N. Tomera, "Selected Predictors of Responsible Environmental Behavior: An Analysis," *Journal of Environmental Education* 17,2 (Winter 1985/86): 31–40; Hines et al., "Analysis and Synthesis," pp. 1–8; Harold R. Hungerford and Trudi L. Volk, "Changing Learner Behavior through Environmental Education," *Journal of Environmental Education* 21,3 (Spring 1990): 8–21; Peter Martin, "A WWF View of Education and the Role of NGOs," in John Huckle and Stephen Sterling, *Education for Sustainability* (London: Earthscan, 1996), pp. 40–51.

67. David W. Orr, *Earth in Mind: On Education, Environment, and the Human Prospect* (Washington, D.C.: Island Press, 1994), pp. 46, 205.

68. See Adolf G. Gundersen, *The Environmental Promise of Democratic Deliberation* (Madison: University of Wisconsin Press, 1995), esp. pp. 190, 205.

69. Pierce et al., *Citizens, Political Communication, and Interest Groups*, p. 109.

70. Gregory A. Smith, *Education and the Environment: Learning to Live with Limits* (Albany: SUNY Press, 1992), p. 94.

71. John Elliott, "Teaching and Learning in the Environment," and Michael Mayer, "Quality Indicators and Innovation in Environmental Education," in Centre for Education Research and Innovation, *Environmental Learning for the 21st Century* (Paris: Organization for Economic Co-operation and Development, 1995), pp. 13–30, 31–46. See also C. A. Bowers and David J. Flinders, *Responsive Teaching: An Ecological Approach to Classroom Patterns of Language, Culture and Thought* (New York: Teachers College Press, 1990), p. 247.

72. C. A. Bowers, *Educating for an Ecologically Sustainable Culture* (Albany: SUNY Press, 1995), p. 199.

73. Debra Friedman and Doug McAdam, "Collective Identity and Activism: Networks, Choices, and the Life of a Social Movement," in *Frontiers in Social Movement Theory*, ed. Aldon Morris and Carol McClurg Mueller (New Haven: Yale University Press, 1992), p. 170.

74. Charles C. Mann and Mark L. Plummer, *Noah's Choice: The Future of Endangered Species* (New York: Alfred A. Knopf, 1995), p. 214.

75. *EDF Letter*, January 1997, p. 7.

76. Barry Commoner, "The Environment," in *Crossroads: Environmental Priorities for the Future* ed. Peter Borrelli, (Washington, D.C.: Island Press, 1988), pp. 121–22.

77. Kirkpatrick Sale, *The Green Revolution: The American Environmental Movement 1962–1992* (New York: Hill and Wang, 1993), p. 90.

78. Ibid., p. 94.

79. Brower, *Let the Mountains Talk*, p. 19.

80. Earth Island Institute, *1993 Annual Report*, p. 5.

81. See Lynton Caldwell, *International Environmental Policy* (Durham: Duke University Press, 1992).

82. Rick Bass, "Last Stand for Wilderness," *Amicus Journal*, Summer 1998, p. 33.

83. Gregg Easterbrook, *A Moment on the Earth: The Coming Age of Environmental Optimism* (New York: Viking, 1995), pp. 79, 82.

84. Peter Dykstra, in *Greenpeace*, January–March 1991, p. 2.

85. *The Planet*, December 1997, p. 3.

86. *American PIE*, Winter 1998, p. 7.

87. *Florida Naturalist*, Spring 1996, p. 2; *Audubon*, July–August 1996, p. 8.

88. *Florida Naturalist*, Winter 1998, p. 14.

89. William Ophuls and A. Stephen Boyan, Jr., *Ecology and the Politics of Scarcity Revisited: The Unraveling of the American Dream* (New York: Freeman, 1992), p. 137.

90. Tom Athanasiou, *Divided Planet: The Ecology of Rich and Poor* (Boston: Little, Brown, 1996), pp. 236–37.

91. According to EPA estimates, as reported in the *Gainesville Sun*, December 15, 1995, p. 12A.

92. *An Environmental Agenda for the Future* (Washington, D.C.: Island Press, 1985), p. 67.

93. Peter Montague, "Earthly Necessities: A New Environmentalism for the 1990s," *The Workbook*. (Southwest Research and Information Center), Summer 1991, p. 59.

94. *Twenty-five Years Defending the Environment* (New York: Natural Resources Defense Council, 1995), pp. 12–13.

95. David Pimentel et al., "Environmental and Economic Impacts of Reducing U.S. Agricultural Pesticide Use," in *Handbook on Pest Management in Agriculture*, ed. D. Pimental (CRC Press, 1991), pp. 679–718.

96. *The Gainesville Sun*, June 19, 1997, 8A.

97. Peter Montague, "After Ten Years: Reason for Hope," in *Ten Years of Triumph* (Falls Church, Va.: Citizens Clearinghouse for Hazardous Waste, 1993), p. 15.

98. Sale, *Green Revolution*, p. 83.

99. Aldo Leopold, *A Sand County Almanac, with Essays on Conservation from Round River* (New York: Balantine Books, 1966), p. 210.

100. Paul Wapner, *Environmental Activism and World Civic Politics* (Albany: SUNY Press, 1996), p. 16.

101. Kai N. Lee, *Compass and Gyroscope: Integrating Science and Politics for the Environment* (Washington, D.C.: Island Press, 1993), p. 200. See also Jim MacNeill, Pieter Winsemius, and Taizo Yakushiji, *Beyond Interdependence: The Meshing of the World's Economy and the Earth's Ecology* (New York: Oxford University Press, 1991), pp. 27–28.

102. Quoted in Stephen Viederman, "Ecological Literacy: Can Colleges Save the World?" EFS Essays, Programs, Second Nature http://www.2nature.org/

103. Scheffer, *The Shaping of Environmentalism in America*, p. 197.

Appendix

1. Max Weber, *The Protestant Ethic and the Spirit of Capitalism*, trans. Talcott Parsons (New York: Scribner's, 1958), p. 98.

2. Ibid., p. 233.

3. Lester Milbrath, *Environmentalists: Vanguard for a New Society* (Albany: SUNY Press, 1984), p. 16.

4. Jennifer Hochschild, *What's Fair: American Beliefs about Distributive Justice* (Cambridge: Harvard University Press, 1981), p. 22.

Selected Bibliography

Publications of Environmental Organizations Consulted

Action Alert, Union of Concerned Scientists
American PIE, American Public Information on the Environment
Amicus Journal, Natural Resources Defense Council
Audubon, The National Audubon Society
Calypso Log, The Cousteau Society
Clearinghouse Bulletin, Carrying Capacity Network
Co-op America Quarterly, Co-op America
The Crane, Alachua Audubon Society
Defenders, Defenders of Wildlife
Earth First!, Earth First!
Earth Island Journal, Earth Island Institute
EDF Letter, Environmental Defense Fund
Everyone's Backyard, CCHW: Center for Health, Environment and Justice (formerly Citizens Clearinghouse for Hazardous Waste)
Florida Fish and Wildlife News, Florida Wildlife Federation
Florida Naturalist, Florida Audubon Society
Florida Population Forum, Floridians for a Sustainable Population
Focus, Carrying Capacity Network
Focus, World Wildlife Fund

Friends of the Earth, Friends of the Earth

Greenpeace Quarterly, Greenpeace

Green Politics, The Greens/Green Party USA

Groundwork (formerly *Green Letter*), The Tides Foundation

IFG News, International Forum on Globalization

International Wildlife, National Wildlife Federation

LCV Insider, League of Conservation Voters

The Monitor, Florida Defenders of the Environment

National Environmental Scorecard, League of Conservation Voters

National Wildlife, National Wildlife Federation

Nature Conservancy, The Nature Conservancy

Nature Conservancy Reporter, The Nature Conservancy

NPG Forum, Negative Population Group

Nucleus, Union of Concerned Scientists

The Pelican, Sierra Club Florida Chapter

The Planet: The Sierra Club Activist Resource, Sierra Club

Popline, Population Institute

Pro Earth Times, Pro Earth Times

Public Citizen, Public Citizen

Sierra, Sierra Club

State of the World, The Worldwatch Institute

Suwannnee–St. Johns Group Sierra Club Newsletter, Suwannnee–
    St. Johns Group Sierra Club

Synthesis/Regeneration, The Greens/Green Party USA

Wildearth, The Wildlands Project

Wilderness, The Wilderness Society

Wilderness America, The Wilderness Society

The Workbook, Southwest Research and Information Center

Annual reports, letters to members, and direct-mail solicitations from these and other environmental organizations were also consulted.

Other Sources

Abbey, Edward. *Desert Solitaire: A Season in the Wilderness*. New York: Ballantine Books, 1968.

Abram, David. *The Spell of the Sensuous: Perception and Language in a More-Than-Human World*. New York: Pantheon Books, 1996.

Adams, John. "Letter from the Executive Director." In *Twenty-five Years Defending the Environment*. New York: Natural Resources Defense Council, 1995.

Agarwal, Anil, and Sunity Narain. "Global Warming in an Unequal World: A Case of Environmental Colonialism," *Earth Island Journal*, Spring 1991, p. 40.

Agrawal, Arun. "Community in Conservation: Beyond Enchantment and Disenchantment." Paper prepared for the Conservation and Development Forum, University of Florida, 1997.

Allen, Frederick, and Gregg Sekscienski. "Greening at the Grassroots: What Polls Say about Americans' Environmental Commitment," *EPA Journal* 18,4 (SeptemberOctober 1992): 52–54.

Allen, Frederick, Gregg Sekscienski, and Roy Popkin. "Environmental Polls: What They Tell Us," *EPA Journal* 14,6 (JulyAugust 1988): 52–53.

Anderson, Leslie E. *The Political Ecology of the Modern Peasant*. Baltimore: Johns Hopkins University Press, 1994.

Anderson, Walter Truett. *To Govern Evolution*. Boston: Harcourt Brace Jovanovich, 1987.

Athanasiou, Tom. *Divided Planet: The Ecology of Rich and Poor*. Boston: Little, Brown, 1996.

Baca, Jim. "People for the West! Challenges and Opportunities." In *Let the People Judge: Wise Use and the Private Property Rights Movement*, ed. John D. Echeverria and Raymond Booth Eby. Washington, D.C.: Island Press, 1995, pp. 53–56.

Barry, John. "Sustainability, Political Judgement, and Citizenship: Connecting Green Politics and Democracy." In *Democracy and Green Political Thought*, ed. Brian Doherty and Marius de Geus. London: Routledge, 1996, pp. 115–31.

Bateson, Gregory. *Steps to an Ecology of Mind*. New York: Ballantine Books, 1972.

Beales, A. F. C. *The History of Peace: A Short Account of the Organized Movements for International Peace*. London: G. Bell and Sons, 1931.

Benedick, Richard Elliot. "Equity and Ethics in a Global Climate Convention." In *Taking Sides: Clashing Views on Controversial Environmental Issues*, 5th edition, ed. Theodore Goldfarb. Guilford, Conn.: Dushkin Publishing, 1993, pp. 306–14.

Benton, L. M. "Selling the Natural or Selling Out? Exploring Environmental Merchandising," *Environmental Ethics* 17 (Spring 1995): 3–23.

Berman, Morris. *The Reenchantment of the World*. Ithaca: Cornell University Press, 1981.

Berry, Thomas. *The Dream of the Earth*. San Francisco: Sierra Club Books, 1988.

Berry, Wendell. *The Gift of Good Land: Further Essays Cultural and Agricultural*. San Francisco: North Point Press, 1981.

———. *Sex, Economy, Freedom and Community*. New York: Pantheon Books, 1993.

————. *The Unsettling of America: Culture and Agriculture.* San Francisco: Sierra Club, 1986.

————. *What Are People For?* San Francisco: North Point Press, 1990.

Bezdek, Roger H. "Environment and Economy: What's the Bottom Line?" *Environment* 35, 7 (September 1993): 7–11, 25–31.

Birch, Charles, and John Cobb, Jr. *The Liberation of Life: From the Cell to the Community.* Cambridge: Cambridge University Press, 1981.

Bookchin, Murray, and Dave Foreman. *Defending the Earth.* Boston: South End Press, 1991.

————. *Remarking Society: Pathways to a Green Future.* Boston: South End Press, 1990.

Bosso, Christopher J. "After the Movement: Environmental Activism in the 1990s." In *Environmental Policy in the 1990s*, 2nd edition, ed. Norman J. Vig and Michael E. Kraft, Washington, D.C.: CQ Press, 1994.

————. "Seizing Back the Day: The Challenge to Environmental Activism in the 1990s." In *Environmental Policy in the 1990s*, 3rd edition, ed. Norman J. Vig and Michael E. Kraft. Washington, D.C.: CQ Press, 1997, pp. 53–74.

Bowers, C. A. *Educating for an Ecologically Sustainable Culture.* Albany: SUNY Press, 1995.

Bowers, C. A., and David J. Flinders. *Responsive Teaching: An Ecological Approach to Classroom Patterns of Language, Culture and Thought.* New York: Teachers College Press, 1990.

Brick, Philip D., and R. McGreggor Cawley. "Taking Back the Rural West." In *Let the People Judge: Wise Use and the Private Property Rights Movement*, ed. John D. Echeverria and Raymond Booth Eby. Washington, D.C.: Island Press, 1995, pp. 61–65.

————, eds. *A Wolf in the Garden: The Land Rights Movement and the New Environmental Debate.* Lanham, Md.: Rowman and Littlefield, 1996.

Bright, Chris. "Tracking the Ecology of Climate Change." In Lester R. Brown et al., *State of the World 1997.* New York: Norton, 1997, pp. 78–94.

Brock, Lothar, "Security through Defending the Environment: An Illusion?" In *New Agendas for Peace Research: Conflict and Security Reexamined*, ed. Elise Boulding. Boulder: Lynne Rienner, 1992, pp. 79–102.

Brower, David. *Let the Mountains Talk, Let the Rivers Run: A Call to Those Who Would Save the Earth.* New York: HarperCollins, 1995.

Brown, Lester, Christopher Flavin, and Sandra Postel. *Saving the Planet: How to Shape an Environmentally Sustainable Global Economy.* New York: Norton, 1991.

Brown, Michael, and John May. *The Greenpeace Story.* New York: Dorling Kindersley, 1991.

Bryant, Bunyan. "Summary." In *Environmental Justice: Issues, Policies, and Solutions*, ed. Bunyan Bryant. Washington, D.C.: Island Press, 1995, pp. 208–19.

Bullard, Robert. *Confronting Environmental Racism: Voices from the Grassroots*, Boston: South End Press, 1993.

———. *Dumping in Dixie: Race, Class and Environmental Quality*. Boulder: Westview Press, 1990.

———. "Environmental Racism and the Environmental Justice Movement." In *Ecology: Key Concepts in Critical Theory*, ed. Carolyn Merchant. Atlantic Highlands, N.J.: Humanities Press, 1994, pp. 254–65.

———. *Unequal Protection: Environmental Justice and Communities of Color*. New York: Random House, 1994.

Burke, Edmund. *Reflections on the Revolution in France*. Garden City: Doubleday, 1961.

Buttel, Frederick H. "Rethinking International Environmental Policy in the Late Twentieth Century." In *Environmental Justice: Issues, Policies, and Solutions*, ed. Bunyan Bryant. Washington, D.C.: Island Press, 1995, pp. 187–207.

Cable, Sherry, and Charles Cable. *Environmental Problems/Grassroots Solutions: The Politics of Grassroots Environmental Conflict*. New York: St. Martin's Press, 1995.

Cairncross, Frances. *Green Inc.: A Guide to Business and the Environment*. Washington, D.C.: Island Press, 1995.

Caldwell, Lynton. *International Environmental Policy*. Durham: Duke University Press, 1992.

Capra, Fritjof. "Systems Theory and the New Paradigm." In *Ecology: Key Concepts in Critical Theory*, ed. Carolyn Merchant. Atlantic Highlands, N.J.: Humanities Press, 1994, pp. 334–41.

———. *The Turning Point*. New York: Simon and Schuster, 1982.

———. *The Web of Life: A New Scientific Understanding of Living Systems*. New York: Doubleday, 1996.

Capra, Fritjof, and Charlene Spretnak. *Green Politics*. New York: Dutton, 1984.

Carroll, John E., Paul Brockelman, and Mary Westfall, eds. *The Greening of Faith: God, the Environment, and the Good Life*. University of New Hampshire Press, 1996.

Carson, Rachel. *Silent Spring*. Boston: Houghton Mifflin, 1962.

Citizens Clearinghouse for Hazardous Waste. *Ten Years of Triumph*. Falls Church, Va.: Citizens Clearinghouse for Hazardous Waste, 1993.

Colborn, Theo, Diane Dumanoski, and John Peterson Myers. *Our Stolen Future*. New York: Dutton Books, 1996.

Commoner, Barry. *The Closing Circle*. New York: Bantam, 1972.

———. "The Environment." In *Crossroads: Environmental Priorities for the Future*, ed. Peter Borrelli. Washington, D.C.: Island Press, 1988, pp. 121–69.

———. "How Poverty Breeds Overpopulation (and Not the Other Way Around)," *Ramparts* (1974): 21–25, 58–59. Excerpted in *Ecology: Key*

Concepts in Critical Theory, ed. Carolyn Merchant. Atlantic Highlands, N.J.: Humanities Press, 1994, pp. 88–95.

————. *Making Peace with the Planet*. New York: Pantheon Books, [1975] 1990.

Conca, Ken, Michael Alberty, and Geoffrey Dabelko, eds. *Green Planet Blues: Environmental Politics from Stockholm to Rio*. Boulder: Westview Press, 1995.

Connors, Donald L., Michael Bliss, and Jack Archer. "The U.S. Environmental Industry and the Global Marketplace for Environmental Goods and Services." In *Let the People Judge: Wise Use and the Private Property Rights Movement*, ed. John D. Echeverria and Raymond Booth Eby. Washington, D.C.: Island Press, 1995, pp. 223–38.

Costanza, Robert. "Three General Policies to Achieve Sustainability." In *Investing in Natural Capital: The Ecological Economics Approach to Sustainability.*, ed. A. Jansson, M. Hammer, C. Folke, and R. Costanza. Washington, D.C.: Island Press, 1994, pp. 392–407.

————, ed. *Ecological Economics*. New York: Columbia University Press, 1991.

Cox, Susan Jane Buck. "No Tragedy on the Commons," *Environmental Ethics* 7 (Spring 1985): 49–61.

Daily, Gretchen C., ed. *Nature's Services: Societal Dependence on Natural Ecosystems*. Washington, D.C.: Island Press, 1997.

Dalton, Russel J. *The Green Rainbow: Environmental Groups in Western Europe*. New Haven: Yale University Press, 1994.

Daly, Herman E. "Farewell Lecture to the World Bank," *Focus* (Carrying Capacity Network) 4, 2 (1994): 9.

————. *Steady-State Economics*. San Francisco: Freeman, 1977.

————. "The Steady-State Economy: Toward a Political Economy of Biophysical Equilibrium and Moral Growth." In ed. Herman Daly *Toward a Steady-State Economy*, San Francisco: Freeman, 1973, pp. 149–174.

Daly, Herman E., and John B. Cobb, Jr. *For the Common Good: Redirecting the Economy toward Community, the Environment, and a Sustainable Future*, 2nd edition. Boston: Beacon Press, 1994.

Davis, Henry Vance. "The Environmental Voting Record of the Congressional Black Caucus." In *Race and the Incidence of Environmental Hazards*, ed. Bunyan Bryant and Paul Mohai. Boulder: Westview Press, 1992, pp. 55–63.

Deslarzes, Luc P. "Strategic Elements of the WWF Environmental Education Programme for the 1990s." In *Environmental Education: An Approach to Sustainable Development,* ed. Hartmut Schneider. Paris: Organization for Economic Co-operation and Development, 1993, pp. 109–16.

Deudney, Daniel. "The Case against Linking Environmental Degradation and National Security," *Millennium: Journal of International Studies* 19,3 (Winter 1990): 461–76.

Devall, Ross, and George Sessions. *Deep Ecology.* Salt Lake City: Gibbs M. Smith, 1985.

Dizard, Jan E. "Going Wild: The Contested Terrain of Nature." In *In the Nature of Things,* ed. Jane Bennett and William Chaloupka. Minneapolis: University of Minnesota Press, 1993, pp. 111–35.

Doherty, Brian, and Marius de Geus, eds. *Democracy and Green Political Thought.* London: Routledge, 1996.

Dowie, Mark. *Losing Ground: American Environmentalism at the Close of the Twentieth Century.* Cambridge: MIT Press, 1995.

Dryzek, John S. *The Politics of the Earth: Environmental Discourses.* Oxford: Oxford University Press, 1997.

———. *Rational Ecology: Environment and Political Economy.* New York: Blackwell, 1987.

Dunlap, Riley, and Angel Mertig, "Trends in Public Opinion toward Environmental Issues: 1965–1990." In *American Environmentalism: The U.S. Environmental Movement 1970–1990,* ed. Riley Dunlap and Angel Mertig. New York: Taylor and Francis, 1992, pp. 89–116.

———, eds. *American Environmentalism: The U.S. Environmental Movement 1970–1990.* New York: Taylor and Francis, 1992.

Dunlap, Riley, Angel Mertig, George H. Gallup, and Alec M. Gallup. "Of Global Concern: Results of the Health of the Planet Survey," *Environment.* 39 (November 1993): 6–23.

Durenberger, David. "A Dissenting Voice," *EPA Journal,* MarchApril 1991. In *Taking Sides: Clashing Views on Controversial Environmental Issues,* 5th edition. ed. Theodore Goldfarb. Guilford, Conn. Dushkin Publishing, 1993, pp. 97–100.

Durning, Alan Thein. "Redesigning the Forest Economy." In *State of the World, 1994,* Lester Brown et al. New York: Norton, 1994, pp. 22–52.

The Earth Works Group. *50 Simple Things You Can Do to Save the Earth.* Berkeley: Earthworks Press, 1989.

Easterbrook, Gregg. *A Moment on the Earth: The Coming Age of Environmental Optimism.* New York: Viking, 1995.

Echeverria, John D. "The Takings Issue." In *Let the People Judge: Wise Use and the Private Property Rights Movement,* ed. John D. Echeverria and Raymond Booth Eby. Washington, D.C.: Island Press, 1995, pp. 143–50.

Ehrenfeld, David. *The Arrogance of Humanism.* New York: Oxford University Press, 1978.

Ehrlich, Paul R., and Peter Raven. "Butterflies and Plants: A Study in Co-evolution," *Evolution* 18 (1965): 586–608.

———. *The Population Bomb.* New York: Ballantine Books, 1968.

Ehrlich, Paul R., Peter Raven, and Ann Ehrlich. *Healing the Planet.* Reading, Mass.: Addison-Wesley, 1991.

Elliott, John. "Teaching and Learning in the Environment." In *Centre for Education Research and Innovation, Environmental Learning for the 21st Century*. Paris: Organization for Economic Co-operation and Development, 1995, pp. 13–30.

Ellis, Richard J., and Fred Thompson. "Culture and the Environment in the Pacific Northwest," *American Political Science Review* 91 (December 1997): 885–97.

An Environmental Agenda for the Future. Washington, D.C.: Island Press, 1985.

Environmental Careers Organization. *Beyond the Green: Redefining and Diversifying the Environmental Movement*. Boston: Environmental Careers Organization, 1992.

Epstein, Richard Allen. *Takings: Private Property and the Power of Eminent Domain*. Cambridge: Harvard University Press, 1985.

Faeth, P. *Paying the Farm Bill*. Washington, D.C.: World Resources Institute, 1991.

Fairfield, Osborn. *Our Plundered Planet*. Boston: Little, Brown, 1948.

Feeney, David, Fikret Berkes, Bonnie J. McCay, and James M. Acheson. "The Tragedy of the Commons: Twenty-two Years Later," *Human Ecology* 18,1 (1990): 1–19.

Ferrier, Grant. "Strategic Overview of the Environmental Industry." In *Let the People Judge: Wise Use and the Private Property Rights Movement*, ed. John D. Echeverria and Raymond Booth Eby. Washington, D.C.: Island Press, 1995, pp. 208–22.

Ferris, Deeohn, and David Hahn-Baker. "Environmentalists and Environmental Justice Policy." In *Environmental Justice: Issues, Policies, and Solutions*, ed. Bunyan Bryant. Washington, D.C.: Island Press, 1995, pp. 66–75

Finger, Matthias. "From Knowledge to Action? Exploring the Relationships between Environmental Experiences, Learning, and Behavior," *Journal of Social Issues* 50,3 (1994):141–60.

Flavin, Christopher, and Odil Tunal. *Climate of Hope: New Strategies for Stabilizing the World's Atmosphere*. Worldwatch Paper 130. Washington, D.C.: Worldwatch Institute, 1996.

Foreman, Dave. *Confessions of an Eco-Warrior*. New York: Crown Trade Paperbacks, 1991.

———. "Wilderness: From Scenery to Nature," *Wild Earth*, Winter 1995/96, pp. 8–16.

Foreman, D. Davis, D. Johns, R. Noss, and M. Soulé. "The Wildlands Project Mission Statement," *Wild Earth*, special issue, 1992.

Foster, John Bellamy. *The Vulnerable Planet: A Short Economic History of the Environment*. New York: Monthly Review Press, 1994.

Fowler, Robert Booth. *The Greening of Protestant Thought*. Chapel Hill: University of North Carolina Press, 1995.

Fox, Stephen. *John Muir and His Legacy: The American Conservation Movement*. Boston: Little, Brown, 1981.

Fox, Warwick. *Toward a Transpersonal Ecology: Developing New Foundations for Environmentalism*. Boston: Shambhala, 1990.

Frank, Jerome, and Earl Nash. "Commitment to Peace Work," *American Journal of Orthopsychiatry* 35 (1965): 115.

French, Hilary F. "Learning from the Ozone Experience." In Lester R. Brown et al., *State of the World 1997*. New York: Norton, 1997, pp. 151–72.

Freudenberg, Nicholas, and Carol Steinsapir. "Not in Our Backyards: The Grassroots Environmental Movement." In *American Environmentalism: The U.S. Environmental Movement 1970–1990*, Riley Dunlap and Angela Mertig. ed. New York: Taylor and Francis, 1992, pp. 27–37.

Friedman, Debra, and Doug McAdam. "Collective Identity and Activism: Networks, Choices, and the Life of a Social Movement." In *Frontiers in Social Movement Theory*, ed. Aldon Morris and Carol McClurg Mueller. New Haven, Conn.: Yale University Press, 1992, pp. 156–73.

Friends of the Earth. *Dirty Little Secrets*. 1995.

Friends of the Earth and the National Taxpayers Union Foundation. *The Green Scissors Report: Cutting Wasteful and Environmentally Harmful Spending and Subsidies*. 1995.

Fuller, R. Buckminster. *Operating Manual for Spaceship Earth*. New York: Dutton, 1971.

Gamson, William A. *Talking Politics*. Cambridge: Cambridge University Press, 1992.

Gelbspan, Ross. "The Heat Is On: The Warming of the World's Climate Sparks a Blaze of Denial," *Harpers Magazine*, December 1995, pp. 31–37.

Gibbs, Lois Marie, and Karen J. Stults. "On Grassroots Environmentalism." In *Crossroads: Environmental Priorities for the Future*, ed. Peter Borrelli. Washington, D.C.: Island Press, 1988, pp. 241–46.

Gillroy, John, and Robert Shapiro. "The Polls: Environmental Protection," *Public Opinion Quarterly* 50 (1986): 270–79.

Gillroy, John Martin. "Public Policy and Environmental Risk: Political Theory, Human Agency and the Imprisoned Rider," *Environmental Ethics* 14 (Fall 1992): 217–37.

Gordon, Josh, and Jane Coppock. "Ecosystem Management and Economic Development." In Marian R. Chertow and Daniel C. Esty, ed. *Thinking Ecologically: The Next Generation of Environmental Policy*. New Haven: Yale University Press, 1997, pp. 37–48.

Gore, Al. *Earth in the Balance: Ecology and the Human Spirit*. Boston: Houghton Mifflin, 1992.

Gottlieb, Robert. *Forcing the Spring: The Transformation of the American Environmental Movement*. Washington, D.C.: Island Press, 1993.

Gottlieb, Roger, ed. *The Ecological Community*. New York: Routledge, 1997.

Gould, Stephen J. *Bully for Brontosaurus: Reflections in Natural History*. New York: Norton, 1991.

Gowdy, John. *Coevolutionary Economics: The Economy, Society and the Environment*. Boston: Kluwer, 1994.

Greenpeace. "Greenpeace Declaration of Interdependence." In *Radical Environmentalism: Philosophy and Tactics*, ed. Peter List. Belmont: Wadsworth Publishing, 1993.

Greeley, Andrew. "Religion and Attitudes toward the Environment," *Journal for the Scientific Study of Religion* 32,1 (March 1993): 19–29.

Guha, Ramachandra. "Radical Environmentalism: A Third-World Critique." In *Ecology: Key Concepts in Critical Theory*, ed. Carolyn Merchant. Atlantic Highlands, N.J.: Humanities Press, 1994, pp. 281–89.

Gundersen, David W. *The Environmental Promise of Democratic Deliberation*. Madison: University of Wisconsin Press, 1995.

Hardin, Garrett. *The Limits of Altruism: An Ecologist's View of Survival*. Bloomington: Indiana University Press, 1977.

———. *Living within Limits: Ecology, Economics, and Population Taboos*. New York: Oxford University Press, 1993.

———. "The Tragedy of the Commons," *Science* 162 (December 13, 1968): 1243–48.

Harrison, Paul. *The Third Revolution: Population, Environment and a Sustainable World*. New York: Penguin, 1993.

Harvey, David. *The Condition of Postmodernity: An Inquiry into the Origins of Cultural Change*. Cambridge: Blackwell Publishers, 1989.

Hawken, Paul. *The Ecology of Commerce: A Declaration of Sustainability*. New York: HarperCollins, 1993.

Hayden, Tom. *The Lost Gospel of the Earth: A Call for Renewing Nature, Spirit and Politics*. San Francisco: Sierra Club Books, 1996.

Hays, Samuel P. *Beauty, Health and Permanence: Environmental Politics in the United States, 1955–1985*. Cambridge: Cambridge University Press, 1987.

———. *Conservation and the Gospel of Efficiency: The Progressive Conservation Movement*. Cambridge: Harvard University Press, 1958.

———. "From Conservation to Environment: Environmental Politics in the United States Since World War II." In *Environmental History: Critical Issues in Comparative Perspective*, ed. Kendall E. Bailes. Lanham, Md.: University Press of America, pp. 198–241.

Heilbroner, Robert L. *An Inquiry into the Human Prospect*. New York: Norton, 1974.

———. *An Inquiry into the Human Prospect: Looked at Again for the 1990s*. New York: Norton, 1991.

Helvarg, David. *The War against the Greens: The "Wise-Use" Movement, the*

New Right, and Anti-environmental Violence. San Francisco: Sierra Club Books, 1994.

Hines, Jody M., Harold R. Hungerford, and Audrey N. Tomera. "Analysis and Synthesis of Research on Responsible Environmental Behavior: A Meta-Analysis," *Journal of Environmental Education* 18 (Winter 1986/87): 1–8.

Hochschild, Jennifer. *What's Fair: American Beliefs about Distributive Justice.* Cambridge: Harvard University Press, 1981.

Homer-Dixon, Thomas F. "Environmental Scarcities and Violent Conflict: Evidence from Cases," *International Security* 19,1 (1994): 5–40.

Howell, Susan, and Shirley Laska. "The Changing Face of the Environmental Coalition," *Environment and Behavior* 24, 1 (January 1992): 134–44.

Hudson, Stewart. "Principles of Basic Tax Reform." NWF monograph. Washington, D.C.: National Wildlife Federation, 1995.

Hungerford, Harold R., and Trudi L. Volk. "Changing Learner Behavior through Environmental Education," *Journal of Environmental Education* 21 (Spring 1990): 8–20.

Inglehart, Ronald. "Public Support for Environmental Protection: Objective Problems and Subjective Values in 43 Societies," *PS: Political Science and Politics* 28 (March 1995): 57–72.

Jansson, A., Hammer, M., Folke, C., and Costanza, R., eds. *Investing in Natural Capital: The Ecological Economics Approach to Sustainability.* Washington D.C.: Island Press, 1994.

Jennings, Cheri Lucas, and Bruce H. Jennings. "Green Fields/Brown Skin: Posting as a Sign of Recognition." In *In the Nature of Things,* ed. Jane Bennett and William Chaloupka. Minneapolis: University of Minnesota Press, 1993, pp. 173–94.

Johnson, Huey D. "Environmental Quality as a National Purpose." In *Crossroads: Environmental Priorities for the Future,* ed. Peter Borrelli. Washington, D.C.: Island Press, 1988, pp. 217–24.

Jordan, Charles, and Donald Snow. "Diversification, Minorities, and the Mainstream Environmental Movement." In *Voices from the Environmental Movement: Perspectives for a New Era,* ed. Donald Snow. Washington, D.C.: Island Press, 1992, pp. 71–109.

Kane, Hal. "Shifting to Sustainable Industries" In Lester R. Brown et al., *State of the World 1996.* New York: Norton, 1996, pp. 152–67.

Kassiola, Joel Jay. *The Death of Industrial Civilization: The Limits to Economic Growth and the Repoliticization of Advanced Industrial Society.* Albany: SUNY Press, 1990.

Kateb, George. *The Inner Ocean: Individualism and Democratic Culture.* Ithaca: Cornell University Press, 1992.

Katz, Eric, and Lauren Oechsli. "Moving beyond Anthropocentrism: Environmental Ethics, Development, and the Amazon," *Environmental Ethics* 15 (1993): 49–60.

Kellert, Stephen, and Edward O. Wilson, eds. *Biophilia Hypothesis*. Washington, D.C.: Island Press, 1993.

Kelman, Steven. "Cost-Benefit Analysis: An Ethical Critique." In *Readings in Risk*, ed. Theodore S. Glickman and Michael Gough. Washington, D.C.: Resources for the Future, 1990, pp. 55–59.

Kempton, Willett, James Boster, and Jennifer Hartley. *Environmental Values in American Culture*. Cambridge: MIT Press, 1995.

Lee, Kai N. *Compass and Gyroscope: Integrating Science and Politics for the Environment*. Washington, D.C.: Island Press, 1993.

Leiss, William. *The Domination of Nature*. New York; George Braziller, 1972.

Lele, Sharachchandra M. "Sustainable Development: A Critical Review," *World Development* 19, 6 (June 1991): 607–21.

Leopold, Aldo. *A Sand County Almanac, with Essays on Conservation from Round River*. New York: Ballantine Books, 1966.

Lewin, R. "In Ecology, Change Brings Stability," *Science* 234 (1986): 1071–73.

Lewis, C. S. *The Abolition of Man*. London: Geoffrey Bles, 1946.

Lewis, Martin W. *Green Delusions*. Durham: Duke University Press, 1992.

Lewis, Thomas. "Cloaked in a Wise Disguise." In *Let the People Judge: Wise Use and the Private Property Rights Movement*, ed. John D. Echeverria and Raymond Booth Eby. Washington, D.C.: Island Press, 1995, pp. 13–20.

Lohmann, Larry. "Whose Common Future," *The Ecologist* 20, 3 (May–June 1990): 82–84.

Lovelock, James. *The Ages of Gaia: A Biography of Our Living Planet*. New York: Norton, 1988.

Luke, Timothy W. *Ecocritique: Contesting the Politics of Nature, Economy, and Culture*. Minneapolis: University of Minnesota Press, 1997.

Lutzenberger, José. "Science, Technology, Economics, Ethics, and Environment." In *Earth Summit Ethics: Toward a Reconstructive Postmodern Philosophy of Environmental Education*, ed. J. Baird Callicott and Fernando J. R. da Rocha Albany: SUNY Press, 1996, pp. 23–46.

MacKenzie, James J., Roger C. Dower, and Don Chen. *The Going Rate: What It Really Costs to Drive*. Washington, D.C.: World Resources Institute, 1992.

MacNeill, Jim, Pieter Winsemius, and Taizo Yakushiji. *Beyond Interdependence: The Meshing of the World's Economy and the Earth's Ecology*. New York: Oxford University Press, 1991.

Mander, Jerry, and Edward Goldsmith. *The Case against the Global Economy— and for a Turn toward the Local*. San Francisco: Sierra Club Books, 1996.

Manes, Christopher. *Green Rage: Radical Environmentalism and the Unmaking of Civilization*. Boston: Little, Brown, 1990.

Mann, Charles, and Mark L. Plummer. *Noah's Choice: The Future of Endangered Species*. New York: Knopf, 1995.

Martin, Peter. "A WWF View of Education and the Role of NGOs." In *Edu-*

cation for Sustainability, ed. John Huckle and Stephen Sterling. London: Earthscan, 1996, pp. 40–51.

Martinez-Alier, Juan. "Environmental Policy and Distributional Conflicts." In *Ecological Economics*, ed. Robert Costanza. New York: Columbia University Press, 1991, pp. 118–36.

Mathews, Freya. "Community and the Ecological Self." In *Ecology and Democracy*, ed. Freya Mathews. London: Frank Cass, 1996, pp. 66–100.

Mayer, Michael. "Quality Indicators and Innovation in Environmental Education." In *Centre for Education Research and Innovation, Environmental Learning for the 21st Century*. Paris: Organization for Economic Cooperation and Development, 1995, pp. 31–46.

Mayhew, A. "Dangers in Using the Idea of Property Rights: Modern Property Rights Theory and the Neoclassical Trap." *Journal of Economic Issues* 19 (1985): 959–66.

McCloskey, Michael. "Twenty Years of Change in the Environmental Movement: An Insider's View." In *American Environmentalism: The U.S. Environmental Movement 1970–1990*, ed. Riley Dunlap and Angel Mertig New York: Taylor and Francis, 1992, pp. 77–88.

McHenry, Robert, and Charles Van Doren, eds. *A Documentary History of Conservation in America*. New York: Praeger, 1972.

McKibben, Bill. *The End of Nature*. New York: Random House, 1989.

Meadows, Donella H., Dennis L. Meadows, Jorgen Randers, and William W. Behrens III. *The Limits to Growth*. New York: Universe Books, 1974.

Melucci, Alberto. "The Symbolic Challenge of Contemporary Movements," *Social Research* 52 (1985): 789–816.

Merchant, Carolyn. *Earthcare*. New York: Routledge, 1995.

Milbrath, Lester W. "Environmental Understanding: A New Concern for Political Socialization." In *Political Socialization: Citizenship, Education and Democracy*. ed. Orit Ichilov. New York: Teachers College Press, 1990, pp. 281–93.

———. *Envisioning a Sustainable Society*. Albany: SUNY Press, 1989.

———. *Learning to Think Environmentally (While There Is Still Time)*. Albany: SUNY Press, 1996.

Moore, Curtis, and Alan Miller. *Green Gold: Japan, Germany, the United States and the Race for Environmental Technology*. Boston: Beacon Press, 1994.

Mpanya, Mutombo. "The Dumping of Toxic Waste in African Countries: A Case of Poverty and Racism." In *Race and the Incidence of Environmental Hazards*, ed. Bunyan Bryant and Paul Mohai. Boulder: Westview Press, 1992, pp. 204–14.

Mueller, Carol McClurg. "Building Social Movement Theory." In *Frontiers in Social Movement Theory*, ed. Aldon Morris and Carol McClurg Mueller. New Haven, Conn.: Yale University Press, 1992, pp. 3–25.

Muir, John. *My First Summer in the Sierra*. Boston: Houghton Mifflin, 1911.

Myers, Norman. *Ultimate Security: The Environmental Basis of Political Stability*. Washington, D.C.: Island Press, 1993.

Naess, Arne. "A Defense of the Deep Ecology Movement," *Environmental Ethics* 6 (1984): 270.

———. "The Shallow and the Deep, Long-Range Ecology Movement: A Summary," *Inquiry* 16 (1973): 95–100. Reprinted in *Ecology: Key Concepts in Critical Theory*, ed. Carolyn Merchant. Atlantic Highlands, N.J.: Humanities Press, 1994, pp. 120–24.

Nash, Roderick. *The Rights of Nature: A History of Environmental Ethics*. Madison: University of Wisconsin Press, 1989.

———. *Wilderness and the American Mind*, revised edition. New Haven, Conn.: Yale University Press, 1973.

National Opinion Research Center. *General Social Survey, 1994–95*. Chicago: National Opinion Research Center, 1995.

Natural Resource Defense Council. *Felling the Myth: The Role of Timber in the Economy of California's Sierra Nevada*. New York: Natural Resource Defense Council, 1995.

Norgaard, Richard B. "Coevolution of Economy, Society and Environment." In Paul Ekins and Manfred Max-Neef, *Real-life Economics*. London: Routledge, 1992, pp. 76–88.

———. *Development Betrayed: The End of Progress and a Coevolutionary Revisioning of the Future*. London: Routledge, 1994.

Norton, Bryan G. *Toward Unity among Environmentalists*. New York: Oxford University Press, 1991.

O'Neill, John. "Time, Narrative, and Environmental Politics." In *The Ecological Community*, ed. Roger Gottlieb. New York: Routledge, 1997.

Ophuls, William, and A. Stephen Boyan, Jr. *Ecology and the Politics of Scarcity Revisited: The Unraveling of the American Dream*. New York: Freeman, 1992.

Oppenheimer, Michael. "Context, Connection and Opportunity in Environmental Problem Solving," *Environment* 37,5 (June 1995): 10–22.

Orr, David W. *Earth in Mind: On Education, Environment, and the Human Prospect*. Washington, D.C.: Island Press, 1994.

Osborn, Fairfield. O.r Plundered Planet. Boston: Little, Brown, 1948.

Ostrom, Elinor. *Governing the Commons: The Evolution of Institutions for Collective Action*. Cambridge: Cambridge University Press, 1990.

Paehlke, Robert C. *Environmentalism and the Future of Progressive Politics*. New Haven, Conn.: Yale University Press, 1989.

Passmore, John. *Responsibility for Nature*. London: Duckworths, 1974.

Payne, Daniel G. *Voices in the Wilderness: American Nature Writing and Environmental Ethics*. Hanover: University Press of New England, 1996.

Pepper, David. *The Roots of Modern Environmentalism*. London: Croom Helm, 1984.

Petulla, Joseph M. *American Environmentalism: Values, Tactics, Priorities*. College Station: Texas A & M University Press, 1980.

Phelan, Shane. "Intimate Distance: The Dislocation of Nature in Modernity." In *In the Nature of Things*, ed. Jane Bennett and William Chaloupka. Minneapolis: University of Minnesota Press, 1993, pp. 44–62.

Pierce, John, Mary Ann Steger, Brent Steel, and Nicholas Lovrich. *Citizens, Political Communication, and Interest Groups: Environmental Organizations in Canada and the United States*. Westport, Conn.: Praeger, 1992.

Pimentel, David, et al. "Environmental and Economic Costs of Soil Erosion and Conservation Benefits." *Science* 267 (February 14, 1995): 1117–23.

Pimm, Stuart. *The Balance of Nature*. Chicago: University of Chicago Press, 1991.

Pinchot, Gifford. *The Fight for Conservation*. New York: Doubleday, Page, 1910.

Ponting, Clive. *A Green History of the World*. New York: Penguin Books, 1991.

Porter, Michael, and Class van der Linde. *The Competitive Advantage of Nations*. London: Macmillan, 1990.

———. "Green and Competitive: Ending the Stalemate." In *Business and the Environment*, ed. Richard Welford and Richard Starkey. Washington, D.C.: Taylor and Francis, 1996, pp. 61–77.

Powers, Charles W., and Marian R. Chertow, "Industrial Ecology:Overcoming Policy Fragmentation." In Marian R. Chertow and Daniel C. Esty, ed. *Thinking Ecologically: The Next Generation of Environmental Policy*. New Haven: Yale University Press, 1997, pp. 19–36.

Powledge, Fred. "A Time of Change—and Promise." In *Gale Environmental Almanac*, ed. Russ Hoyle. Detroit: Gale Research, 1993, pp. 173–74.

Prefecto, Ivette "Pesticide Exposure of Farm Workers and the International Connection." In *Race and the Incidence of Environmental Hazards*, ed. Bunyan Bryant and Paul Mohai. Boulder: Westview Press, 1992, pp. 177–203.

President's Council on Sustainable Development. *Sustainable Development: A New Consensus*. Internet, 1995. Available http://www.whitehouse.gov/PCSD.

Press, Daniel. *Democratic Dilemmas in the Age of Ecology: Trees and Toxics in the American West*. Durham: Duke University Press, 1994.

Rainforest Action Network. *Ten Years of Rainforest Action*. 1995.

Redford, Kent H., and Allyn M. Stearman. "Forest Dwelling Native Amazonians and the Conservation of Biodiversity: Interests in Common or in Collision?" *Conservation Biology* 7, 2 (1993): 248–55.

Renner, Michael. "Transforming Security." In Lester R. Brown et al., *State of the World 1997*. New York: Norton, 1997, pp. 115–31.

Repetto, Robert. *Jobs, Competitiveness, and Environmental Regulation*. Washington, D.C.: World Resources Institute, 1995.

Repetto, Robert, Dale S. Rothman, Paul Faeth, and Duncan Austin. *Has Envi-*

ronmental Protection Really Reduced Productivity Growth? Washington, D.C.: World Resources Institute, 1996.

Rolston, Holmes, III. *Environmental Ethics: Duties to and Values in the Natural World.* Philadelphia: Temple University Press, 1988.

Roodman, David Malin. *Paying the Piper: Subsidies, Politics, and the Environment.* Worldwatch Paper 133. Washington, D.C.: Worldwatch Institute, 1995.

Rosecrance, Richard. *The Rise of the Trading State: Commerce and Conquest in the Modern World.* New York: Basic Books, 1986.

Roszak, Theodore. *Ecopsychology: Restoring the Earth, Healing the Mind.* San Francisco: Sierra Club Books, 1995.

Roush, G. Jon. "Introduction" and "Conservation's Hour." In *Voices from the Environmental Movement: Perspectives for a New Era,* ed. Donald Snow. Washington, D.C.: Island Press, 1992, pp. 21–40.

Rubin, Charles T. *The Green Crusade: Rethinking the Roots of Environmentalism.* New York: The Free Press, 1994.

Sachs, Aaron. *Eco-Justice: Linking Human Rights and the Environment.* Worldwatch Paper 127. Washington, D.C.: Worldwatch Institute, 1995.

———. "Upholding Human Rights and Environmental Justice." In Lester R. Brown et al., *State of the World 1996* New York: Norton, 1996, pp. 133–51.

Sachs, Wolfgang "Global Ecology and the Shadow of 'Development.' " In *Deep Ecology for the 21st Century,* George Sessions. Boston: Shambala, 1995, pp. 428–44.

Sagoff, Mark. *The Economy of the Earth: Philosophy, Law and the Environment.* Cambridge: Cambridge University Press, 1988.

Sale, Kirkpatrick. *The Green Revolution: The American Environmental Movement 1962–1992.* New York: Hill and Wang, 1993.

Scheffer, Victor B. *The Shaping of Environmentalism in America.* Seattle: University of Washington Press, 1991.

Schmidheiny, Stephan. *Changing Course: A Global Business Perspective on Development and the Environment.* Cambridge: MIT Press, 1992.

Schnaiberg, Allan. *The Environment: From Surplus to Scarcity.* New York: Oxford University Press, 1980.

Schumacher, E. F. *Small Is Beautiful: Economics as if People Mattered.* New York: Harper and Row, 1973.

Schwab, Jim. *Deeper Shades of Green: The Rise of Blue-Collar and Minority Environmentalism in America.* San Francisco: Sierra Club Books, 1994.

Scott, David, and Fern K. Willets. "Environmental Attitudes and Behavior," *Environment and Behavior* 26,2 (March 1994): 239–62.

Seager, Joni. *Earth Follies: Coming to Feminist Terms with the Global Environmental Crisis.* New York: Routledge, 1993.

Serageldein, Ismail, and Richard Barnett, eds. *Ethics and Spiritual Values: Pro-*

moting Environmentally Sustainable Development. Washington, D.C.: The World Bank, 1996.

Shabecoff, Philip. *A Fierce Green Fire: The American Environmental Movement*. New York: Hill and Wang, 1993.

Shaffer, Mark. *Beyond the Endangered Species Act: Conservation in the 21st Century*. Washington, D.C.: The Wilderness Society, 1992.

Shnaiberg, Allan, and Kenneth Alan Gould. *Environment and Society: The Enduring Conflict*. New York: St. Martin's Press, 1994.

Shrader-Frechette, K. S. *Risk and Rationality: Philosophical Foundations for Populist Reforms*. Berkeley: University of California Press, 1991.

———. *Science Policy, Ethics, and Economic Methodology: Some Problems of Technology Assessment and Environmental-Impact Analysis*. Boston: Reidel, 1985.

Sia, Archibald P., Harold R. Hungerford, and Audrey N. Tomera. "Selected Predictors of Responsible Environmental Behavior: An Analysis," *Journal of Environmental Education* 17 (Winter 1985/86): 31–40.

Sikorski, Wade. "Building Wilderness." In *In the Nature of Things*, ed. Jane Bennett and William Chaloupka. Minneapolis: University of Minnesota Press, 1993, pp. 24–43.

Singer, Peter. *Animal Liberation: A New Ethics for Our Treatment of Animals*. New York: Random House, 1976.

Skrentny, John D. "Concern for the Environment: A Cross National Perspective," *International Journal of Public Opinion Research* 5,4 (Winter 1993): 335–52.

Smith, Gregory A. *Education and the Environment: Learning to Live with Limits*. Albany: SUNY Press, 1992.

Smith, Tom W. *Environmental and Scientific Knowledge around the World*. GSS Cross-National Report No. 16. Chicago: University of Chicago, National Opinion Research Center, 1996.

Snodgrass, Ronald. "The Endangered Species Act: A Commitment Worth Keeping." In *Let the People Judge: Wise Use and the Private Property Rights Movement*, ed. John D. Echeverria and Raymond Booth Eby. Washington, D.C.: Island Press, 1995, pp. 278–84.

Snow, Donald. *Inside the Environmental Movement: Meeting the Leadership Challenge*. Washington, D.C.: Island Press, 1992.

Soulé, Michael, and Bruce Wilcox, eds. *Conservation Biology: An Evolutionary Ecological Perspective*. Sunderland, Mass.: Sinauer, 1980.

Spash, Clive L. "Economics, Ethics, and Long-Term Environmental Damages," *Environmental Ethics* 15 (Summer 1993): 117–33.

Spretnak, Charlene. *The Spiritual Dimensions of Green Politics*. Santa Fe/Bear and Co., 1986.

Starke, Linda. "Foreword." In Alan Thein Durning, *How Much Is Enough? The Consumer Society and the Future of the Earth*. New York: Norton, 1992.

Steverson, Brian K. "Contextualism and Norton's Convergence Hypothesis," *Environmental Ethics* 17 (Summer 1995): 135–51.

Sugameli, Glenn P. "Environmentalism: The Real Movement to Protect Property Rights." In Philip D. Brick and R. McGreggor Cawley, *A Wolf in the Garden: The Land Rights Movement and the New Environmental Debate.* Lanham, Md.: Rowman and Littlefield, 1996, pp. 59–72.

Switzer, Jacqueline Vaughn. *Environmental Politics: Domestic and Global Dimensions.* New York: St. Martin's Press, 1994.

———. *Green Backlash: The History and Politics of Environmental Opposition in the U.S.* Boulder: Lynne Rienner, 1997.

Switzer, Jacqueline Vaughn, with Gary Bryner. *Environmental Politics: Domestic and Global Dimensions*, 2nd edition. New York: St. Martin's Press, 1998.

Taylor, Dorceta. "Can the Environmental Movement Attract and Maintain the Support of Minorities?" In *Race and the Incidence of Environmental Hazards*, ed. Bunyan Bryant and Paul Mohai. Boulder: Westview Press, 1992, pp. 28–54.

Thoreau, Henry David. *Great Short Works of Henry David Thoreau.* New York: Harper and Row, 1982.

Tokar, Brian. *Earth for Sale: Reclaiming Ecology in the Age of Corporate Greenwash.* Boston: South End Press, 1997.

Trocki, Linda. "Science, Technology, Environment, and Competitiveness in a North American Context." In *Let the People Judge: Wise Use and the Private Property Rights Movement*, ed. John D. Echeverria and Raymond Booth Eby. Washington, D.C.: Island Press, 1995, pp. 239–52.

Varner, Gary E. "Can Animal Rights Activists Be Environmentalists?" In *Environmental Philosophy and Environmental Activism*, ed. Don E. Marietta, Jr., and Lester Embree. Lanham, Md.: Rowman and Littlefield, 1995, pp. 169–201.

Vitousek, Peter, Paul Ehrlich, Anne Ehrlich, and Pamela Matson, "Human Appropriation of the Products of Photosynthesis." *BioScience* 36 (June 1986): 368–73.

Wapner, Paul. *Environmental Activism and World Civic Politics.* Albany: SUNY Press, 1996.

———. "Politics beyond the State: Environmental Activism and World Civic Politics." *World Politics* 47 (April 1995): 311–41.

Weber, Max. *The Protestant Ethic and the Spirit of Capitalism*, trans. Talcott Parsons. New York: Scribner's, 1958.

Index